Green Antimicrobials

Green Antimicrobials

Editor

Helena Prado Felgueiras

MDPI • Basel • Beijing • Wuhan • Barcelona • Belgrade • Manchester • Tokyo • Cluj • Tianjin

Editor
Helena Prado Felgueiras
University of Minho
Guimarães, Portugal

Editorial Office
MDPI
St. Alban-Anlage 66
4052 Basel, Switzerland

This is a reprint of articles from the Special Issue published online in the open access journal *Antibiotics* (ISSN 2079-6382) (available at: https://www.mdpi.com/journal/antibiotics/special_issues/green_antibiotics).

For citation purposes, cite each article independently as indicated on the article page online and as indicated below:

LastName, A.A.; LastName, B.B.; LastName, C.C. Article Title. *Journal Name* **Year**, *Volume Number*, Page Range.

ISBN 978-3-0365-8238-2 (Hbk)
ISBN 978-3-0365-8239-9 (PDF)

Contents

antibiotics

Editorial

Green Antimicrobials

Helena P. Felgueiras

Centre for Textile Science and Technology (2C2T), University of Minho, Campus de Azurém, 4800-058 Guimarães, Portugal; helena.felgueiras@2c2t.uminho.pt

In the last couple of years, the awareness of climate change and high pollution levels have raised our sense of ecological responsibility. Pharmaceutical industries play a major role in these issues; as such, alternatives must be found. New environmentally friendly approaches to deal with the growing concern associated with antimicrobial-resistant bacteria are also in great demand. The excessive consumption and misuse of these agents have accelerated the rise of such pathogens responsible for compromising global health—not only the health of humans, but also the health of all living systems. Considering our natural resources are in great danger, finding green and less environmentally impactful alternatives for fighting these resistant microbials is imperative. From green chemistries, natural extracts and waste products, the sources of these alternate antimicrobial agents can be immense, and their implications are of great impact for future generations [1–3].

Fernández-Fernández et al. reported the great biodiversity and biomass of microorganisms found in soil microbiota that could be used in the production of new antimicrobial agents. Through the MicroMundo project, they illustrated the relevant link between science and education and the benefits of implementing service-learning methodologies to raise awareness of the antimicrobial resistance problem. They collected soil samples from different areas and analyzed their content for antimicrobial-producing bacteria. In total, 132 potentially producing bacteria were identified, from which 18 isolates were deemed low producers of antimicrobial agents, 12 were recognized as medium producers, and 2 were considered highly antimicrobial-producing isolates [4]. Aside from soil, agricultural wastes can also serve as sources of antimicrobial agents. Arumugam et al. explored the abilities of crude solvent extracts of both groundnut shells and black gram pods to reduce aerolysin formation and biofilm matrix formation by *Aeromonas hydrophila*, an opportunistic bacterium, responsible for many diseases in humans and animals, particularly aquatic species. Twelve potent metabolites extracted from these wastes showed interactions with aerolysin during molecular docking analysis, confirming their potential for uses in pharmacological solutions for treating *A. hydrophila*-induced infections in aquaculture [5].

Foodborne pathogens can pose great risks to human health. Torres Neto et al. investigated the antimicrobial potency of the essential oils of oregano, thyme, lemongrass and their blends against three foodborne pathogens, *Salmonella enterica* serotype Enteritidis, *Escherichia coli* and *Staphylococcus aureus*. The data highlighted the ability of the blends containing thyme to inhibit and kill each bacterium individually [6]. Adnan et al. focused their efforts on finding effective agents against the foodborne pathogen *Listeria monocytogenes*, responsible for the disease listeriosis. Here, active biosurfactants and their potential targets were tested against the bacterium, revealing their ability in regulating several bacterium pathways responsible for the bacterium's pathogenic nature, evidencing their pharmacological potency [7]. Stavropoulou et al. demonstrated that honey samples obtained from different geographical locations in Greece and with diverse pollen origins were influenced by the pollinizing animals. In fact, they detected 335 distinct taxa in honey microbiota, native to the gut microbiota of melliferous bees and microbiota of their flowering plants, including beneficial bacteria like probiotic strains. These could work as indicators of the authenticity of honey, facilitating the identification of honey samples

Citation: Felgueiras, H.P. Green Antimicrobials. *Antibiotics* **2023**, *12*, 1128. https://doi.org/10.3390/antibiotics12071128

Received: 20 June 2023
Accepted: 28 June 2023
Published: 29 June 2023

with potent antimicrobial profiles [8]. β-Glucan is an important natural prebiotic abundant in lentinan, cereals and the yeast cell wall. Zhou et al. evaluated the influence of this natural-origin component on growth performance and intestinal epithelium functions and established β-glucan as an attenuator of intestinal damage by suppressing the secretion of inflammatory cytokines, enhancing serum immunoglobulins and improving intestinal epithelium functions and microbiota [9].

Skin disorders have benefited immensely from natural-origin antimicrobials. Diabetic foot ulcers, for instance, have been shown to accelerate their healing in the presence of antimicrobial peptides derived from green, eco-friendly processes [10]. In chronic wound care, many antiseptic agents have also been highlighted, namely octenidine and polyhexamide, as beneficial for wound care without inducing any cytotoxic effects or raising significant environmental concerns [11]. Ranjutha et al. identified synergisms between ceftriaxone and *Polyalthia longifolia* methanol leaf extract against methicillin-resistant *Staphylococcus aureus*, supporting their use as an etiological agent for skin disease therapies. The gene responsible for the bacterium resistance was significantly suppressed by the antimicrobial agents' combination, and the same occurred with the concentration required to eliminate the bacterium [12]. Matias et al. disclosed the seaweed *Gelidium corneum* as a sustainable source of antimicrobial ingredients for new dermatological formulations, highlighting its potential to be explored in a circular economy context. For this purpose, seaweed fractions were compiled and examined against common pathogenic bacteria. The data showed that the antimicrobial effects of the fractions were manifested in the form of membrane hyperpolarization and DNA damage [13].

Innovations in silver nanoparticle production have also taken advantage of green processes. For instance, Padmanabhan et al. reported the formation of clusters containing biphasic calcium phosphate-modified silver nanoparticles for improved antimicrobial and cytocompatibility performances and, thus, engineering of bone tissue replacements [14]. Sharma et al. explored the use of silver nanoparticles embedded into polymeric films to prevent and mitigate pathogen transmission in biomedical surfaces. It was shown that the incorporation of silver nanoparticles inhibited the growth of Gram-positive and Gram-negative bacteria, generating a polymeric substrate that is both cost-effective and highly scalable, by means of simple production methodologies (sputter deposition) [15].

Funding: This research was funded by the Portuguese Foundation for Science and Technology (FCT) via grant UID/CTM/00264/2020 of 2C2T Strategic Project 2020–2023. H.P.F. also acknowledges the FCT for Auxiliary Researcher contract 2021.02720.CEECIND.

Conflicts of Interest: The author declares no conflict of interest.

References

1. Felgueiras, H.P. Emerging Antimicrobial and Immunomodulatory Fiber-Based Scaffolding Systems for Treating Diabetic Foot Ulcers. *Pharmaceutics* **2023**, *15*, 258. [CrossRef] [PubMed]
2. Barcellos, L.; Pham, C.K.; Menezes, G.; Bettencourt, R.; Rocha, N.; Carvalho, M.; Felgueiras, H.P. A concise review on the potential applications of Rugulopteryx okamurae macroalgae. *Mar. Drugs* **2023**, *21*, 40. [CrossRef] [PubMed]
3. Felgueiras, H.P. An insight into biomolecules for the treatment of skin infectious diseases. *Pharmaceutics* **2021**, *13*, 1012. [CrossRef] [PubMed]
4. Fernández-Fernández, R.; Robredo, B.; Navajas, E.; Torres, C. Citizen Contribution for Searching for Alternative Antimicrobial Activity Substances in Soil. *Antibiotics* **2023**, *12*, 57. [CrossRef]
5. Arumugam, M.; Manikandan, D.B.; Marimuthu, S.K.; Muthusamy, G.; Kari, Z.A.; Téllez-Isaías, G.; Ramasamy, T. Evaluating Biofilm Inhibitory Potential in Fish Pathogen, Aeromonas hydrophila by Agricultural Waste Extracts and Assessment of Aerolysin Inhibitors Using In Silico Approach. *Antibiotics* **2023**, *12*, 891. [CrossRef] [PubMed]
6. Torres Neto, L.; Monteiro, M.L.G.; Machado, M.A.M.; Galvan, D.; Conte Junior, C.A. An Optimization of Oregano, Thyme, and Lemongrass Essential Oil Blend to Simultaneous Inactivation of Relevant Foodborne Pathogens by Simplex–Centroid Mixture Design. *Antibiotics* **2022**, *11*, 1572. [CrossRef] [PubMed]
7. Adnan, M.; Siddiqui, A.J.; Noumi, E.; Hannachi, S.; Ashraf, S.A.; Awadelkareem, A.M.; Snoussi, M.; Badraoui, R.; Bardakci, F.; Sachidanandan, M. Integrating Network Pharmacology Approaches to Decipher the Multi-Target Pharmacological Mechanism of Microbial Biosurfactants as Novel Green Antimicrobials against Listeriosis. *Antibiotics* **2022**, *12*, 5. [CrossRef] [PubMed]

8. Stavropoulou, E.; Remmas, N.; Voidarou, C.C.; Vrioni, G.; Konstantinidis, T.; Ntougias, S.; Tsakris, A. Microbial Community Structure among Honey Samples of Different Pollen Origin. *Antibiotics* **2023**, *12*, 101. [CrossRef]
9. Zhou, Y.; Luo, Y.; Yu, B.; Zheng, P.; Yu, J.; Huang, Z.; Mao, X.; Luo, J.; Yan, H.; He, J. Effect of β-Glucan supplementation on growth performance and intestinal epithelium functions in weaned pigs challenged by enterotoxigenic *Escherichia coli*. *Antibiotics* **2022**, *11*, 519. [CrossRef] [PubMed]
10. Teixeira, I.D.; Carvalho, E.; Leal, E.C. Green Antimicrobials as Therapeutic Agents for Diabetic Foot Ulcers. *Antibiotics* **2023**, *12*, 467. [CrossRef] [PubMed]
11. Barrigah-Benissan, K.; Ory, J.; Sotto, A.; Salipante, F.; Lavigne, J.-P.; Loubet, P. Antiseptic agents for chronic wounds: A systematic review. *Antibiotics* **2022**, *11*, 350. [CrossRef] [PubMed]
12. Ranjutha, V.; Chen, Y.; Al-Keridis, L.A.; Patel, M.; Alshammari, N.; Adnan, M.; Sahreen, S.; Gopinath, S.C.; Sasidharan, S. Synergistic Antimicrobial Activity of Ceftriaxone and Polyalthia longifolia Methanol (MEPL) Leaf Extract against Methicillin-Resistant Staphylococcus aureus and Modulation of mecA Gene Presence. *Antibiotics* **2023**, *12*, 477. [CrossRef] [PubMed]
13. Matias, M.; Pinteus, S.; Martins, A.; Silva, J.; Alves, C.; Mouga, T.; Gaspar, H.; Pedrosa, R. Gelidiales Are Not Just Agar—Revealing the Antimicrobial Potential of Gelidium corneum for Skin Disorders. *Antibiotics* **2022**, *11*, 481. [CrossRef] [PubMed]
14. Padmanabhan, V.P.; Sivashanmugam, P.; Kulandaivelu, R.; Sagadevan, S.; Sridevi, B.; Govindasamy, R.; Thiruvengadam, M. Biosynthesised Silver Nanoparticles Loading onto Biphasic Calcium Phosphate for Antibacterial and Bone Tissue Engineering Applications. *Antibiotics* **2022**, *11*, 1780. [CrossRef] [PubMed]
15. Sharma, P.; Fialho, L.; Figueiredo, N.M.; Serra, R.; Cavaleiro, A.; Carvalho, S. Antimicrobial Polymeric Surfaces Using Embedded Silver Nanoparticles. *Antibiotics* **2023**, *12*, 207. [CrossRef] [PubMed]

antibiotics

MDPI

Article

Evaluating Biofilm Inhibitory Potential in Fish Pathogen, *Aeromonas hydrophila* by Agricultural Waste Extracts and Assessment of Aerolysin Inhibitors Using *In Silico* Approach

Manikandan Arumugam [1], Dinesh Babu Manikandan [1], Sathish Kumar Marimuthu [2], Govarthanan Muthusamy [3], Zulhisyam Abdul Kari [4,5], Guillermo Téllez-Isaías [6] and Thirumurugan Ramasamy [1,*]

1 Laboratory of Aquabiotics/Nanoscience, Department of Animal Science, School of Life Sciences, Bharathidasan University, Tiruchirappalli 620024, India
2 Department of Pharmaceutical Technology, University College of Engineering, Bharathidasan Institute of Technology (BIT) Campus, Anna University, Tiruchirappalli 620024, India
3 Department of Environmental Engineering, Kyungpook National University, Daegu 41566, Republic of Korea
4 Department of Agricultural Sciences, Faculty of Agro-Based Industry, Jeli Campus, Universiti Malaysia Kelantan, Jeli 17600, Malaysia
5 Advanced Livestock and Aquaculture Research Group, Faculty of Agro-Based Industry, Jeli Campus, Universiti Malaysia Kelantan, Jeli 17600, Malaysia
6 Department of Poultry Science, University of Arkansas, Fayetteville, AR 72701, USA
* Correspondence: ramthiru72@bdu.ac.in

Abstract: *Aeromonas hydrophila*, an opportunistic bacteria, causes several devastating diseases in humans and animals, particularly aquatic species. Antibiotics have been constrained by the rise of antibiotic resistance caused by drug overuse. Therefore, new strategies are required to prevent appropriate antibiotic inability from antibiotic-resistant strains. Aerolysin is essential for *A. hydrophila* pathogenesis and has been proposed as a potential target for inventing drugs with anti-virulence properties. It is a unique method of disease prevention in fish to block the quorum-sensing mechanism of *A. hydrophila*. In SEM analysis, the crude solvent extracts of both groundnut shells and black gram pods exhibited a reduction of aerolysin formation and biofilm matrix formation by blocking the QS in *A. hydrophila*. Morphological changes were identified in the extracts treated bacterial cells. Furthermore, in previous studies, 34 ligands were identified with potential antibacterial metabolites from agricultural wastes, groundnut shells, and black gram pods using a literature survey. Twelve potent metabolites showed interactions between aerolysin and metabolites during molecular docking analysis, in that H-Pyran-4-one-2,3 dihydro-3,5 dihydroxy-6-methyl (−5.3 kcal/mol) and 2-Hexyldecanoic acid (−5.2 kcal/mol) showed promising results with potential hydrogen bond interactions with aerolysin. These metabolites showed a better binding affinity with aerolysin for 100 ns in molecular simulation dynamics. These findings point to a novel strategy for developing drugs using metabolites from agricultural wastes that may be feasible pharmacological solutions for treating *A. hydrophila* infections for the betterment of aquaculture.

Keywords: aerolysin; agri-waste; antimicrobial metabolites; molecular docking and dynamics; quorum sensing

Citation: Arumugam, M.; Manikandan, D.B.; Marimuthu, S.K.; Muthusamy, G.; Kari, Z.A.; Téllez-Isaías, G.; Ramasamy, T. Evaluating Biofilm Inhibitory Potential in Fish Pathogen, *Aeromonas hydrophila* by Agricultural Waste Extracts and Assessment of Aerolysin Inhibitors Using *In Silico* Approach. *Antibiotics* **2023**, *12*, 891. https://doi.org/10.3390/antibiotics12050891

Academic Editors: Helena P. Felgueiras and Jarl Bøgwald

Received: 23 March 2023
Revised: 27 April 2023
Accepted: 8 May 2023
Published: 11 May 2023

1. Introduction

Although aquaculture is one of the industries producing food with the greatest growth rate, bacterial fish infections result in large output losses every year [1]. The disease is a significant negative socioeconomic consequence for those dependent on aquaculture and a major barrier to aquaculture operations [2]. With the growth of aquaculture activities, stress conditions also increase, encouraging the frequent incidence and development of pathogens [3,4]. Furthermore, approximately 10 to 50% of output loss is brought on by epizootics, which severely hampers the efforts to increase productivity [5,6]. *Aeromonas hydrophila* is a freshwater chemoorganoheterotrophic, facultatively anaerobic, gram-negative pathogenic bacteria that mostly affects fish, mammals, birds, amphibians, and reptiles. It causes infections such as gastroenteritis, necrotizing fasciitis, and septicemia in the species mentioned above [7,8]. Since the condition is driven by several virulence factors, including cytotoxins, adhesions, hemolysins, proteases, lipases, and biofilm development, *A. hydrophila* is multifaceted in its pathogenicity [9,10].

Aerolysin causes symptoms of many sorts of infections, including hemorrhagic and ulcerative lesions on the skin and other organs [11,12]. Aerolysin has numerous effects such as hemolytic, enterotoxin, and cytotoxic activities [13–15]. Aerolysin can enter the target cell's membrane after producing seven oligomeric subunits with a transmembrane pore [16–18]. The channel pore breaches the cellular membrane permeability barrier, resulting in cell death [19]. One of the main virulence factors in developing fish disorders linked to *A. hydrophila* is the gene aerolysin, a destructive pore-forming enterotoxin [20]. Tetramycin and romet-30 are the most widely and frequently used antimicrobial drugs against *A. hydrophila* contamination in freshwater aquaculture [2]. The key pathogenic factors assisted by *Aeromonas* sp. are surface polysaccharides, extracellular proteins, iron-binding systems, and exotoxins, which are crucial in the pathogenic mechanisms. These virulence factors have antibiotic resistance that might develop in aquaculture systems; nevertheless, these antimicrobial drugs are used indiscriminately [21–23]. These antibiotics are mostly administered directly to the aquaculture system by feed or submersion. According to the scientific literature, between 70 and 80% of prescribed antibiotics eventually enter water [24–27].

As an outcome, the aquaculture systems have been designated as "biological hubs" for bacterial transduction, conjugation, and transformation of antibiotic-resistant genes [28,29]. Therefore, as our reliance on aquaculture increases, it is vital to investigate appropriate antibiotic substitutes that feedstuffs may deliver, lower the risk of antimicrobial resistance emerging, and increase the fish immune system [30–32]. Various approaches have been recently proposed to combat the rise of antibiotic resistance, including the use of plant metabolites to improve and stimulate the fish's immune characteristics in aquaculture [33]. Plants and their secondary metabolites have a wide range of activities, which raises the possibility that they could be used as antimicrobial agents. In particular, the main natural plant species come under the *Apiaceae*, *Anacardiaceae*, *Burseraceae*, *Cupressaceae*, *Dracenaceae*, *Euphorbiaceae*, *Fabaceae*, *Palmaceae*, and *Pinaceae* families of plants [34,35]. Tannins are the phenolic polymers in all plants that tend to inhibit the bacterial matrices, outer membranes, and protein transport in bacterial cells and may prevent several hydrolytic enzymes such as α-amylase which is essential for cell growth [36]. Several plant natural products or secondary metabolites have been shown to positively impact disease virulence factors *in vivo* and *in vitro* [37].

Quorum sensing (QS) system controls the expression of aerolysin and numerous other virulence factors and biofilm development [38,39]. Blocking the action of aerolysin and biofilm by inhibiting QS has been shown in prior research to reduce the pathogenicity of *A. hydrophila* [40]. Additionally, developing novel and quick molecular docking techniques has enhanced molecular simulations with critical applications for screening and drug discovery [41–43]. A useful approach in drug design and compound screening for the development of natural drugs is the study of molecular docking of protein–ligand interactions [44,45]. It is possible to anticipate the conformations and binding affinities of the putative phytoconstituents from the extracts. This research will focus on the successful development of new medications by screening agricultural waste-derived metabolites for the diseases caused by *A. hydrophila* in aquaculture industries.

2. Results

2.1. Scanning Electron Microscopy

SEM images demonstrated a decrease in the biofilm development of *A. hydrophila* when treated with agricultural waste extracts. A maximum cell size and shape reduction was seen at the treatment concerning the calculated minimal inhibitory concentration (MIC) when control images revealed a cell matrix. Streptomycin (50 µg/mL) was used as a positive control and it exhibited dispersed cells without biofilm formation and morphological changes of the *A. hydrophila* cells (Figures 1 and 2).

2.2. FT-IR Analysis of Bacterial Biomass

The FT-IR spectra of bacterial biomass treated with black gram pod extracts and groundnut shell extracts (Figures 3 and 4). The peak at 3350–3450 cm^{-1} indicates the existence of the OH group, which contains carbohydrates, proteins, and polyphenols and is classified as an alcoholic group, and a minor intensity peak at 1700–1715 cm^{-1} indicates the presence of fatty acid groups. Certain peaks at 3000 cm^{-1} show the presence of C-H alkenes and aromatic rings as a result of the interaction of the metabolites present in the ethyl acetate, acetone, methanol, and ethanol extracts of both groundnut shells and black gram pods. In contrast, non-polar solvents such as petroleum ether and hexane did not show any clear bands.

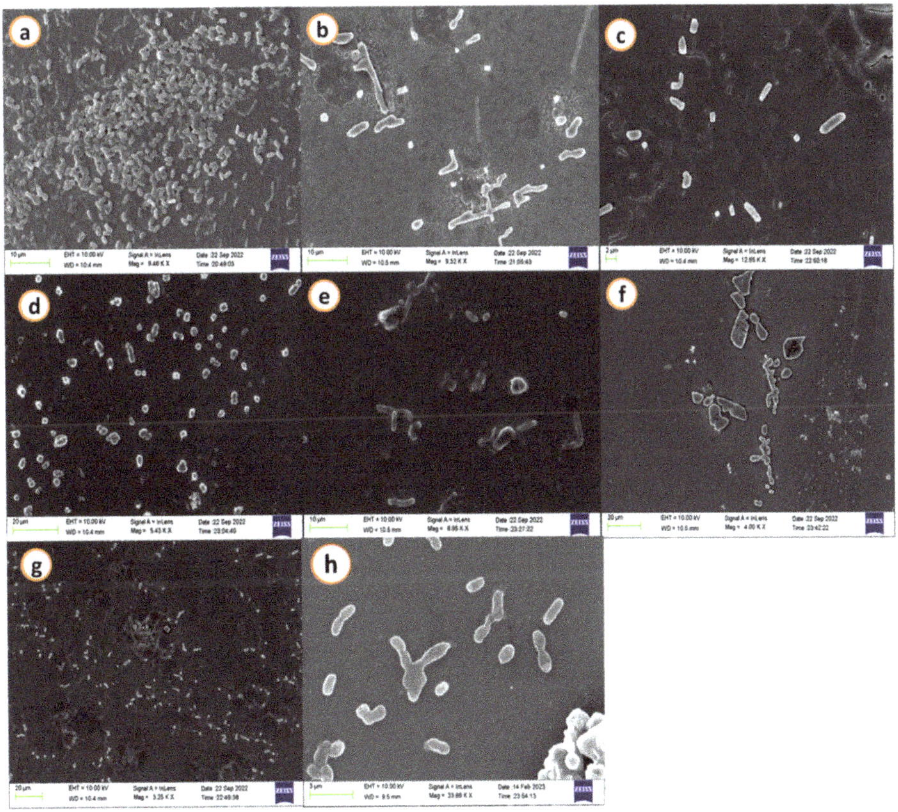

Figure 1. Scanning electron microscopic images of *A. hydrophila* biofilm matrix. (**a**) Negative control: shows dense and thick adherence of biofilm; treated with groundnut shell extracts (**b**) methanol, (**c**) ethanol, (**d**) acetone, (**e**) ethyl acetate, (**f**) hexane, (**g**) petroleum ether, and (**h**) positive control: streptomycin (50 µg/mL) exhibits dispersed biofilm.

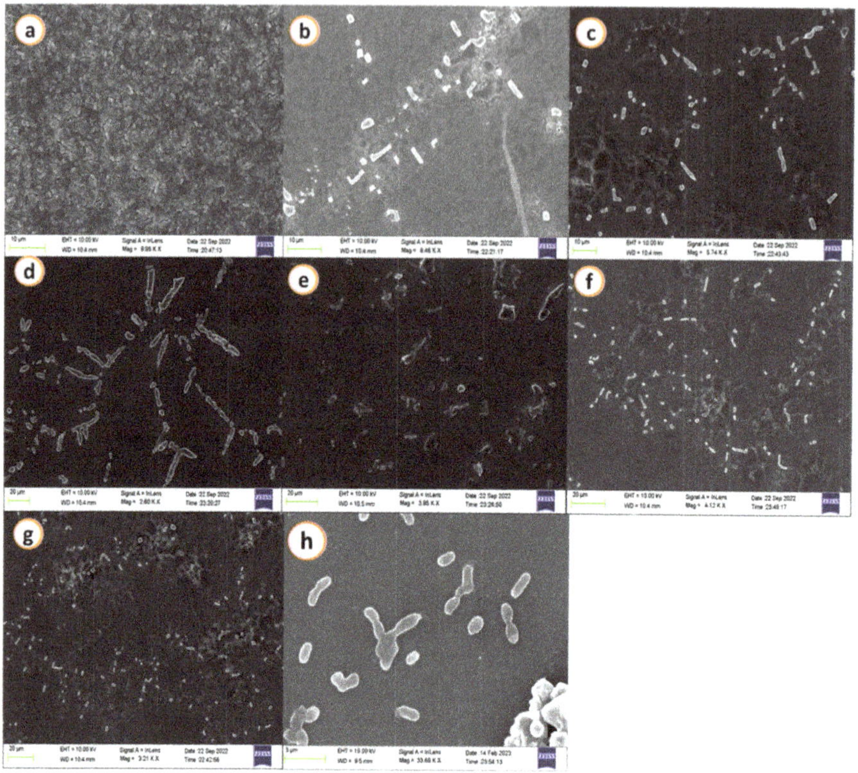

Figure 2. Scanning electron microscopic images of *A. hydrophila* biofilm matrix. (**a**) Negative control: shows dense and thick adherence of biofilm; treated with black gram pod extracts (**b**) methanol, (**c**) ethanol, (**d**) acetone, (**e**) ethyl acetate, (**f**) hexane (**g**) petroleum ether, and (**h**) positive control: streptomycin (50 µg/mL) exhibits dispersed biofilm.

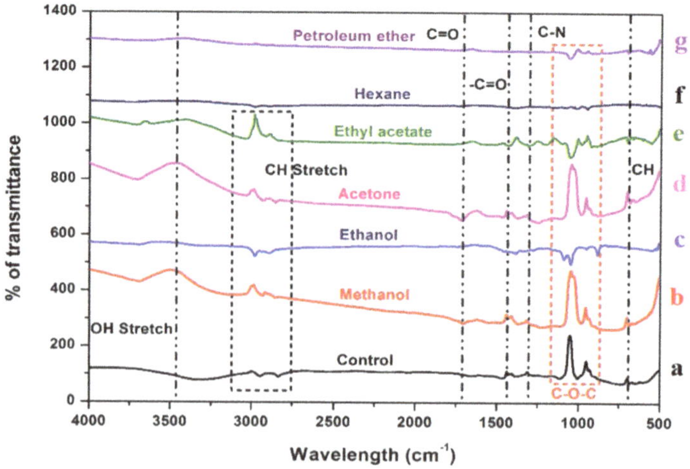

Figure 3. FT-IR spectra of the *Aeromonas hydrophila* biomass treated against various extracts of groundnut shell. Red dotted lines indicate the presence of polysaccharides. Untreated: (**a**) control. Treated: (**b**) methanol, (**c**) ethanol, (**d**) acetone, (**e**) ethyl acetate, (**f**) hexane, and (**g**) petroleum ether.

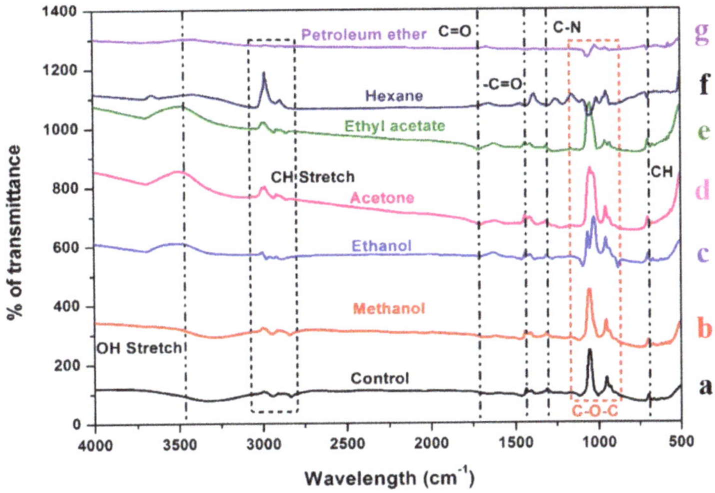

Figure 4. FT-IR spectra of the *Aeromonas hydrophila* biomass treated against various extracts of black gram pods. Red dotted lines indicate the presence of polysaccharides. Untreated: (**a**) control. Treated: (**b**) methanol, (**c**) ethanol, (**d**) acetone, (**e**) ethyl acetate, (**f**) hexane, and (**g**) petroleum ether.

2.3. Homology Modelling of AhEUS112 Aerolysin

Homology (comparative) modelling is typically considered the most reliable *in silico* technique for predicting accurate 3D protein models using amino acid sequences [46,47]. The best AhEUS112 aerolysin sequence model (Figure 5a) with the lowest DOPE (discrete optimized protein energy) score (Figure 5b) is chosen.

a

MMNKIITANLALLASSLMLVQAQAAEPVYPDQVKWFGLGEGVCAS
YRPLTRDEAMSIKGNLVSRMGQWQITGLANRWVIMGPGYNGEIKQGT
AGETWCYPNSPISGEIPSLSAWNIPAGDEVDVQWRMVHDKDYFIKPVS
YLAHYLGYAWVGGNHSPYVGEDMDVTRLSDGWLIKGNNDGGCSGY
RCGEKSSIKVSNFSYTLEPDSFSHGQVTESGKQLIKTITANATNYTDLPQ
QVVVTLKYDKATNWSKTDTYSLSEKVGIKKTFQIPQVSSTEYSVEISST
QSWAEQKGGATTETVSVEARPTVPPHSSVPVRVALYKSNISYPYEFKAE
VNYDLTMKGFLRWGGNAWYTHPTNRPTWEHTFAVGPFRDKASSIRYQ
WDKRYIPGEVKWWDWNWTIGEYGLSTMQNNLGRVLRPTRSAVTGDF
YAESQFAGDIEIGQPQTRAAAAVQQRSASAEGVALTALELDQDTLASE
GFGNVTLTITPAQ

b

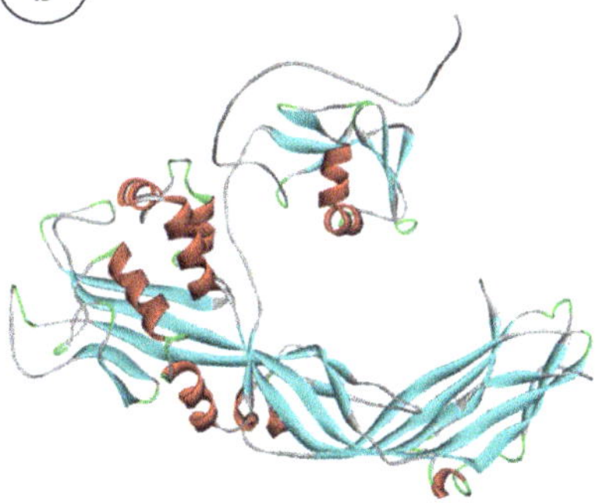

Figure 5. (**a**) Sequence of the AhEUS112 aerolysin. (**b**) Predicted three-dimensional structure of aerolysin from *A. hydrophila*.

Using PROCHECK, the modelled structure is validated using a Ramachandran plot using the RAMPAGE server [48]. The Ramachandran plot of the modelled protein represents 89.1% (366 aa) of the total residues in the most favoured regions. In comparison, 10.7% (44 aa) are in further and generously allowed regions, and only 0.2% of residues are found in the disallowed region. Based on the Ramachandran plot, the modelled structure indicates a good quality model (Figure 6). As a result, the predicted structure is chosen for molecular docking and molecular dynamics simulations.

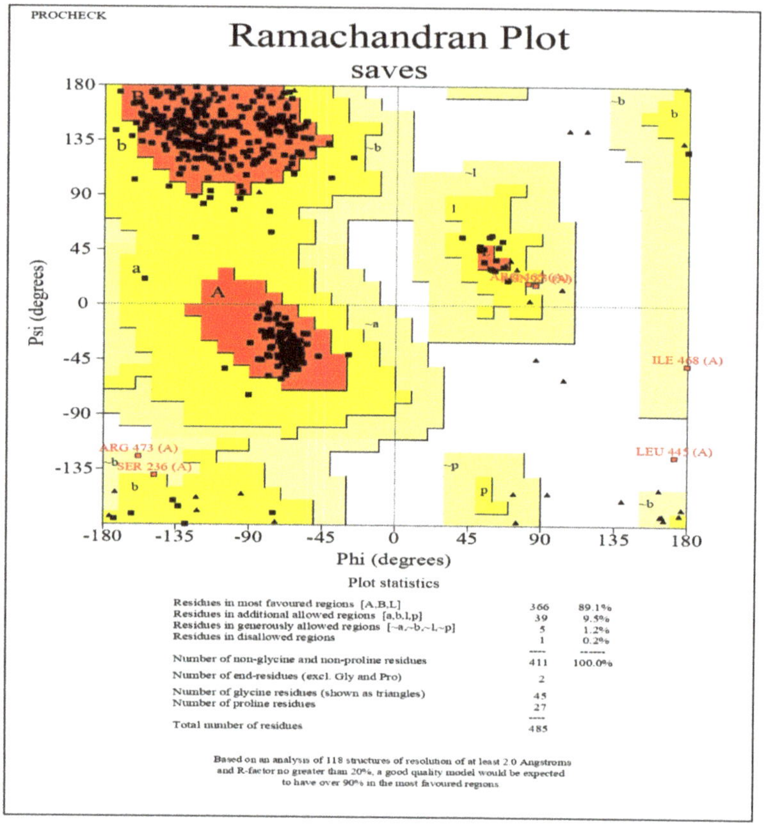

Figure 6. The stereochemical spatial arrangement of amino acid residues in the preferred area of the Ramachandran plot of the modelled 3D structure of aerolysin (Red coloured squares indicate residues in most favoured regions, dark yellow-coloured squares indicate the residues in additional allowed regions and pale yellow coloured square indicates residues in the generously allowed regions).

2.4. Phylogenetic Analysis of the Aerolysin

The AhEUS112 aerolysin shared 90–96% of its identity with other bacterial aerolysin when their multiple sequence alignment was analyzed and aerolysin from other *Aeromonas* sp. (Figure 7). MEGAX software was used to construct the distance matrix of the aerolysin sequence obtained from the different species [49].

2.5. Molecular Docking

The main goal of *in silico* docking analysis of this study was to identify the optimal binding conformations between aerolysin and metabolites from the agri-wastes that blocks the function and membrane potential. To interpret the optimal binding position for the ligands and the drugs developed, molecular docking was used to assess the great affinity for the aerolysin active site residues of the *A. hydrophila*. Based on this, several positions were created and evaluated. Crude extracts have both active and inactive chemical compounds in their mixture of diverse chemical molecules that exhibit high affinity and complementarity to the target protein. The capacity of the ligands to interact with the target protein *in vivo*, which impacts the outcomes of molecular docking studies, relates to the substances in crude extracts. However, the molecular docking study may only foresee a potent binding relationship, if the active components in the crude extract have low affinities or can efficiently access the target region in the protein [50,51]. The metabolites chosen from the different extracts of BGP and GNS for the interactions of aerolysin are shown in Table 1.

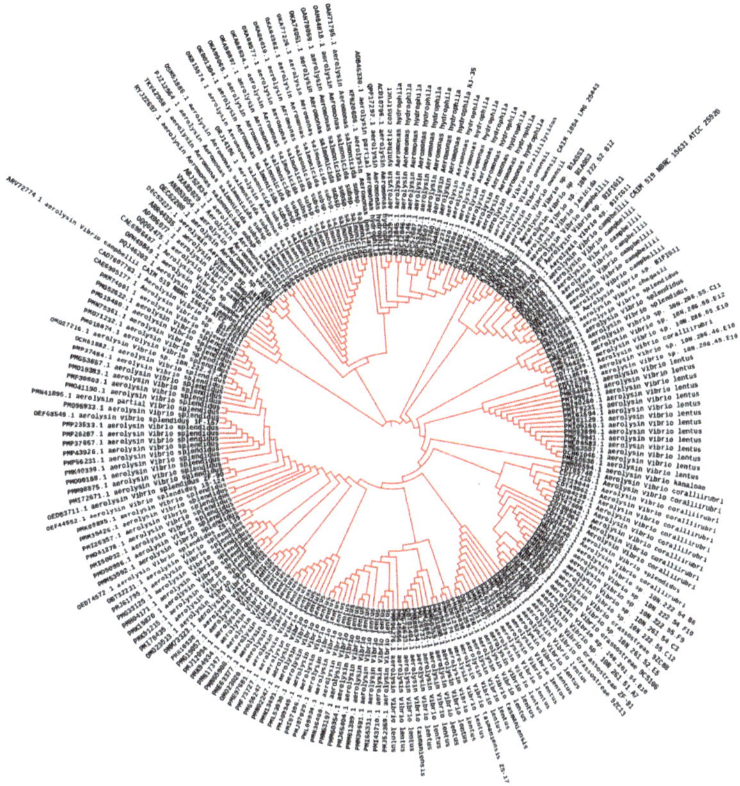

Figure 7. Phylogenetic tree of aerolysin from various species of *Aeromonas hydrophila*.

Table 1. List of compounds for molecular docking with aerolysin from black gram pods and groundnut shells.

Black Gram Pods		Groundnut Shells	
S. No.	**Name of the Compound**	**S. No.**	**Name of the Compound**
1.	1-Hexadecene	1.	2-Hexyldecanoic acid
2.	1-Isopropoxy-2-propanol	2.	2-Pentanone, 5-methoxy
3.	2,2-Difluorocycloheptan-1-one	3.	3-O-Methyl-d-glucose
4.	3-7-11-15-Tetramethyl-2-hexadecen-1-O	4.	4H-Pyran-4-one- 2-3-dihydro-3-5-dihydroxy-6-methyl
5.	5-Hydroxymethylfurfural	5.	Cyclohexanone
6.	Azulene	6.	Eicosane
7.	Butyronitrile	7.	Ethyl linoleate
8.	Cholesterol propionate	8.	Hexatriacontane
9.	Cholesterol	9.	Methyl alpha-D-glucopyranoside
10.	Cyanoacetic acid	10.	Octadecane
11.	Diacetone alcohol	11.	Palmitic acid
12.	Dodecanel	12.	Pentadecane- 2-6-10-13-tetramethyl
13.	Heptadecane	13.	Stearic acid
14.	Hexadecane	14.	Tetracosane
15.	Methyl palmitate		
16.	Methyl propyl ether		
17.	Naphthalene		
18.	Tetracontane		
19.	Tetratetracontane		
20.	Z-5-Nonadecene		

The number of hydrogen bonds that interacted with the aerolysin and the residues involved in the interactions were given in Table 2. The H-Pyran-4-one-2,3 dihydro-3,5 dihydroxy-6-methyl showed the strongest affinity with aerolysin possessing binding energy of −5.3 (kcal/mol), followed by 2-Hexyldecanoic acid and 2,2-Difluorocycloheptan-1-one (−5.2 kcal/mol), Methyl alpha-D-glucopyranoside (−5.1 kcal/mol), 5-Hydroxymethylfurfural (−5.0 kcal/mol), Methyl-d-glucose and Palmitic acid (−4.9 kcal/mol), Ethyl linoleate (−4.6 kcal/mol), Pentanone-5-methoxy (−4.3 kcal/mol), Diacetone alcohol (−4.1 kcal/mol), Methyl palmitate (−3.9 kcal/mol), and Cyanoacetic acid (−3.6 kcal/mol), respectively. The number of hydrogen bonds found during the interactions of metabolites with the aerolysin was represented in (Figures 8 and 9).

2.6. Simulation Dynamics

Molecular dynamic simulation (MDS) was used to determine the precise interaction of the ligand candidates with the protein under investigation. A methodology involving molecular docking, molecular dynamics, and free energy computing was used to identify the properties of specific natural compounds in a solvation state. In the current study, a 100 ns MDS was used to determine the best-docked molecule of H-Pyran-4-one-2,3 dihydro-3,5 dihydroxy-6-methyl, and 2-Hexyldecanoic acid to the aerolysin based on binding affinity and conformational stability (Figure 10a–d).

A significant RMS fluctuation was found between aerolysin and the other two ligands up to 30 residues, then showed similar fluctuation with all three complexes throughout the protein residues. The root-mean-square deviation (RMSD) estimate of backbone atoms varied from 0.25 nm to a maximum of 1.5 nm across the whole simulation. The RMSD value of the protein aerolysin was increased to 1.5 nm (10 ns), then showed at 0.75 nm (22.5 ns), and maintained steadily at 1 nm (up to 100 ns). However, aerolysin interacted with H-Pyran-4-one-2,3 dihydro-3,5 dihydroxy-6-methyl exhibited at 1 nm (45 ns) and then maintained at 0.75 nm (until 100 ns). Similarly, aerolysin interacted with 2-Hexyldecanoic acid and possessed an RMSD value of 0.75 nm at 10 ns, and it was gradually increased and maintained at 0.75 nm (until 100 ns). This data showed that the aerolysin formed a stable complex with the H-Pyran-4-one-2,3 dihydro-3,5 dihydroxy-6-methyl and 2-Hexyldecanoic acid at the range of 0.75 nm steadily (from 25 ns to 100 ns).

Table 2. Molecular docking of aerolysin with metabolites identified from both black gram pod and groundnut shell extracts. (Arg-Arginine, Pro-Proline, Tyr-Tyrosine, Asn-Asparagine, Met-Methionine, Val-Valine, Ile-Isoleucine, Ser-Serine, Phe-Phenylalanine, Gly-Glycine, Lys-Lysine).

S. No.	Compound	Binding Energy (kcal/mol)	Hydrogen Bond Interactions	Residues Involved During Interactions
1	2,2-Difluorocycloheptan-1-one	−5.2	2	Arg 414, Pro 139, Tyr 146, Asn 411, Tyr 142
2	5-Hydroxymethylfurfural	−5.0	5	Arg 417, Met 129, Tyr 135, Arg 414, Val 415, Tyr 337
3	Cyanoacetic acid	−3.6	1	Tyr 337
4	Diacetone alcohol	−4.1	2	Asn 411, Tyr 146
5	Methyl palmitate	−3.9	1	Tyr 146
6	2-Hexyldecanoic acid	−5.2	3	Phe 371, Tyr 380, Ile 378, Ser 377, Arg 379, Val 368
7	Pentanone-5-methoxy	−4.3	4	Tyr 337, Arg 417, Arg 414, Tyr 135
8	Methyl-d-glucose	−4.9	2	Gly 369, Ile 378, Ser 377, Arg 379, Ala 369
9	H-Pyran-4-one-2,3 dihydro-3,5 dihydroxy-6-methyl	−5.3	4	Val 335, Thr 419, Val 415, Pro 418
10	Ethyl linoleate	−4.6	1	Arg 414, Pro 139, Tyr 142, Lys 138
11	Methyl alpha-D-glucopyranoside	−5.1	4	Leu 416, Thr 419, Glu 334
12	Palmitic acid	−4.9	3	Arg 379, Ile 378, Ser 377, Phe 371, Tyr 380

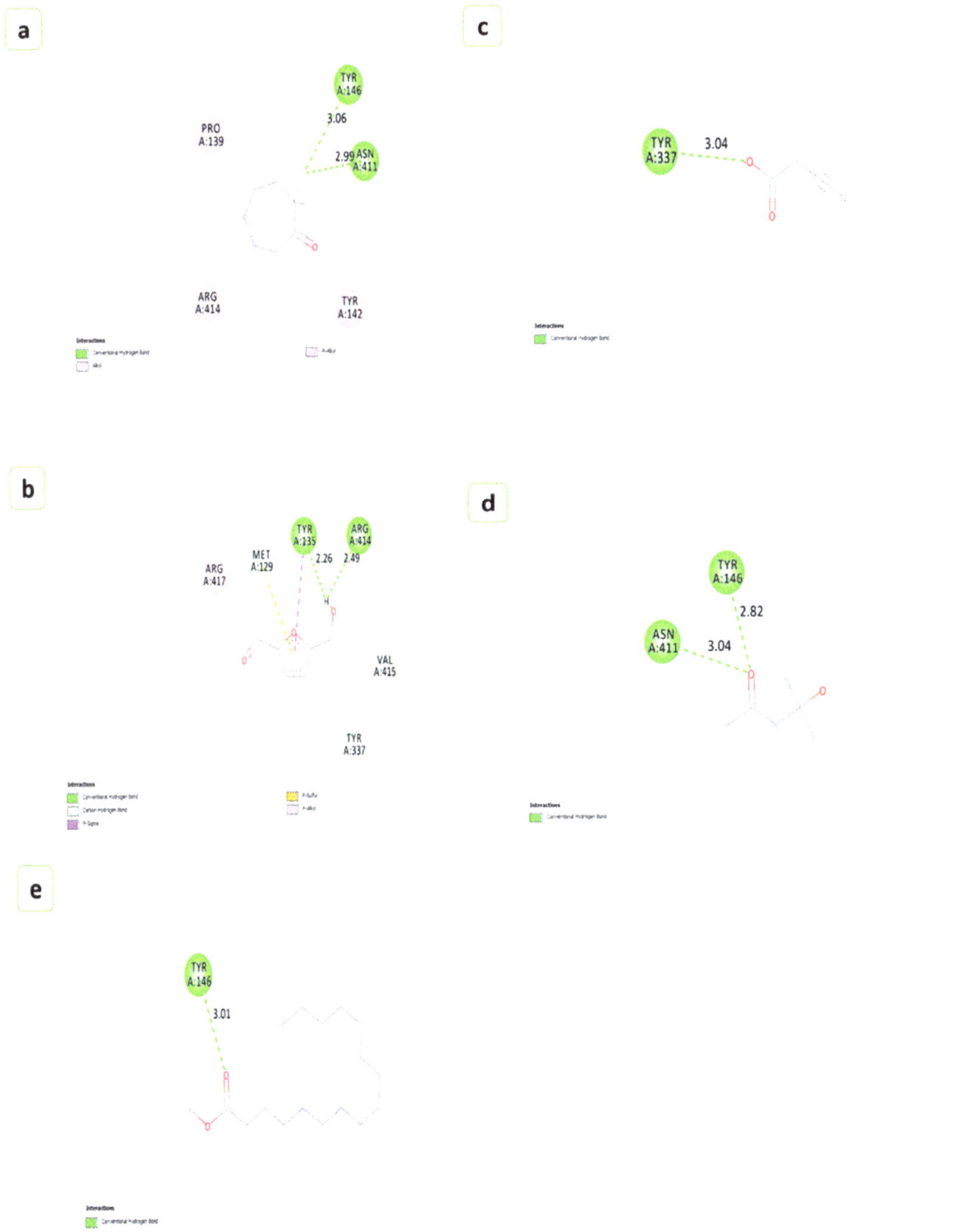

Figure 8. A 2D view of the interactions of the compounds extracted from black gram pods. (**a**) 2,2-Difluorocycloheptan-1-one, (**b**) 5-Hydroxymethylfurfural, (**c**) Cyanoacetic acid, (**d**) Diacetone alcohol, (**e**) Methyl palmitate.

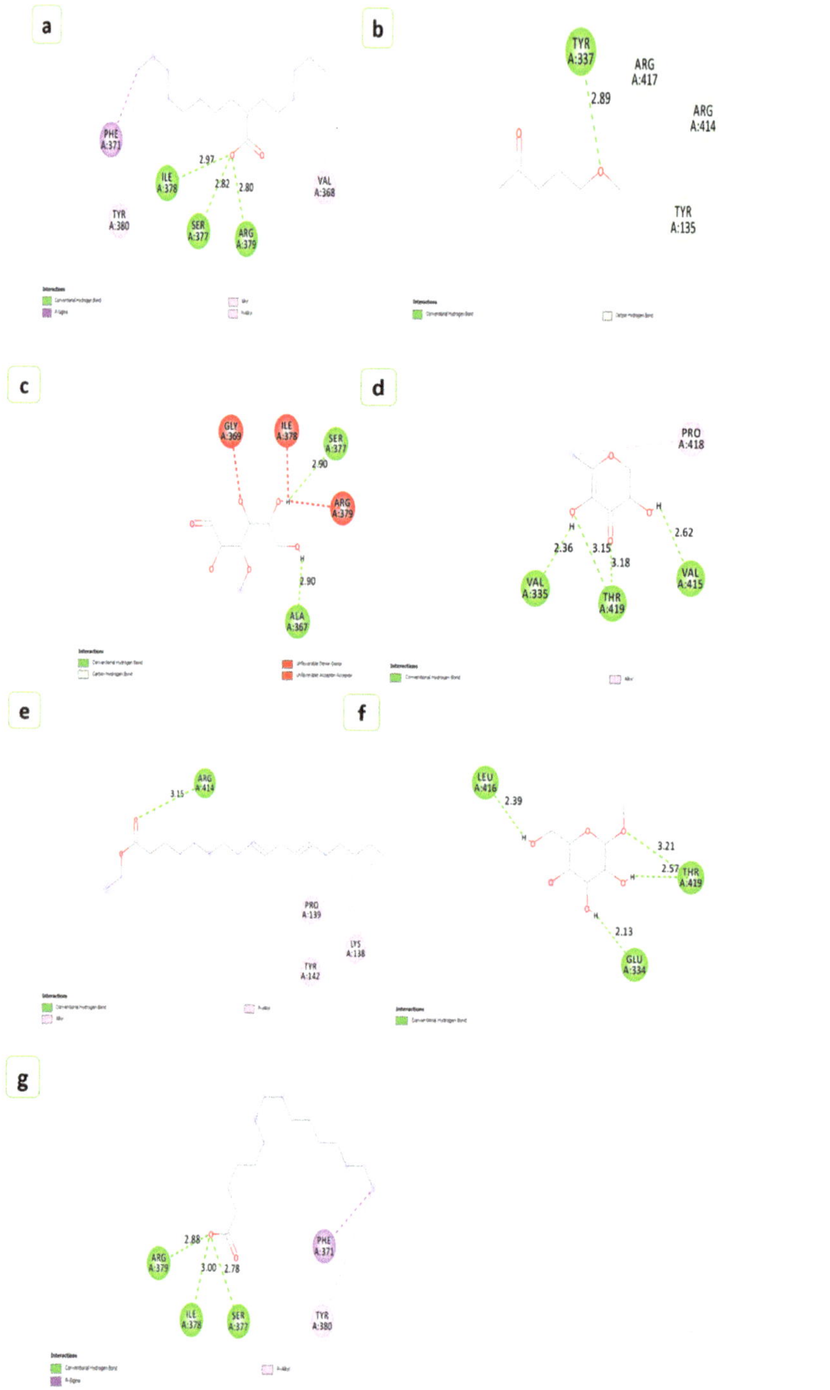

Figure 9. A 2D view of the interactions of the compounds from groundnut shells. (**a**) 2-Hexyldecanoic acid, (**b**) Pentanone-5-methoxy, (**c**) Methyl-d-glucose, (**d**) H-Pyran-4-one-2,3 dihydro-3,5 dihydroxy-6-methyl, (**e**) Ethyl linoleate, (**f**) Methyl alpha-D-glucopyranoside, (**g**) Palmitic acid.

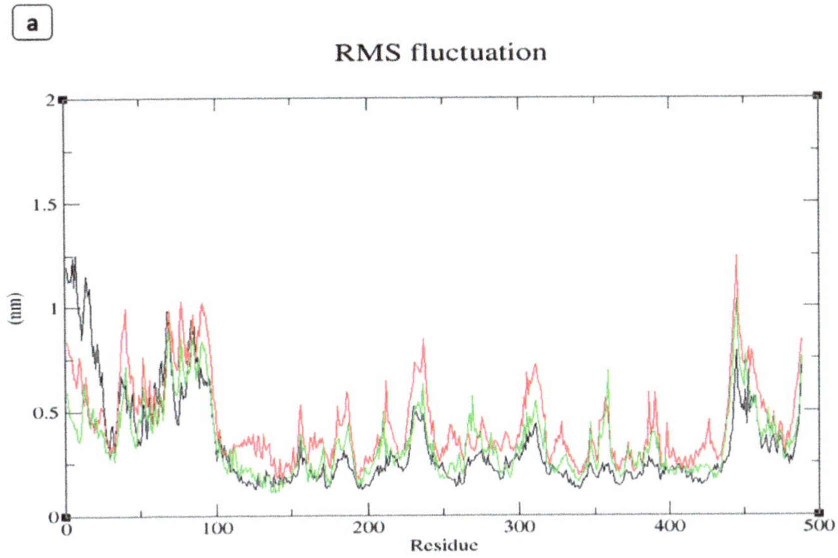

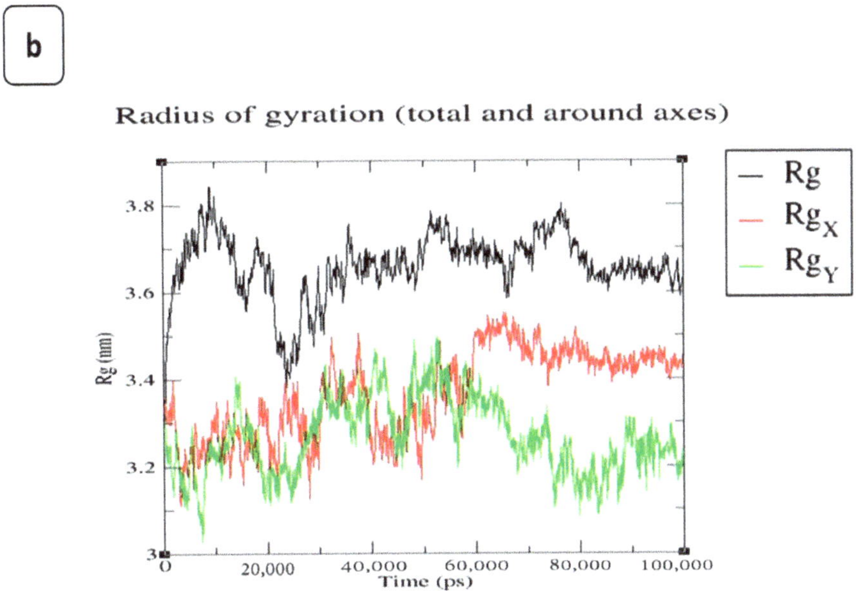

Figure 10. *Cont.*

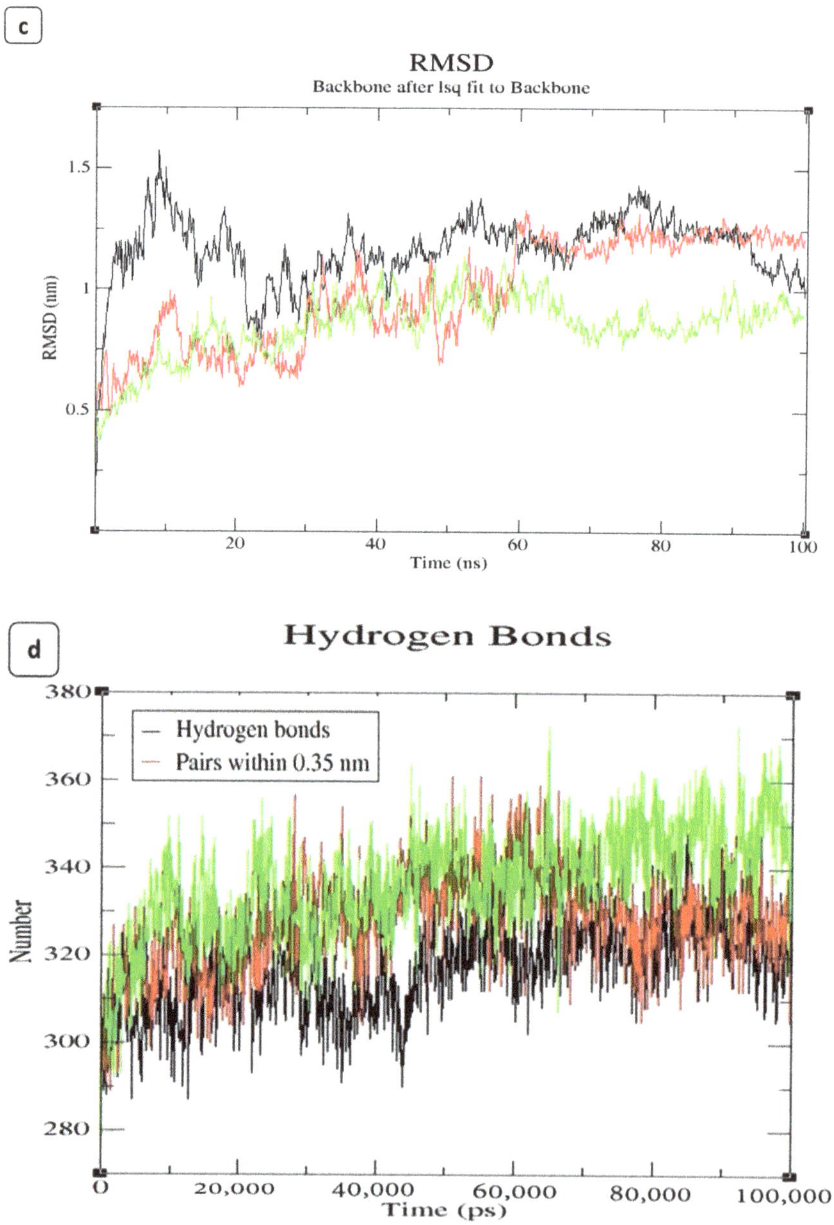

Figure 10. Graphical representation of 100 ns MD simulation analysis of aerolysin (in black colour), a protein with H-Pyran-4-one-2,3 dihydro-3,5 dihydroxy-6-methyl (in green colour), and protein with 2-Hexyldecanoic acid (in red colour). (**a**) RMSF values of the backbone atoms, (**b**) radius of gyration of the backbone atoms, (**c**) RMSD values of Cα atoms in the protein, and (**d**) hydrogen bonds stability of the protein and complexes.

3. Discussion

The extraction of phenolic compounds depends on the nature of the solvents. Polar solvents have lower electrostatic interactions that easily interact with the compounds present in the plant extracts that interchange their functional groups [52]. However, non-polar solvents can easily penetrate bacterial cells due to their lower charge [53]. Gram-negative bacteria such as *A. hydrophila* have rigid cell membranes that prevent the entry of the compounds into the cytoplasm [54]. They also have lipopolysaccharides that limit the penetration of hydrophobic compounds [55]. Based on our previous results, both GNS and BGP solvent extracts possess phenols and tannins [56,57]. These primary bioactive compounds are the major cause that exhibits better antioxidant and antibacterial properties [58,59]. The metabolites from the polar solvents also tend to diffuse the fatty acid composition of the rigid layer of *A. hydrophila* [60]. The metabolites of plant extracts may act on reducing the colonization of body surfaces and different epithelial layers, certain inorganic and organic molecules, along with other micro and macronutrients which are necessary for cell growth also promotes cell adhesion [61]. After a 48 h treatment with the extracts of both GNS and BGP, *A. hydrophila* cells were shrunken. They underwent splitting due to metabolites such as palmitic acid, methyl linoleate, H-Pyran-4-one-2,3 dihydro-3,5 dihydroxy-6-methyl, and 2-Hexyldecanoic acid [62–65]. These metabolites adhered to the lipopolysaccharides of the cell membrane, thus altering the bacterial cell morphology [66]. The metabolites from the extracts may inhibit nutrient availability that paved the way for bacterial cell growth [67]. The formation of the matrix by bacterial cells was separated due to inhibiting quorum-sensing signals from one cell to another [39,68]. This QS controls the synthesis of exopolysaccharide (EPS) by the *A. hydrophila* [69,70]. These polysaccharides, proteins, and nucleic acids played a crucial role in preventing the entry of antimicrobial agents and antibiotic exposure [71,72]. These exopolysaccharides play a vital role in cell detachment, colonization, and safeguarding forces of bacterial cells. The approaches to developing the new drug to combat multi-drug resistance and tolerance by polysaccharide lyases, a key enzyme which targeting the production of exopolysaccharides. Reduction in the exopolysaccharide production affects the QS signals between the cells [73,74]. According to Pellock et al. [75], *expr* is the major gene that maintains the quorum-sensing mechanism, and it is a homologue to *lux* receptors that leads to controls the production of exopolysaccharides. However, gram-negative bacteria such as *A. hydrophila* had autoinducers that tend to diffuse in and out of the cell [76]. These autoinducers, such as acyl-homoserine lactones (AHLs) synthesized by S-adenosylmethionine, bind to the cytoplasmic receptors and regulate the quorum-sensing gene expression [77–79]. Interfering with the synthesis, transport, or identification of autoinducers can be used to prevent quorum sensing. The key strategy is to utilize quorum-sensing inhibitors, which imitate or interfere with autoinducer binding to their receptors. These metabolites can impair quorum sensing in several bacterial species and limit biofilm development [80]. In gram-negative bacteria such as *A. hydrophila*, LuxR-type cytoplasmic receptors interact with another cell by detecting the AHLs; this complex transfers the quorum signals [81]. Additionally, fatty acids inhibit energy generation and cell lysis by interfering with components and preventing food intake [82]. Several studies have investigated the effects of fatty acids on mixed culture biofilms in the presence of natural conditions that may affect microbial signal production and reception [83].

In the FT-IR spectrum, the intensity peak at 1120–1160 cm^{-1} indicates the presence of polysaccharides in both control and treated groups due to bacterial biofilm formation [84,85]. Peaks obtained in the 2800–2600 cm^{-1} confirm the presence of aldehydes in the extract-treated biomass compared to the control [86]. The fatty acid groups in the polar and mid-polar extracts interact with the electron transport chain of bacteria. It involves the ATP transfer, which inhibits the bacterial enoyl-acyl reductase and leads to bacterial death [87]. In microbes, the electron acceptor is oxygen; when it demands, the organism tends to find an alternative to accept in the form of oxidized metals or non-metals [88]. During oxygen depletion, *A. hydrophila* utilizes iron (III) as an electron acceptor [89,90]. Carbon dioxide formed during the reactions will generate electrons which are accepted by iron (III) [91,92]. In our study, it is suggested that the metabolites present in the extracts of the GNS and BGP inhibit electron transfer by directly inhibiting iron reductase in the complex reactions. The antibacterial nature of the metabolites is based on solvent extraction [93]. Non-polar solvents are chemically inert and do not mix with water, so the microorganism can easily grow in the watery phase [94]. Essentially, polyphenols and bacteria interact in a non-specific manner, relying on the hydrogen group and hydrophobic effects that may have a significant influence owing to lipophilic interactions and the creation of covalent bonds [95]. Phenolic compounds present in the extracts may directly interact with the bacterial cell membrane, which causes intracellular leakage and ROS generation [96]. FT-IR analysis of *A. hydrophila* biomass can reveal important information about the bacterium's chemical composition, such as the presence of proteins, lipids, and fatty acids [97]. This knowledge can help us understand the structure and function of the bacterial cell, as well as create ways to prevent or treat *A. hydrophila* infections. An ideal tree was generated by utilizing the neighbour-joining method to analyze the evolutionary history of the aerolysin of *A. hydrophila* [98]. The existence of several branches representing the different architectural structures of a protein was evident in the phylogenetic tree created from the multiple sequence alignment of the AhEUS112 aerolysin with aerolysin from 200 different bacterial species [99]. It is a feasible approach to find protein areas that have been conserved during evolution by comparing the sequences of various species [100]. These conserved regions of the protein may be critical for protein function and might be targeted for drug development or other purposes.

A mixture of hydrophobic and van der Waals interactions with active site residues also stabilized the ligand configurations [101,102]. To emphasize, amino acid characteristics impact the functional activities of certain residues based on the physicochemical restrictions to variation of amino acid position/alignment [103]. The data analysis showed that aerolysin had common interaction residues with most test compounds.

Molecular docking and homology modelling were unique and useful tools for characterizing protein–ligand interaction patterns in configuration [104]. Due to the strong covalent bonds, weak intermolecular linkages encompassed a variety of interactions that did not involve the exchange of electrons. Still, hydrogen bonds played a vital role in the interaction of proteins and ligands [102]. GRID detects favourable sites for ligand binding with protein [105]. The binding nature between the ligand and protein depends on the length and orientation [101,106]. These protein–ligand interactions formed due to the cavity shape, size, and energy level of pocket formation [107,108]. Ligands are compounds that can control the activity of a protein or enzyme by binding to specific sites on the target protein or enzyme. In the case of aerolysin, ligands can be employed to prevent the production of toxin aggregates, which can injure host cells and tissues [109]. Ligands can bind to different sites of aerolysin, such as hydrophobic regions on the toxin's surface, particular spots on the pore-forming domain, and other sections of the molecule [110]. Flavonoids and polyphenols have been demonstrated to suppress the production of aerolysin aggregates. These bioactive compounds can attach to particular sites on the toxin and prevent it from building huge complexes that can damage host cells. This kind of *in vitro* approach is beneficial in decreasing aerolysin toxicity [111]. Our docking studies showed that the key residues of the aerolysin protein's binding pocket, such as Tyr 337, Arg 417, Arg 414, and Tyr 135 interacted with pentanone-5-methoxy via traditional hydrogen bonding and hydrophobic interactions [112]. These findings stated that pentanone-5-methoxy might reduce quorum sensing by decreasing the expression of aerolysin, which then affects other virulence-associated genes.

Palmitic acid reacted significantly with aerolysin, with a binding energy of −4.3 kcal/mol and a two-hydrogen bonding interaction (Arg 379, Ile 378, Ser 377, Phe 371, Tyr 380); these results agree with studies that reported the palmitic acid inhibiting the virulence factors associated with biofilm [113,114]. According to Dong et al. [115], heptamer formation was controlled by the ARG 414 and ARG 417 residual movements. This was the basic action mechanism behind the inhibition of aerolysin by the ligand H-Pyran-4-one-2,3 dihydro-3,5 dihydroxy-6-methyl. However, 2-Hexyldecanoic acid is bound with the ASP360 and does not involve forming heptamer [19]. Aerolysin often had the propensity to form a heptamer after entering the host cell membrane [116]. This heptamer had a transmembrane pore that affected the permeability of the host cell membrane and caused cell death [117]. The flexible portion of a protein or the parts of structures that change concerning the overall structure was evaluated by the root-mean-square fluctuation (RMSF) [118].

The simulation's dynamics give scientists a unique perspective on the structural and functional changes that occur during ligand binding by allowing them to watch the movement of specific atoms in the protein and the ligand over time [119]. The radius of gyration of aerolysin, protein complex with 2-Hexyldecanoic acid, and protein complex with the H-Pyran-4-one-2,3 dihydro-3,5 dihydroxy-6-methyl was determined. Using thermodynamic concepts, the radius of gyration indicated the protein's compactness with protein folding and unfolding [120]. The radius of gyration cannot be precisely measured because of diverse samples [121]. The Rg values were obtained in the range of 2.75–3.0 nm, whereas the aerolysin was maintained at 3.2 nm, and the aerolysin complexed with 2-Hexyldecanoic acid lay at 3.2 nm, respectively, which gradually increased and maintained at 3.4 nm from 40–100 ns. However, the aerolysin with H-Pyran-4-one-2,3 dihydro-3,5 dihydroxy-6-methyl showed an Rg value at 3.4 nm initially and it held at 3.6 nm. With this evidence, the examination of dynamics' mean radius of gyration fell within the range of random-coil statistics, confirming the protein folding in the presence of residual structure [122]. According to studies, aerolysin pores are fairly far from the host membrane surface and are shown as nanodisc-entrapped pores compatible with the absence of hydrophobicity [123]. The molecular dynamics trajectories for the whole examined protein–ligand complex is typically stable and within acceptable limits for the 100 ns simulation period, according to the RMSD fluctuation analysis [124]. According to the findings, the inhibitor attaches to a particular site of the protein and stabilizes it in a closed conformation, preventing the formation of the opening in the membrane [125]. Overall, the results of the molecular dynamics simulation study imply the stability of the protein–ligand complex of aerolysin with metabolites from agricultural waste.

4. Materials and Methods

4.1. Maintenance of Bacterial Strain and Culture Media Preparation

A fish pathogen, *A. hydrophila* (glycerol stock preserved at Laboratory of Aquabiotics/Nanoscience, Bharathidasan University, Tiruchirappalli, Tamil Nadu, India) (35 ± 2 °C/24 h), and bacterial cultures were maintained and grown in Tryptic soy agar or broth (TSA/TSB) containing Tryptone 1%, yeast extract 0.5%, and sodium chloride 0.5% with 1.2% agar.

4.2. Preparation of Extract

Groundnut shells (GNS) and black gram pods (BGP) were collected based on their detailed experimental procedures [56,57]. The collected agri-wastes were shade dried at 37 °C, ground into a coarse powder, sieved using 0.2 mm sieve plates, and then stored at −20 °C for subsequent examination in an airtight container. Cold maceration was used to elution the extracts from powdered agri-wastes using six solvents (10:90 w/v): ethyl acetate, petroleum ether, methanol, ethanol, hexane, and acetone. Additionally, the solvents employed for this study is based on the polarity which ranges from least polar to most polar [126]. The extracts were concentrated at roughly 40 °C in a rotating vacuum evaporator under decreasing pressure until agglomerates were formed; before that, the filtrate was collected using Whatman No. 1 filter paper. The extracts were dried to remove excess solvents and kept at 4 °C for future research. For experimental purposes, dry extracts were reconstituted with DMSO (0.1%) [127].

4.3. Scanning Electron Microscopy

The inhibition and deterioration of *A. hydrophila* biofilm using the various extracts of GNS and BGP were visualized using scanning electron microscopy (SEM) with a slight modification of the detailed protocol by Zhou et al. [128]. In short, biofilms of *A. hydrophila* grown on glass coverslips (18 mm) submerged in nutrient broth with determined minimum inhibitory concentration (MIC) were given in Table 3. Based on this, various extracts (acetone, methanol, hexane, ethanol, ethyl acetate, and petroleum ether) of groundnut shells and black gram pods were poured into six-well plates, and the untreated (without extracts) acted as the negative control. Streptomycin (50 μg/mL) was used as a positive control. The treated and untreated plates were incubated for 48 h at 37 °C, and then gently washed with miliQ to extract adherent bacterial cells. Samples were kept in 2.5% glutaraldehyde for 15 min and dehydrated with 25–95% gradient ethanol for 10 min. The dried biofilms were gold coated and examined under a scanning electron microscope (SEM, TESCAN, Czech Republic, and Vega 3).

Table 3. Determination of minimum inhibitory concentration (MIC) of groundnut shell and black gram pods for *A. hydrophila* (μg/mL).

S. No..	Extraction Solvents	Minimum Inhibitory Concentration (MIC) for *Aeromonas hydrophila* (μg/mL)	
		Groundnut Shell	Black Gram Pod
1.	Methanol	250	250
2.	Ethanol	250	250
3.	Acetone	500	500
4.	Ethyl acetate	500	500
5.	Hexane	500	500
6.	Petroleum ether	500	500

4.4. FT-IR Analysis of Bacterial Biomass

The bacterial cells of *A. hydrophila* treated with the extracts of agricultural wastes were collected through centrifugation at 10,000 rpm for 10 min and washed with phosphate buffer, pH 7.0, then made into a die using a desiccator at 45 °C. The KBr crystals were vacuum-dried as described [129]; 1000 mg of KBr and 2.5 mg of bacterial biomass were finely powdered and homogenized. The KBr beta press was used to form 100 mg of this bulk mixture into a single pellet. With the bacterial biomass abundant as a clear pellet within the KBr beta press barrel, the barrel was put on the sample holder in the FT-IR chamber. Fourier transform infrared (FT-IR) spectrophotometer (Perkin Elmer, Waltham, MA, USA) (4000–500 cm^{-1}) scans were then performed, and the FT-IR chamber was carefully modified until water vapour peaks were eliminated [130].

4.5. Ligand Screening for Molecular Docking

This ligand screening is based on the metabolites identified through GC-MS analysis from various solvent extracts of groundnut shells and black gram pods from our previous study [56,57]. A total of 325 metabolites were identified, of which 14 compounds from groundnut shells and 20 compounds from black gram pods with potential antibacterial efficacy were chosen for this study based on the earlier literature to analyze their interaction with the aerolysin (Table 1). The chemical structure of each drug/compound was retrieved in structure-data file (SDF) format from the PubChem database (https://pubchem.ncbi.nlm.nih.gov/; accesed on 9 March 2021), and Open Babel was used to convert SDF to mol2 format [131].

4.6. Phylogenetic Analysis of Aerolysin

A multiple sequence alignment with the AhEUS112 aerolysin amino acid sequence was performed on the amino acid sequences obtained from 200 various bacterial species (*Aeromonas* sp. and other related bacteria). The evolutionary analysis was constructed with a neighbour-joining (NJ) algorithm using the MEGAX maximum likelihood method [98,132]. In the bootstrap test, the numbers next to the branches indicated the fraction of duplicate trees in which the related taxa were clustered together (100 repetitions) [133]. The tree was built using the maximum likelihood method and visualized using iTOL (Interactive Tree of Life) (https://itol.embl.de/; accessed on 24 April 2023). The phylogenetic tree's branch lengths were shown to scale and correspond to the evolutionary distances. The number of amino acid changes per site was used to calculate evolutionary distances using the P-distance approach [134]. All unclear places for each sequence pair were eliminated, leaving a final data set of 523 positions that were utilized for analysis.

4.7. Structural Analysis of Aerolysin

The AhEUS112 aerolysin's amino acid sequence was analyzed using BLAST-P to find the most appropriate template for homology modelling (accession no. MT491733) [135]. Following a similarity search for the best-aligned aerolysin crystal structures published in the Protein Data Bank (PDB), 1HWG (PDB ID) was selected for the modelling template. MODELLER software was used to construct the 3D model of the target sequence and structure validated with the Ramachandran plot using the SAVES server.

4.8. Molecular Docking and Simulation Dynamics

The potential ligands identified from GNS and BGP with aerolysin were used for molecular docking through AutoDock software [136,137]. The compounds with the highest binding affinity were chosen for the best-docked complexes. Using Webgro (online server), the modelled protein aerolysin was subjected to a molecular dynamics simulation. Using OPLS forcefield, MD simulation of both ligands and protein was analyzed for 100 ns [138,139].

5. Conclusions

This is a kind, sensible, and effective tactic in anti-virulence treatment, which involves employing different aerolysin inhibitors or substances that lead to preventing QS in bacterial pathogens such as *A. hydrophila*. Extracts from agricultural waste, such as groundnut shells and black gram pods, have been evident in SEM micrographs to block QS signals and disrupt the growth of biofilms. Additionally, the 3D structure of aerolysin has been generated, and it plays a major role in causing septicemia. Using an *in silico* technique in this study, H-Pyran-4-one-2,3 dihydro-3,5 dihydroxy-6-methyl and 2-Hexyldecanoic acid are shown to be more effective in inhibiting aerolysin oligomerization of *A. hydrophila*. This protein homology implies that a different potential pharmacological target could possibly work to restrict the activity of aerolysin in other pathogenic bacteria to form biofilms. It also provides novel insight that limits the broad usage of pharmaceutical drugs for *in vitro* testing. Thus, the agricultural waste extracts could be used as an appropriate medicine to block aerolysin activity by *A. hydrophila*, and they may aid in treating hemorrhagic septicemia. The outcome of this study enlightens the aquafarmer and the country's economy by overcoming the major disease outbreak in aquaculture by *A. hydrophila*.

Author Contributions: M.A.; investigation, methodology, formal analysis, writing—original draft, writing—review and editing. D.B.M.; investigation, resources, data curation. S.K.M.; formal analysis, data curation, writing—review and editing. G.M.; formal analysis, review and editing. Z.A.K.: formal analysis, review and editing. G.T.-I.; formal analysis, review and editing. T.R.; conceptualization, project administration, supervision, validation, visualization, writing—review and editing. All authors have read and agreed to the published version of the manuscript.

Funding: This research received no external funding. And the APC was funded by USDA-NIFA Sustainable Agriculture Systems, Grant No. 2019-69012-29905. Title of project: "Empowering US Broiler Production for Transformation and Sustainability USDA-NIFA (Sustainable Agriculture Systems)": No. 2019-69012-29905.

Institutional Review Board Statement: Not applicable.

Informed Consent Statement: Not applicable.

Data Availability Statement: Data will be made available on reasonable request.

Acknowledgments: The first author, Manikandan Arumugam, is grateful to Bharathidasan University for providing a University Research Fellowship (Ref. No. 026525/URF/DIR-RES/2020 dt: 04.01.2020). The authors are thankful to UGC-SAP-DRS-II (F.3-9/2013[SAP-II], the Department of Science and Technology-Fund for Improvement of Science and Technology Infrastructure (DST-FIST) Level-I (stage-II) (Ref. No. SR/FST/LSI-647/2015(C) Date.11.08.2016), and the Department of Science and Technology Promotion of University Research and Scientific Excellence (DST PURSE Phase II) (Ref. No. SR/PURSE PHASE 2/16(G)/& 16(C) Date. 21 February 2017) of the Department of Animal Science, Bharathidasan University, for the instrumentation facility. The authors also thank "RUSA, 2.0-Biological Sciences, Bharathidasan University" Tiruchirappalli.

Conflicts of Interest: The authors declared that there are no conflicts of interest.

References

1. Foysal, M.J.; Momtaz, F.; Ali, M.H.; Siddik, M.A.; Chaklader, M.R.; Rahman, M.M.; Prodhan, M.S.; Cole, A. Molecular characterization and interactome analysis of aerolysin (aer) gene from fish pathogen *Aeromonas veronii*: The pathogenicity inferred from sequence divergence and linked to histidine kinase (cheA). *J. Fish Dis.* **2019**, *42*, 465–475. [CrossRef] [PubMed]
2. Watts, J.E.; Schreier, H.J.; Lanska, L.; Hale, M.S. The rising tide of antimicrobial resistance in aquaculture: Sources, sinks and solutions. *Mar. Drugs* **2017**, *15*, 158. [CrossRef] [PubMed]
3. Bhat, R.A.; Rehman, S.; Tandel, R.S.; Dash, P.; Bhandari, A.; Ganie, P.A.; Shah, T.K.; Pant, K.; Yousuf, D.J.; Bhat, I.A.; et al. Immunomodulatory and Antimicrobial potential of ethanolic extract of Himalayan *Myrica esculanta* in *Oncorhynchus mykiss*: Molecular modelling with *Aeromonas hydrophila* functional proteins. *Aquaculture* **2021**, *533*, 736213. [CrossRef]
4. Tandel, R.S.; Dash, P.; Bhat, R.A.; Sharma, P.; Kalingapuram, K.; Dubey, M.; Sarma, D. Morphological and molecular characterization of *Saprolegnia spp.* from Himalayan snow trout, *Schizothorax richardsonii*: A case study report. *Aquaculture* **2021**, *531*, 735824. [CrossRef]
5. Jayasankar, P. Present status of freshwater aquaculture in India—A review. *Indian J. Fish.* **2018**, *65*, 157–165. [CrossRef]
6. Bagum, N.; Monir, M.S.; Khan, M.H. Present status of fish diseases and economic losses due to incidence of disease in rural freshwater aquaculture of Bangladesh. *J. Innov. Dev. Strategy* **2013**, *7*, 48–53.
7. Rasmussen-Ivey, C.R.; Figueras, M.J.; McGarey, D.; Liles, M.R. Virulence factors of *Aeromonas hydrophila*: In the wake of reclassification. *Front. Microbiol.* **2016**, *7*, 1337. [CrossRef]
8. Aguilera-Arreola, M.G.; Hernández-Rodríguez, C.; Zúñiga, G.; Figueras, M.J.; Castro-Escarpulli, G. *Aeromonas hydrophila* clinical and environmental ecotypes as revealed by genetic diversity and virulence genes. *FEMS Microbiol. Lett.* **2005**, *242*, 231–240. [CrossRef]
9. Pessoa, R.B.; de Oliveira, W.F.; Marques, D.S.; dos Santos Correia, M.T.; de Carvalho, E.V.; Coelho, L.C. The genus Aeromonas: A general approach. *Microb. Pathog.* **2019**, *130*, 81–94. [CrossRef]
10. Biscardi, D.; Castaldo, A.; Gualillo, O.; De Fusco, R. The occurrence of cytotoxic *Aeromonas hydrophila* strains in Italian mineral and thermal waters. *Sci. Total Environ.* **2002**, *292*, 255–263. [CrossRef]
11. Bücker, R.; Krug, S.M.; Rosenthal, R.; Günzel, D.; Fromm, A.; Zeitz, M.; Chakraborty, T.; Fromm, M.; Epple, H.J.; Schulzke, J.D. Aerolysin from *Aeromonas hydrophila* perturbs tight junction integrity and cell lesion repair in intestinal epithelial HT-29/B6 cells. *Int. J. Infect. Dis.* **2011**, *204*, 1283–1292. [CrossRef] [PubMed]
12. Ran, C.; Qin, C.; Xie, M.; Zhang, J.; Li, J.; Xie, Y.; Wang, Y.; Li, S.; Liu, L.; Fu, X.; et al. *Aeromonas veronii* and aerolysin are important for the pathogenesis of motile aeromonad septicemia in cyprinid fish. *Environ. Microbiol.* **2018**, *20*, 3442–3456. [CrossRef] [PubMed]
13. Zhang, L.; Ma, L.; Yang, Q.; Liu, Y.; Ai, X.; Dong, J. Sanguinarine Protects Channel Catfish against *Aeromonas hydrophila* Infection by Inhibiting Aerolysin and Biofilm Formation. *Pathogens* **2022**, *11*, 323. [CrossRef] [PubMed]
14. Banerji, R.; Karkee, A.; Kanojiya, P.; Saroj, S.D. Pore-forming toxins of foodborne pathogens. *Compr. Rev. Food Sci. Food Saf.* **2021**, *20*, 2265–2285. [CrossRef]
15. Wang, G.; Clark, C.G.; Liu, C.; Pucknell, C.; Munro, C.K.; Kruk, T.M.; Caldeira, R.; Woodward, D.L.; Rodgers, F.G. Detection and characterization of the hemolysin genes in *Aeromonas hydrophila* and *Aeromonas sobria* by multiplex PCR. *J. Clin. Microbiol.* **2003**, *41*, 1048–1054. [CrossRef] [PubMed]
16. Lata, K.; Singh, M.; Chatterjee, S.; Chattopadhyay, K. Membrane dynamics and remodelling in response to the action of the membrane-damaging pore-forming toxins. *J. Membr. Biol.* **2022**, *255*, 161–173. [CrossRef]
17. Kulma, M.; Anderluh, G. Beyond pore formation: Reorganization of the plasma membrane induced by pore-forming proteins. *Cell. Mol. Life Sci.* **2021**, *78*, 6229–6249. [CrossRef]
18. Van der Goot, F.G.; Pattus, F.; Wong, K.R.; Buckley, J.T. Oligomerization of the channel-forming toxin aerolysin precedes insertion into lipid bilayers. *Biochemistry* **1993**, *32*, 2636–2642. [CrossRef]
19. Dong, J.; Zhang, L.; Liu, Y.; Xu, N.; Zhou, S.; Yang, Y.; Yang, Q.; Ai, X. Luteolin decreases the pathogenicity of *Aeromonas hydrophila* via inhibiting the activity of aerolysin. *Virulence* **2021**, *12*, 165–176. [CrossRef]

20. Casabianca, A.; Orlandi, C.; Barbieri, F.; Sabatini, L.; Di Cesare, A.; Sisti, D.; Pasquaroli, S.; Magnani, M.; Citterio, B. Effect of starvation on survival and virulence expression of *Aeromonas hydrophila* from different sources. *Arch. Microbiol.* **2015**, *197*, 431–438. [CrossRef]
21. Klase, G.; Lee, S.; Liang, S.; Kim, J.; Zo, Y.G.; Lee, J. The microbiome and antibiotic resistance in integrated fishfarm water: Implications of environmental public health. *Sci. Total Environ.* **2019**, *649*, 1491–1501. [CrossRef]
22. Manyi-Loh, C.; Mamphweli, S.; Meyer, E.; Okoh, A. Antibiotic use in agriculture and its consequential resistance in environmental sources: Potential public health implications. *Molecules* **2018**, *23*, 795. [CrossRef] [PubMed]
23. Petit, F. Spread of antibiotic resistance in water: A public health and environmental issue. *Environ. Risques St.* **2018**, *17*, 40–46. [CrossRef]
24. Wu, J.; Liu, D.F.; Li, H.H.; Min, D.; Liu, J.Q.; Xu, P.; Li, W.W.; Yu, H.Q.; Zhu, Y.G. Controlling pathogenic risks of water treatment biotechnologies at the source by genetic editing means. *Environ. Microbiol.* **2021**, *23*, 7578–7590. [CrossRef]
25. Vanderhaeghen, W.; Dewulf, J. Antimicrobial use and resistance in animals and human beings. *Lancet Planet. Health* **2017**, *1*, e307–e308. [CrossRef] [PubMed]
26. McEwen, S.A. Human health importance of use of antimicrobials in animals and its selection of antimicrobial resistance. In *Antimicrobial Resistance in the Environment*; Wiley-Blackwell: Hoboken, NJ, USA, 2012; pp. 389–422.
27. Taylor, D.J. Antimicrobial use in animals and its consequences for human health. *Clin. Microbiol. Infect.* **1999**, *5*, 119–124. [CrossRef]
28. Jiang, X.; Ellabaan, M.M.; Charusanti, P.; Munck, C.; Blin, K.; Tong, Y.; Weber, T.; Sommer, M.O.; Lee, S.Y. Dissemination of antibiotic resistance genes from antibiotic producers to pathogens. *Nat. Commun.* **2017**, *8*, 15784. [CrossRef]
29. Peterson, E.; Kaur, P. Antibiotic resistance mechanisms in bacteria: Relationships between resistance determinants of antibiotic producers, environmental bacteria, and clinical pathogens. *Front. Microbiol.* **2018**, *9*, 2928. [CrossRef]
30. Monteiro, S.H.; Andrade, G.M.; Garcia, F.; Pilarski, F. Antibiotic residues and resistant bacteria in aquaculture. *Pharmaceut. Chem. J.* **2018**, *5*, 127–147.
31. Pepi, M.; Focardi, S. Antibiotic-resistant bacteria in aquaculture and climate change: A challenge for health in the Mediterranean Area. *Int. J. Environ. Res. Public Health* **2021**, *18*, 5723. [CrossRef]
32. Reverter, M.; Sarter, S.; Caruso, D.; Avarre, J.C.; Combe, M.; Pepey, E.; Pouyaud, L.; Vega-Heredía, S.; De Verdal, H.; Gozlan, R.E. Aquaculture at the crossroads of global warming and antimicrobial resistance. *Nat. Commun.* **2020**, *11*, 1870. [CrossRef] [PubMed]
33. Schar, D.; Klein, E.Y.; Laxminarayan, R.; Gilbert, M.; Van Boeckel, T.P. Global trends in antimicrobial use in aquaculture. *Sci. Rep.* **2020**, *10*, 21878. [CrossRef] [PubMed]
34. Natarajan, D.; Srinivasan, R.; Shivakumar, M.S. *Phyllanthus wightianus* Müll. Arg.: A potential source for natural antimicrobial agents. *BioMed Res. Int.* **2014**, *2014*, 135082. [CrossRef] [PubMed]
35. Termentzi, A.; Fokialakis, N.; Leandros Skaltsounis, A. Natural resins and bioactive natural products thereof as potential anitimicrobial agents. *Curr. Pharm. Des.* **2011**, *17*, 1267–1290. [CrossRef]
36. Makarewicz, M.; Drożdż, I.; Tarko, T.; Duda-Chodak, A. The Interactions between polyphenols and microorganisms, especially gut microbiota. *Antioxidants* **2021**, *10*, 188. [CrossRef] [PubMed]
37. Barbieri, R.; Coppo, E.; Marchese, A.; Daglia, M.; Sobarzo-Sánchez, E.; Nabavi, S.F.; Nabavi, S.M. Phytochemicals for human disease: An update on plant-derived compounds antibacterial activity. *Microbiol. Res.* **2017**, *196*, 44–68. [CrossRef]
38. Kiran, G.S.; Sajayan, A.; Priyadharshini, G.; Balakrishnan, A.; Prathiviraj, R.; Sabu, A.; Selvin, J. A novel anti-infective molecule nesfactin identified from sponge associated bacteria *Nesterenkonia* sp. MSA31 against multidrug resistant *Pseudomonas aeruginosa*. *Microb. Pathog.* **2021**, *157*, 104923. [CrossRef]
39. Rutherford, S.T.; Bassler, B.L. Bacterial quorum sensing: Its role in virulence and possibilities for its control. *Cold Spring Harb. Perspect. Med.* **2012**, *2*, a012427. [CrossRef]
40. Patel, B.; Kumari, S.; Banerjee, R.; Samanta, M.; Das, S. Disruption of the quorum sensing regulated pathogenic traits of the biofilm-forming fish pathogen *Aeromonas hydrophila* by tannic acid, a potent quorum quencher. *Biofouling* **2017**, *33*, 580–590. [CrossRef]
41. Lakshmanan, D.K.; Murugesan, S.; Rajendran, S.; Ravichandran, G.; Elangovan, A.; Raju, K.; Prathiviraj, R.; Pandiyan, R.; Thilagar, S. *Brassica juncea* (L.) Czern. leaves alleviate adjuvant-induced rheumatoid arthritis in rats via modulating the finest disease targets-IL2RA, IL18 and VEGFA. *J. Biomol. Struct Dyn.* **2021**, *40*, 8155–8168. [CrossRef]
42. Pinzi, L.; Rastelli, G. Molecular docking: Shifting paradigms in drug discovery. *Int. J. Mol. Sci.* **2019**, *20*, 4331. [CrossRef] [PubMed]
43. Torres, P.H.; Sodero, A.C.; Jofily, P.; Silva, F.P., Jr. Key topics in molecular docking for drug design. *Int. J. Mol. Sci.* **2019**, *20*, 4574. [CrossRef] [PubMed]
44. Murugan, A.; Prathiviraj, R.; Mothay, D.; Chellapandi, P. Substrate-imprinted docking of *Agrobacterium tumefaciens* uronate dehydrogenase for increased substrate selectivity. *Int. J. Biol. Macromol.* **2019**, *140*, 1214–1225. [CrossRef] [PubMed]
45. Vilar, S.; Sobarzo-Sanchez, E.; Santana, L.; Uriarte, E. Molecular docking and drug discovery in β-adrenergic receptors. *Curr. Med. Chem.* **2017**, *24*, 4340–4359. [CrossRef] [PubMed]
46. Muhammed, M.T.; Aki-Yalcin, E. Homology modeling in drug discovery: Overview, current applications, and future perspectives. *Chem. Biol. Drug Des.* **2019**, *93*, 12–20. [CrossRef]

47. Tasleem, M.; Alrehaily, A.; Almeleebia, T.M.; Alshahrani, M.Y.; Ahmad, I.; Asiri, M.; Alabdallah, N.M.; Saeed, M. Investigation of antidepressant properties of yohimbine by employing structure-based computational assessments. *Curr. Issues Mol. Biol.* **2021**, *43*, 1805–1827. [CrossRef]
48. Lovell, S.C.; Davis, I.W.; Arendall, W.B., III; De Bakker, P.I.; Word, J.M.; Prisant, M.G.; Richardson, J.S.; Richardson, D.C. Structure validation by Cα geometry: ϕ, ψ and Cβ deviation. *Proteins Struct. Funct. Genet.* **2003**, *50*, 437–450. [CrossRef]
49. Marimuthu, S.K.; Nagarajan, K.; Perumal, S.K.; Palanisamy, S.; Subbiah, L. In silico alpha-helical structural recognition of temporin antimicrobial peptides and its interactions with Middle East respiratory syndrome-coronavirus. *Int. J. Pept. Res. Ther.* **2020**, *26*, 1473–1483. [CrossRef]
50. Aliye, M.; Dekebo, A.; Tesso, H.; Abdo, T.; Eswaramoorthy, R.; Melaku, Y. Molecular docking analysis and evaluation of the antibacterial and antioxidant activities of the constituents of *Ocimum cufodontii*. *Sci. Rep.* **2021**, *11*, 10101. [CrossRef]
51. Mir, W.R.; Bhat, B.A.; Rather, M.A.; Muzamil, S.; Almilaibary, A.; Alkhanani, M.; Mir, M.A. Molecular docking analysis and evaluation of the antimicrobial properties of the constituents of *Geranium wallichianum* D. Don ex Sweet from Kashmir Himalaya. *Sci. Rep.* **2022**, *12*, 12547. [CrossRef]
52. Miralrio, A.; Espinoza Vázquez, A. Plant extracts as green corrosion inhibitors for different metal surfaces and corrosive media: A review. *Processes* **2020**, *8*, 942. [CrossRef]
53. Mushtaq, Z.; Khan, U.; Seher, N.; Shahid, M.; Shahzad, M.T.; Bhatti, A.A.; Sikander, T. Evaluation of antimicrobial, antioxidant and enzyme inhibition roles of polar and non-polar extracts of *Clitoria ternatea* seeds. *JAPS J. Anim. Plant Sci.* **2021**, *31*, 1405–1418. [CrossRef]
54. Belyagoubi-Benhammou, N.; Belyagoubi, L.; Gismondi, A.; Di Marco, G.; Canini, A.; Atik Bekkara, F. GC/MS analysis, and antioxidant and antimicrobial activities of alkaloids extracted by polar and apolar solvents from the stems of *Anabasis articulata*. *Med. Chem. Res.* **2019**, *28*, 754–767. [CrossRef]
55. Vaara, M. Lipopolysaccharide and the permeability of the bacterial outer membrane. In *Endotoxin in Health and Disease*; CRC Press: Boca Raton, FL, USA, 2020; pp. 31–38.
56. Arumugam, M.; Manikandan, D.B.; Sridhar, A.; Palaniyappan, S.; Jayaraman, S.; Ramasamy, T. GC–MS Based Metabolomics Strategy for Cost-Effective Valorization of Agricultural Waste: Groundnut Shell Extracts and Their Biological Inhibitory Potential. *Waste Biomass Valorization* **2022**, *13*, 4179–4209. [CrossRef]
57. Arumugam, M.; Manikandan, D.B.; Mohan, S.; Sridhar, A.; Veeran, S.; Jayaraman, S.; Ramasamy, T. Comprehensive metabolite profiling and therapeutic potential of black gram (*Vigna mungo*) pods: Conversion of biowaste to wealth approach. *Biomass Convers. Biorefin.* **2022**, 1–32. [CrossRef]
58. Jiang, Y.; Fang, Z.; Leonard, W.; Zhang, P. Phenolic compounds in Lycium berry: Composition, health benefits and industrial applications. *J. Funct. Foods* **2021**, *77*, 104340. [CrossRef]
59. Prasathkumar, M.; Raja, K.; Vasanth, K.; Khusro, A.; Sadhasivam, S.; Sahibzada, M.U.K.; Gawwad, M.R.A.; Al Farraj, D.A.; Elshikh, M.S. Phytochemical screening and *in vitro* antibacterial, antioxidant, anti-inflammatory, anti-diabetic, and wound healing attributes of *Senna auriculata* (L.) Roxb. leaves. *Arab. J. Chem.* **2021**, *14*, 103345. [CrossRef]
60. Alav, I.; Kobylka, J.; Kuth, M.S.; Pos, K.M.; Picard, M.; Blair, J.M.; Bavro, V.N. Structure, assembly, and function of tripartite efflux and type 1 secretion systems in gram-negative bacteria. *Chem. Rev.* **2021**, *121*, 5479–5596. [CrossRef]
61. Sandasi, M.; Leonard, C.M.; Viljoen, A.M. The *in vitro* antibiofilm activity of selected culinary herbs and medicinal plants against *Listeria monocytogenes*. *Lett. Appl. Microbiol.* **2010**, *50*, 30–35. [CrossRef]
62. Ghezzal, S.; Postal, B.G.; Quevrain, E.; Brot, L.; Seksik, P.; Leturque, A.; Thenet, S.; Carriere, V. Palmitic acid damages gut epithelium integrity and initiates inflammatory cytokine production. *Biochim. Biophys. Acta BBA Mol. Cell Biol. Lipids* **2020**, *1865*, 158530. [CrossRef]
63. El-anssary, A.A.; Raoof, G.F.; Saleh, D.O.; El-Masry, H.M. Bioactivities, physicochemical parameters and GC/MS profiling of the fixed oil of *Cucumis melo* L seeds: A focus on anti-inflammatory, immunomodulatory, and antimicrobial activities. *J. Herb. Med. Pharmacol.* **2021**, *10*, 476–485. [CrossRef]
64. El-Benawy, N.M.; Abdel-Fattah, G.M.; Ghoneem, K.M.; Shabana, Y.M. Antimicrobial activities of *Trichoderma atroviride* against common bean seed-borne *Macrophomina phaseolina* and *Rhizoctonia solani*. *Egypt. J. Basic Appl. Sci.* **2020**, *7*, 267–280. [CrossRef]
65. Lin, H.; Meng, L.; Sun, Z.; Sun, S.; Huang, X.; Lin, N.; Zhang, J.; Lu, W.; Yang, Q.; Chi, J.; et al. Yellow wine polyphenolic compound protects against doxorubicin-induced cardiotoxicity by modulating the composition and metabolic function of the gut microbiota. *Circ. Heart Fail.* **2021**, *14*, e008220. [CrossRef] [PubMed]
66. Asghar, M.A.; Asghar, M.A. Green synthesized and characterized copper nanoparticles using various new plants extracts aggravate microbial cell membrane damage after interaction with lipopolysaccharide. *Int. J. Biol. Macromol.* **2020**, *160*, 1168–1176. [CrossRef] [PubMed]
67. Famuyide, I.M.; Aro, A.O.; Fasina, F.O.; Eloff, J.N.; McGaw, L.J. Antibacterial and antibiofilm activity of acetone leaf extracts of nine under-investigated south African Eugenia and Syzygium (Myrtaceae) species and their selectivity indices. *BMC Complement. Altern. Med.* **2019**, *19*, 141. [CrossRef]
68. Lynch, M.J.; Swift, S.; Kirke, D.F.; Keevil, C.W.; Dodd, C.E.; Williams, P. The regulation of biofilm development by quorum sensing in *Aeromonas hydrophila*. *Environ. Microbiol.* **2008**, *4*, 18–28. [CrossRef]

69. Defoirdt, T.; Bossier, P.; Sorgeloos, P.; Verstraete, W. The impact of mutations in the quorum sensing systems of *Aeromonas hydrophila, Vibrio anguillarum* and *Vibrio harveyi* on their virulence towards gnotobiotically cultured *Artemia franciscana*. *Environ. Microbiol.* **2005**, *7*, 1239–1247. [CrossRef]
70. Sun, B.; Luo, H.; Jiang, H.; Wang, Z.; Jia, A. Inhibition of Quorum Sensing and Biofilm Formation of Esculetin on *Aeromonas hydrophila*. *Front. Microbiol.* **2021**, *12*, 737626. [CrossRef]
71. Ramanathan, S.; Ravindran, D.; Arunachalam, K.; Arumugam, V.R. Inhibition of quorum sensing-dependent biofilm and virulence genes expression in environmental pathogen Serratia marcescens by petroselinic acid. *Antonie Van Leeuwenhoek* **2018**, *111*, 501–515. [CrossRef]
72. Shivaprasad, D.P.; Taneja, N.K.; Lakra, A.; Sachdev, D. *In vitro* and in situ abrogation of biofilm formation in *E. coli* by vitamin C through ROS generation, disruption of quorum sensing and exopolysaccharide production. *Food Chem.* **2021**, *341*, 128171. [CrossRef]
73. Singh, S.; Datta, S.; Narayanan, K.B.; Rajnish, K.N. Bacterial exo-polysaccharides in biofilms: Role in antimicrobial resistance and treatments. *J. Genet. Eng. Biotechnol.* **2021**, *19*, 140. [CrossRef]
74. Geng, Y.F.; Yang, C.; Zhang, Y.; Tao, S.N.; Mei, J.; Zhang, X.C.; Sun, Y.J.; Zhao, B.T. An innovative role for luteolin as a natural quorum sensing inhibitor in *Pseudomonas aeruginosa*. *Life Sci.* **2021**, *274*, 119325. [CrossRef] [PubMed]
75. Pellock, B.J.; Teplitski, M.; Boinay, R.P.; Bauer, W.D.; Walker, G.C. A LuxR homolog controls production of symbiotically active extracellular polysaccharide II by *Sinorhizobium meliloti*. *J. Bacteriol. Res.* **2002**, *184*, 5067–5076. [CrossRef]
76. Sahreen, S.; Mukhtar, H.; Imre, K.; Morar, A.; Herman, V.; Sharif, S. Exploring the function of quorum sensing regulated biofilms in biological wastewater treatment: A review. *Int. J. Mol. Sci.* **2022**, *23*, 9751. [CrossRef] [PubMed]
77. Li, Y.H.; Tian, X. Quorum sensing and bacterial social interactions in biofilms. *Sensors* **2012**, *12*, 1195–1205. [CrossRef] [PubMed]
78. Papenfort, K.; Bassler, B.L. Quorum sensing signal–response systems in Gram-negative bacteria. *Nat. Rev. Microbiol.* **2016**, *14*, 576–588. [CrossRef]
79. Zhou, J.; Lin, Z.J.; Cai, Z.H.; Zeng, Y.H.; Zhu, J.M.; Du, X.P. Opportunistic bacteria use quorum sensing to disturb coral symbiotic communities and mediate the occurrence of coral bleaching. *Environ. Microbiol.* **2020**, *22*, 1944–1962. [CrossRef]
80. Shaaban, M.; Elgaml, A.; Habib, E.S.E. Biotechnological applications of quorum sensing inhibition as novel therapeutic strategies for multidrug resistant pathogens. *Microb. Pathog.* **2019**, *127*, 138–143. [CrossRef]
81. Silpe, J.E.; Bassler, B.L. Phage-encoded LuxR-type receptors responsive to host-produced bacterial quorum-sensing autoinducers. *MBio* **2019**, *10*, e00638-19. [CrossRef] [PubMed]
82. Kumar, P.; Lee, J.H.; Beyenal, H.; Lee, J. Fatty acids as antibiofilm and antivirulence agents. *Trends Microbiol.* **2020**, *28*, 753–768. [CrossRef]
83. Defoirdt, T. Quorum-sensing systems as targets for antivirulence therapy. *Trends Microbiol.* **2018**, *26*, 313–328. [CrossRef] [PubMed]
84. Mayers, J.J.; Flynn, K.J.; Shields, R.J. Rapid determination of bulk microalgal biochemical composition by Fourier-Transform Infrared spectroscopy. *Bioresour. Technol.* **2013**, *148*, 215–220. [CrossRef]
85. Pradhan, S.; Nautiyal, V.; Dubey, R.C. Antioxidant potential and molecular docking of bioactive compound of *Camellia sinensis* and *Camellia assamica* with cytochrome P450. *Arch. Microbiol.* **2022**, *204*, 350. [CrossRef]
86. Yuen, C.W.; Ku, S.K.; Choi, P.S.; Kan, C.W.; Tsang, S.Y. Determining functional groups of commercially available ink-jet printing reactive dyes using infrared spectroscopy. *Res. J. Text. Appar.* **2005**, *9*, 26–38. [CrossRef]
87. Pradhan, J.; Das, S.; Das, B.K. Antibacterial activity of freshwater microalgae: A review. *Afr. J. Pharm. Pharmacol.* **2014**, *8*, 809–818. [CrossRef]
88. Ayangbenro, A.S.; Babalola, O.O. A new strategy for heavy metal polluted environments: A review of microbial biosorbents. *Int. J. Environ. Res. Public Health* **2017**, *14*, 94. [CrossRef]
89. Knight, V.; Blakemore, R. Reduction of diverse electron acceptors by *Aeromonas hydrophila*. *Arch. Microbiol.* **1998**, *169*, 239–248. [CrossRef]
90. Castro, L.; Vera, M.; Muñoz, J.Á.; Blázquez, M.L.; González, F.; Sand, W.; Ballester, A. *Aeromonas hydrophila* produces conductive nanowires. *Res. Microbiol.* **2014**, *165*, 794–802. [CrossRef] [PubMed]
91. Woźnica, A.; Dzirba, J.; Mańka, D.; Łabużek, S. Effects of electron transport inhibitors on iron reduction in *Aeromonas hydrophila* strain KB1. *Anaerobe* **2003**, *9*, 125–130. [CrossRef] [PubMed]
92. Holmes, D.E.; Rotaru, A.E.; Ueki, T.; Shrestha, P.M.; Ferry, J.G.; Lovely, D.R. Electron and proton flux for carbon dioxide reduction in *Methanosarcina barkeri* during direct interspecies electron transfer. *Front. Microbiol.* **2018**, *9*, 3109. [CrossRef]
93. Abu-Ghannam, N.; Rajauria, G. Antimicrobial activity of compounds isolated from algae. In *Functional Ingredients from Algae for Foods and Nutraceuticals*; Woodhead Publishing: Sawston, UK, 2013; pp. 287–306. [CrossRef]
94. Dyrda, G.; Boniewska-Bernacka, E.; Man, D.; Barchiewicz, K.; Słota, R. The effect of organic solvents on selected microorganisms and model liposome membrane. *Mol. Biol. Rep.* **2019**, *46*, 3225–3232. [CrossRef]
95. Górniak, I.; Bartoszewski, R.; Króliczewski, J. Comprehensive review of antimicrobial activities of plant flavonoids. *Phytochem. Rev.* **2019**, *18*, 241–272. [CrossRef]
96. Zhang, L.L.; Zhang, L.F.; Xu, J.G. Chemical composition, antibacterial activity and action mechanism of different extracts from hawthorn (*Crataegus pinnatifida* Bge.). *Sci. Rep.* **2020**, *10*, 8876. [CrossRef]
97. Srinivasan, R.; Devi, K.R.; Santhakumari, S.; Kannappan, A.; Chen, X.; Ravi, A.V.; Lin, X. Anti-quorum sensing and protective efficacies of naringin against *Aeromonas hydrophila* infection in *Danio rerio*. *Front. Microbiol.* **2020**, *11*, 600622. [CrossRef] [PubMed]

98. Saitou, N.; Nei, M. The neighbor-joining method: A new method for reconstructing phylogenetic trees. *Mol. Biol. Evol.* **1987**, *4*, 406–425. [CrossRef]
99. Hossain, S.; De Silva, B.C.; Dahanayake, P.S.; De Zoysa, M.; Heo, G.J. Phylogenetic characteristics, virulence properties and antibiogram profile of motile *Aeromonas spp.* isolated from ornamental guppy (*Poecilia reticulata*). *Arch. Microbiol.* **2020**, *202*, 501–509. [CrossRef]
100. Young, A.D.; Gillung, J.P. Phylogenomics—Principles, opportunities and pitfalls of big-data phylogenetics. *Syst. Entomol.* **2020**, *45*, 225–247. [CrossRef]
101. Du, X.; Li, Y.; Xia, Y.L.; Ai, S.M.; Liang, J.; Sang, P.; Ji, X.L.; Liu, S.Q. Insights into protein–ligand interactions: Mechanisms, models, and methods. *Int. J. Mol. Sci.* **2016**, *17*, 144. [CrossRef]
102. Brylinski, M. Aromatic interactions at the ligand–protein interface: Implications for the development of docking scoring functions. *Chem. Biol. Drug Des.* **2018**, *91*, 380–390. [CrossRef] [PubMed]
103. Abriata, L.A.; Bovigny, C.; Dal Peraro, M. Detection and sequence/structure mapping of biophysical constraints to protein variation in saturated mutational libraries and protein sequence alignments with a dedicated server. *BMC Bioinform.* **2016**, *17*, 242. [CrossRef]
104. Ferreira, L.G.; Dos Santos, R.N.; Oliva, G.; Andricopulo, A.D. Molecular docking and structure-based drug design strategies. *Molecules* **2015**, *20*, 13384–13421. [CrossRef]
105. Harigua-Souiai, E.; Cortes-Ciriano, I.; Desdouits, N.; Malliavin, T.E.; Guizani, I.; Nilges, M.; Blondel, A.; Bouvier, G. Identification of binding sites and favorable ligand binding moieties by virtual screening and self-organizing map analysis. *BMC Bioinform.* **2015**, *16*, 93. [CrossRef]
106. Ruepp, M.D.; Wei, H.; Leuenberger, M.; Lochner, M.; Thompson, A.J. The binding orientations of structurally-related ligands can differ; A cautionary note. *Neuropharmacology* **2017**, *119*, 48–61. [CrossRef]
107. Gao, M.; Skolnick, J. The distribution of ligand-binding pockets around protein-protein interfaces suggests a general mechanism for pocket formation. *Proc. Natl. Acad. Sci. USA* **2012**, *109*, 3784–3789. [CrossRef]
108. Stank, A.; Kokh, D.B.; Fuller, J.C.; Wade, R.C. Protein binding pocket dynamics. *Acc. Chem. Res.* **2016**, *49*, 809–815. [CrossRef]
109. MacKenzie, C.R.; Hirama, T.; Buckley, J.T. Analysis of receptor binding by the channel-forming toxin aerolysin using surface plasmon resonance. *J. Appl. Biol. Chem.* **1999**, *274*, 22604–22609. [CrossRef]
110. Podobnik, M.; Kisovec, M.; Anderluh, G. Molecular mechanism of pore formation by aerolysin-like proteins. *Philos. Trans. R. Soc. Lond. B Biol. Sci.* **2017**, *372*, 20160209. [CrossRef] [PubMed]
111. Chakraborty, N.; Das, B.K.; Das, A.K.; Manna, R.K.; Chakraborty, H.J.; Mandal, B.; Bhattacharjya, B.K.; Raut, S.S. Antibacterial prophylaxis and molecular docking studies of ketone and ester compounds isolated from *Cyperus rotundus* L. against *Aeromonas veronii*. *Aqua. Res.* **2022**, *53*, 1363–1377. [CrossRef]
112. Yang, L.; Wei, Z.; Li, S.; Xiao, R.; Xu, Q.; Ran, Y.; Ding, W. Plant secondary metabolite, daphnetin reduces extracellular polysaccharides production and virulence factors of *Ralstonia solanacearum*. *Pestic. Biochem. Physiol.* **2021**, *179*, 104948. [CrossRef] [PubMed]
113. Garba, L.; Yussoff, M.A.; Abd Halim, K.B.; Ishak, S.N.; Ali, M.S.; Oslan, S.N.; Rahman, R.N. Homology modeling and docking studies of a Δ9-fatty acid desaturase from a Cold-tolerant *Pseudomonas* sp. AMS8. *PeerJ* **2018**, *6*, e4347. [CrossRef] [PubMed]
114. Selvaraj, A.; Valliammai, A.; Premika, M. *Sapindus mukorossi* Gaertn. and its bioactive metabolite oleic acid impedes methicillin-resistant *Staphylococcus aureus* biofilm formation by down regulating adhesion genes expression. *Microbiol. Res.* **2021**, *242*, 126601. [CrossRef]
115. Dong, J.; Liu, Y.; Xu, N.; Yang, Q.; Ai, X. Morin protects channel catfish from *Aeromonas hydrophila* infection by blocking aerolysin activity. *Front. Microbiol.* **2018**, *9*, 2828. [CrossRef] [PubMed]
116. Knapp, O.; Stiles, B.; Popoff, M. The aerolysin-like toxin family of cytolytic, pore-forming toxins. *Toxicol. Open Access* **2010**, *3*, 53–68. [CrossRef]
117. Li, Y.; Li, Y.; Mengist, H.M.; Shi, C.; Zhang, C.; Wang, B.; Li, T.; Huang, Y.; Xu, Y.; Jin, T. Structural basis of the pore-forming toxin/membrane interaction. *Toxins* **2021**, *13*, 128. [CrossRef]
118. Lobanov, M.Y.; Bogatyreva, N.S.; Galzitskaya, O.V. Radius of gyration as an indicator of protein structure compactness. *Mol. Biol.* **2008**, *42*, 623–628. [CrossRef]
119. Salo-Ahen, O.M.; Alanko, I.; Bhadane, R.; Bonvin, A.M.; Honorato, R.V.; Hossain, S.; Juffer, A.H.; Kabedev, A.; Lahtela-Kakkonen, M.; Larsen, A.S.; et al. Molecular dynamics simulations in drug discovery and pharmaceutical development. *Processes* **2020**, *9*, 71. [CrossRef]
120. Moharana, M.; Das, A.; Sahu, S.N.; Pattanayak, S.K.; Khan, F. Evaluation of binding performance of bioactive compounds against main protease and mutant model spike receptor binding domain of SARS-CoV-2: Docking, ADMET properties and molecular dynamics simulation study. *J. Indian Chem. Soc.* **2022**, *99*, 100417. [CrossRef]
121. Bondos, S.E. Methods for measuring protein aggregation. *Curr. Anal. Chem.* **2006**, *2*, 157–170. [CrossRef]
122. Sorokina, I.; Mushegian, A.R.; Koonin, E.V. Is protein folding a thermodynamically unfavorable, active, energy-dependent process? *Int. J. Mol. Sci.* **2022**, *23*, 521. [CrossRef]
123. Degiacomi, M.T.; Iacovache, I.; Pernot, L.; Chami, M.; Kudryashev, M.; Stahlberg, H.; Van Der Goot, F.G.; Dal Peraro, M. Molecular assembly of the aerolysin pore reveals a swirling membrane-insertion mechanism. *Nat. Chem. Biol.* **2013**, *9*, 623–629. [CrossRef]

124. Fuglebakk, E.; Echave, J.; Reuter, N. Measuring and comparing structural fluctuation patterns in large protein datasets. *Bioinformatics* **2012**, *28*, 2431–2440. [CrossRef]
125. Loschwitz, J.; Olubiyi, O.O.; Hub, J.S.; Strodel, B.; Poojari, C.S. Computer simulations of protein–membrane systems. *Prog. Mol. Biol. Transl. Sci.* **2020**, *170*, 273–403. [CrossRef] [PubMed]
126. Abubakar, A.R.; Haque, M. Preparation of medicinal plants: Basic extraction and fractionation procedures for experimental purposes. *J. Pharm. Bioallied. Sci.* **2020**, *12*, 1–10. [CrossRef]
127. Tanhaeian, A.; Nazifi, N.; Shahriari Ahmadi, F.; Akhlaghi, M. Comparative study of antimicrobial activity between some medicine plants and recombinant Lactoferrin peptide against some pathogens of cultivated button mushroom. *Arch. Microbiol.* **2020**, *202*, 2525–2532. [CrossRef]
128. Zhou, J.; Bi, S.; Chen, H.; Chen, T.; Yang, R.; Li, M.; Fu, Y.; Jia, A.Q. Anti-biofilm and antivirulence activities of metabolites from *Plectosphaerella cucumerina* against *Pseudomonas aeruginosa*. *Front. Microbiol.* **2017**, *8*, 769. [CrossRef]
129. Kamnev, A.A.; Dyatlova, Y.A.; Kenzhegulov, O.A.; Vladimirova, A.A.; Mamchenkova, P.V.; Tugarova, A.V. Fourier transform infrared (FTIR) spectroscopic analyses of microbiological samples and biogenic selenium nanoparticles of microbial origin: Sample preparation effects. *Molecules* **2021**, *26*, 1146. [CrossRef] [PubMed]
130. Fung-Khee, F. FTIR Analysis of Bacteria Biomass-Mineral Interactions in Soils. 2020. Available online: https://academicworks.cuny.edu/cgi/viewcontent.cgi?article=1934&context=cc_etds_theses (accessed on 1 August 2020).
131. O'Boyle, N.M.; Banck, M.; James, C.A.; Morley, C.; Vandermeersch, T.; Hutchison, G.R. Open Babel: An open chemical toolbox. *J. Cheminform.* **2011**, *3*, 33. [CrossRef]
132. Kumar, S.; Stecher, G.; Li, M.; Knyaz, C.; Tamura, K. MEGA X: Molecular evolutionary genetics analysis across computing platforms. *Mol. Biol. Evol.* **2018**, *35*, 1547. [CrossRef]
133. Felsenstein, J. Confidence limits on phylogenies: An approach using the bootstrap. *Evolution* **1985**, *39*, 783–791. [CrossRef]
134. Nei, M.; Kumar, S. *Molecular Evolution and Phylogenetics*; Oxford University Press: New York, NY, USA, 2000.
135. Altschul, S.F.; Madden, T.L.; Schäffer, A.A.; Zhang, J.; Zhang, Z.; Miller, W.; Lipman, D.J. Gapped BLAST and PSI-BLAST: A new generation of protein database search programs. *Nucleic Acids Res.* **1997**, *25*, 3389–3402. [CrossRef] [PubMed]
136. Forli, S.; Huey, R.; Pique, M.E.; Sanner, M.F.; Goodsell, D.S.; Olson, A.J. Computational protein–ligand docking and virtual drug screening with the AutoDock suite. *Nat. Protoc.* **2016**, *11*, 905–919. [CrossRef] [PubMed]
137. Morris, G.M.; Huey, R.; Lindstrom, W.; Sanner, M.F.; Belew, R.K.; Goodsell, D.S.; Olson, A.J. AutoDock4 and AutoDockTools4: Automated docking with selective receptor flexibility. *J. Comput. Chem.* **2009**, *30*, 2785–2791. [CrossRef] [PubMed]
138. Idoko, V.O.; Sulaiman, M.A.; Adamu, R.M.; Abdullahi, A.D.; Tajuddeen, N.; Mohammed, A.; Inuwa, H.M.; Ibrahim, M.A. Evaluating Khaya senegalensis for dipeptidyl peptidase–IV inhibition using *in vitro* analysis and molecular dynamic simulation of identified bioactive compounds. *Chem. Biodivers.* **2022**, *20*, e202200909. [CrossRef]
139. Singh, M.B.; Vishvakarma, V.K.; Lal, A.A.; Chandra, R.; Jain, P.; Singh, P.A. A comparative study of 5-fluorouracil, doxorubicin, methotrexate, paclitaxel for their inhibition ability for Mpro of nCoV: Molecular docking and molecular dynamics simulations. *J. Indian Chem. Soc.* **2022**, *99*, 100790. [CrossRef]

antibiotics

Article

Synergistic Antimicrobial Activity of Ceftriaxone and *Polyalthia longifolia* Methanol (MEPL) Leaf Extract against Methicillin-Resistant *Staphylococcus aureus* and Modulation of *mecA* Gene Presence

Valiappan Ranjutha [1], Yeng Chen [2], Lamya Ahmed Al-Keridis [3,*], Mitesh Patel [4], Nawaf Alshammari [5], Mohd Adnan [5], Sumaira Sahreen [1], Subash C. B. Gopinath [6,7,8] and Sreenivasan Sasidharan [1,*]

1 Institute for Research in Molecular Medicine (INFORMM), Universiti Sains Malaysia (USM), Gelugor 11800, Pulau Pinang, Malaysia
2 Department of Oral & Craniofacial Sciences, Faculty of Dentistry, University of Malaya, Kuala Lumpur 50603, Selangor, Malaysia
3 Department of Biology, College of Science, Princess Nourah bint Abdulrahman University, P.O. Box 84428, Riyadh 11671, Saudi Arabia
4 Department of Biotechnology, Parul Institute of Applied Sciences and Centre of Research for Development, Parul University, Vadodara 391760, India
5 Department of Biology, College of Science, University of Hail, P.O. Box 2440, Hail 81451, Saudi Arabia
6 Faculty of Chemical Engineering & Technology, Universiti Malaysia Perlis (UniMAP), Arau 02600, Perlis, Malaysia
7 Institute of Nano Electronic Engineering, Universiti Malaysia Perlis (UniMAP), Kangar 01000, Perlis, Malaysia
8 Micro System Technology, Centre of Excellence (CoE), Universiti Malaysia Perlis (UniMAP), Pauh Campus, Arau 02600, Perlis, Malaysia
* Correspondence: laalkeridis@pnu.edu.sa (L.A.A.-K.); srisasidharan@yahoo.com (S.S.)

Citation: Ranjutha, V.; Chen, Y.; Al-Keridis, L.A.; Patel, M.; Alshammari, N.; Adnan, M.; Sahreen, S.; Gopinath, S.C.B.; Sasidharan, S. Synergistic Antimicrobial Activity of Ceftriaxone and *Polyalthia longifolia* Methanol (MEPL) Leaf Extract against Methicillin-Resistant *Staphylococcus aureus* and Modulation of *mecA* Gene Presence. *Antibiotics* **2023**, *12*, 477. https://doi.org/10.3390/antibiotics12030477

Academic Editor: William N. Setzer

Received: 10 January 2023
Revised: 22 February 2023
Accepted: 23 February 2023
Published: 27 February 2023

Abstract: Medicinal plants are an essential source of traditional curatives for numerous skin diseases. *Polyalthia longifolia* (Sonn.) Thwaites (Annonaceae family) *is a medicinal plant* used to cure skin illnesses. *P. longifolia* is usually applied in folkloric therapeutical systems to treat skin diseases. The methicillin-resistant *Staphylococcus aureus* (MRSA) bacteria is among the essential bacteria contributing to skin diseases. Hence, to verify the traditional medicinal claim of *P. longifolia* usage in skin disease treatment, the current research was performed to study the synergistic antibacterial activity of standardized *Polyalthia longifolia* methanol leaf extract (MEPL) against MRSA bacteria. The synergistic antimicrobial activity result of ceftriaxone, when mixed with MEPL, against MRSA was investigated by the disc diffusion method, broth microdilution method, checkerboard dilution test, and modulation of *mecA* gene expression by multiplex polymerase chain reaction (multiplex PCR). The MEPL extract exhibited good *synergistic* antimicrobial activity against MRSA. Using the checkerboard method, we confirmed the synergistic effect of MEPL from *P. longifolia* and ceftriaxone (2:1) for MRSA with a marked reduction of the MIC value of the ceftriaxone from 8000 μg/mL to 1000 μg/mL. Moreover, the combination of MEPL with ceftriaxone significantly ($p < 0.05$) inhibited the presence of the resistant *mecA* gene in the tested strain. The LC–ESI–MS/MS analysis identified compounds that were reported to exhibit antimicrobial activity. Conclusively, the MEPL extract, an important etiological agent for skin diseases, showed worthy synergistic antimicrobial action against MRSA bacteria, thus supporting the traditional use of *P. longifolia*.

Keywords: ceftriaxone; methicillin-resistant *Staphylococcus aureus*; *Polyalthia longifolia*; synergistic effect; *mecA* gene; multiplex PCR; gene expression; skin diseases

1. Introduction

Microbial infectious diseases have become the third most crucial reason for mortality and morbidity worldwide. The contagion instigated by methicillin-resistant *Staphylococcus aureus* (MRSA) has contributed significantly to deadly infections and diseases [1]. There is growing proof that *S. aureus* is becoming resistant to all the standard antibiotics. Ceftriaxone belongs to a class of drugs identified as cephalosporin antibiotics and is extensively used to treat resistant bacterial strains, including *S. aureus* infection [2]. Nevertheless, disturbingly, the emergence of ceftriaxone-resistance MRSA bacteria was reported in the literature [3]. Moreover, a genetic mutation was involved in the development of resistance to the antibiotic. The attainment of the *mecA* gene by horizontal transmission by conjugation was the leading cause of antibiotic resistance in *S. aureus* [4]. This important *mecA* gene has contributed to methicillin resistance in *S. aureus* strains, which encodes a novel penicillin-binding protein 2A (PBP2A) [5]. Therefore, new alternative strategies are needed to address this issue by developing new antimicrobial agents, modifying the existing antibiotic activity with a combination of plant extracts as resistance modifying agents, or using the plant extract combined with existing antibiotics against resistant bacteria to suppress the expression of the *mecA* gene in MRSA bacteria. Consequently, the increasing incidence of MRSA bacterial infection has drawn the pharmaceutical and scientific community's attention to studies on the potential antimicrobial activity of plant-derived substances used in traditional medicine in different countries. Scientists from divergent fields are investigating medicinal plants regarding their antimicrobial usefulness. Hence, the development of a new antibacterial against MRSA is of crucial importance.

Consequently, the search for drugs derived from medicinal plants by scientists has accelerated in recent years worldwide. The medicinal plant, a famous healthcare agent, is used daily by billions of people globally for their primary healthcare. The medicinal plant was considered a panacea with various curative values in traditional medicine, including anti-infectious activity. One crucial medicinal plant with multiple curative values is *Polyalthia longifolia* var. angustifolia Thw. (Annonaceae). *P. longifolia* is a medicinal plant with linear–lanceolate leaves found in Sri Lanka, India's tropical parts, and Malaysia. This tree is normally planted along roadsides and gardens due to its beautiful appearance. *P. longifolia* is one of the most important traditional indigenous medicinal plants commonly used in folk medicine to treat skin diseases, fever, hypertension, helminthiasis, and diabetes [6]. The MRSA bacteria is also one of the important bacteria contributing to skin and soft tissue infection [7], which leads to major illness and death [8], comprising endocarditis, septic shock, bacteremia, and pneumonia [9]. Hence, to verify the traditional medicinal practitioner's claims on the contribution of *P. longifolia* to skin disease treatment, the present research studied the synergistic antimicrobial action of *P. longifolia* leaf extract and ceftriaxone antibiotic against MRSA bacteria.

Until 2019, there was limited experimental evidence of the synergistic activity between *P. longifolia* leaf extract and synthetic antibiotics against MRSA. Previous experiments have demonstrated the in vitro interaction of ampicillin and *P. longifolia* leaf ethyl acetate fraction (PLEAF) by checkerboard and microscopic techniques against MRSA [10,11]. That previous study showed that the PLEAF fraction worked synergistically with ampicillin to kill MRSA's local resistance strain. Moreover, the PLEAF fraction also exhibited excellent antioxidant activity. The combination of the PLEAF fraction with ampicillin also increased Vero cell viability. This critical finding showed the non-toxic nature of ampicillin in the presence of PLEAF in combinational therapy. Further study was also conducted to observe the in situ synergistic antimicrobial effects between PLEAF and ampicillin against a local MRSA isolate using modern scanning electron microscopy (SEM) observation [11]. PLEAF and ampicillin combination exhibited significant antibacterial activity against MRSA by killing the resistant MRSA bacteria, as observed via SEM analysis. However, as a further study in understanding multidrug-resistant bacteria's challenges, *P. longifolia* leaf extract antibacterial activity, antibiotic modifying activity, and mutagenic effects combined with different first-line antibiotics commonly used against infectious agents should be investigated. Investigating the synergistic antimicrobial effects of the *P. longifolia* leaf methanolic extract combined with β-lactam antibiotics, such as ceftriaxone, will enhance the understanding of the synergistic antimicrobial effects of *P. longifolia* leaf extract, which has never been studied in detail before. In addition, the synergistic effect of ceftriaxone and *P. longifolia* methanol leaf extract in combination against MRSA bacteria and the *mecA* gene is still unclear, and few studies were conducted in this line. Therefore, the objective of the current research was to study the action of MEPL from *P. longifolia* on the regulation of *mecA* gene presence in the MRSA strain and study the synergistic effect of ceftriaxone and MEPL in this bacterium.

2. Results

2.1. Ceftriaxone and MEPL Antibacterial Activity against MRSA Isolates

Antimicrobial susceptibility of MRSA isolates shows complete resistance to the standard dosage strengths (8 µg/mL, 16 µg/mL, 32 µg/mL, and 64 µg/mL) of ceftriaxone, and no diameter of zone of inhibition was produced by all the different ceftriaxone dosages tested in this study (Table 1). Conversely, the tested MEPL exhibited significant antibacterial activity against MRSA by producing a clear zone of inhibition between 21 mm and 34 mm (Table 1). The negative control 5% dimethyl sulfoxide (DMSO) did not produce any zone of inhibition.

Table 1. Antimicrobial activity of MEPL against MRSA.

Concentration of Ceftriaxone (µg/mL)	Diameter of Zone of Inhibition (mm)	Concentration of MEPL (mg/mL)	Diameter of Zone of Inhibition (mm)
8	0	1	21 ± 2
16	0	2	24 ± 1
32	0	3	26 ± 2
64	0	4	28 ± 2
		5	29 ± 2
		6	31 ± 1
		7	32 ± 1
		8	34 ± 1

2.2. Determination of the MIC and MBC Concentration of Ceftriaxone and MEPL against the MRSA Isolate

The antibiotic MIC value is an essential aid in evaluating bacterial resistance. According to 2022 CLSI interpretive measures, MRSA is susceptible to ceftriaxone when the MIC value is ≤8 µg/mL, and MRSA is susceptible to ceftriaxone with a MIC value of 32 µg/mL. The MIC of ceftriaxone was obtained using the broth dilution method, and the ceftriaxone MIC value was 8000 µg/mL, visibly inhibiting MRSA growth in the broth. While the ceftriaxone MBC value, where the lowest concentration showed zero growth on sterile NA, was found at 8000 µg/mL. The MIC result demonstrated MRSA growth in a concentration of ≥62.5 µg/mL (the breaking point of ceftriaxone is ≤16 to ≥64 µg/mL). This proves that the MRSA used in this study was highly resistant towards ceftriaxone. The MEPL recorded the MIC value of 16,000 µg/mL. On the other hand, when a volume of 100 µL of inoculum from each tube was plated on fresh sterile NA, the lowest concentration of MEPL where no MRSA growth was observed was at a concentration of 16,000 µg/mL, therefore, indicating the MBC value of the MEPL to be also 16,000 µg/mL. It should be noted that the MIC and MBC results for MEPL against the MRSA strain showed a larger value than ceftriaxone.

2.3. Synergistic Activity of Antibiotic with MEPL

The interrelation effects between ceftriaxone and MEPL against MRSA were tested using the checkerboard technique in association with the MIC value. Ceftriaxone and MEPL combination treatment enhanced the antimicrobial effect and exhibited synergistic activity on MRSA (Table 2). In the combination treatment, the MIC values of ceftriaxone and MEPL against MRSA were reduced to eight times lower (1000 µg/mL and 2000 µg/mL). As predicted, unique antibacterial activity with a lower MIC value was demonstrated by ceftriaxone in the presence of the MEPL in the combination therapy. Coherently, it resulted in a synergistic antibacterial effect against the tested MRSA via the combination therapy of ceftriaxone and MEPL extract.

Table 2. Synergistic effect of MEPL and ceftriaxone sodium was determined by the checkerboard test.

		MEPL (μg/mL)								
		16,000	8000	4000	2000	1000	500	250	125	62.5
Ceftriaxone sodium (μg/mL)	8000	No growth	No growth	No growth	No growth	No growth	Mild growth	Heavy Growth	Heavy Growth	Heavy Growth
	4000	No growth	No growth	No growth	No growth	Mild growth	Mild growth	Heavy Growth	Heavy Growth	Heavy Growth
	2000	No growth	No growth	No growth	No growth	Mild growth	Heavy Growth	Heavy Growth	Heavy Growth	Heavy Growth
	1000	No growth	No growth	No growth	No growth	Mild growth	Heavy Growth	Heavy Growth	Heavy Growth	Heavy Growth
	500	No growth	Mild Growth	Mild Growth	Heavy Growth	Heavy Growth	Heavy Growth	Heavy Growth	Heavy Growth	Heavy Growth
	250	No growth	Mild Growth	Heavy Growth	Heavy Growth	Heavy Growth	Heavy Growth	Heavy Growth	Heavy Growth	Heavy Growth
	125	Mild Growth	Heavy Growth	Heavy Growth	Heavy Growth	Heavy Growth	Heavy Growth	Heavy Growth	Heavy Growth	Heavy Growth
	62.5	Heavy Growth	Heavy Growth	Heavy Growth	Heavy Growth	Heavy Growth	Heavy Growth	Heavy Growth	Heavy Growth	Heavy Growth

Calculation of the FIC index of MEPL and ceftriaxone to determine the synergistic effect:

MIC of ceftriaxone alone = 8000 μg/mL
MIC of ceftriaxone in combination = 1000 μg/mL
MIC of MEPL alone = 16,000 μg/mL
MIC of MEPL in combination = 2000 μg/mL
$\mathbf{FIC}_{\text{ceftriaxone}}$ = 1000 μg/mL ÷ 8000 μg/mL = **0.125**
$\mathbf{FIC}_{\text{MEPL}}$ = 2000 μg/mL ÷ 16000 μg/mL = **0.125**

The sum of FIC (**ΣFIC**) is calculated as follows:

$$\begin{aligned} \mathbf{\Sigma FIC} &= \mathbf{FIC}_{\text{ceftriaxone}} + \mathbf{FIC}_{\text{MEPL}} \\ &= 0.125 + 0.125 \\ &= \mathbf{0.25} \end{aligned}$$

In brief, the MIC value of the MEPL and ceftriaxone in the checkerboard test were 2000 μg/mL and 1000 μg/mL, respectively. The FIC index of the combination of MEPL and ceftriaxone was 0.25, which indicates a significant synergistic antimicrobial activity against the MRSA bacteria. The combination is considered synergistic when the ΣFIC index is ≤0.5, and indifference is indicated by an FIC index > 0.5 to ≤4, while antagonism is when the ΣFIC is >4. In addition, the initial MIC values of MEPL (16,000 μg/mL) and ceftriaxone (8000 μg/mL) were found to reduce to 2000 μg/mL for MEPL and 1000 μg/mL for ceftriaxone ($p < 0.05$), respectively, in the checkerboard test against the MRSA bacteria.

2.4. *Presence of the mecA Gene in MRSA Treated with Different Combinations of MEPL and Ceftriaxone*

2.4.1. Purity of Genomic DNA

DNA concentration, purity, and contamination are the three factors that can affect the multiplex PCR test. Nucleic acids are typically quantified (at an absorption ratio of 260 nm/280 nm) to obtain an average DNA concentration and purity necessary to be considered when carrying out PCR (Table S1). All the DNA extracted demonstrated a purity ratio value of 1.8, which indicates low protein contamination. The result was analyzed by electrophoresis on a 0.8% agarose gel followed by ethidium bromide staining to confirm an adequate amount of the DNA present with a clear band for further amplification with primary and targeted band detection. As shown in Figure 1, bands of genomic DNA can be seen on top of the gel. The absence of smearing indicates that the DNA is intact and not degraded.

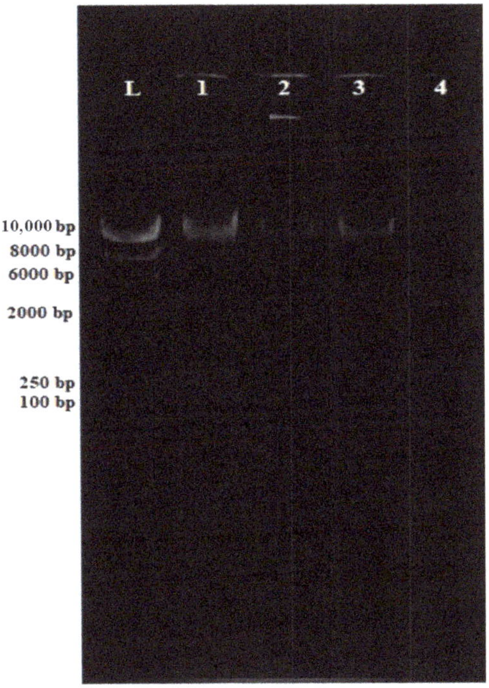

Figure 1. Electrophoresis gel (0.8% agarose) of the extracted genomic DNA from MRSA isolates. All genomic MRSA DNA (Lane 1–Lane 4) are intact for downstream applications.

2.4.2. Optimization of *mecA* Gene Amplification

The gradient amplification was performed to obtain the optimum annealing temperature for the multiplex PCR. Isolate 20 with a purity value of 1.7 and DNA concentration of 33.5 ng/µL (Table S1), was used throughout optimization since this isolate shows an enhanced DNA band in extracted product during electrophoresis observation. Four specific temperatures at 55.0 °C, 56.6 °C, 60.0 °C, and 61.0 °C were selected for the gradient PCR. As shown in Figure 2, a clear thick band was visible using the annealing temperature of 60.0 °C. Table S2 shows the relative intensity of the PCR amplicons on the 3% agarose gel. This finding proved that the DNA band present at 60.0 °C annealing temperature was the perfect band for *mecA* gene amplification via PCR.

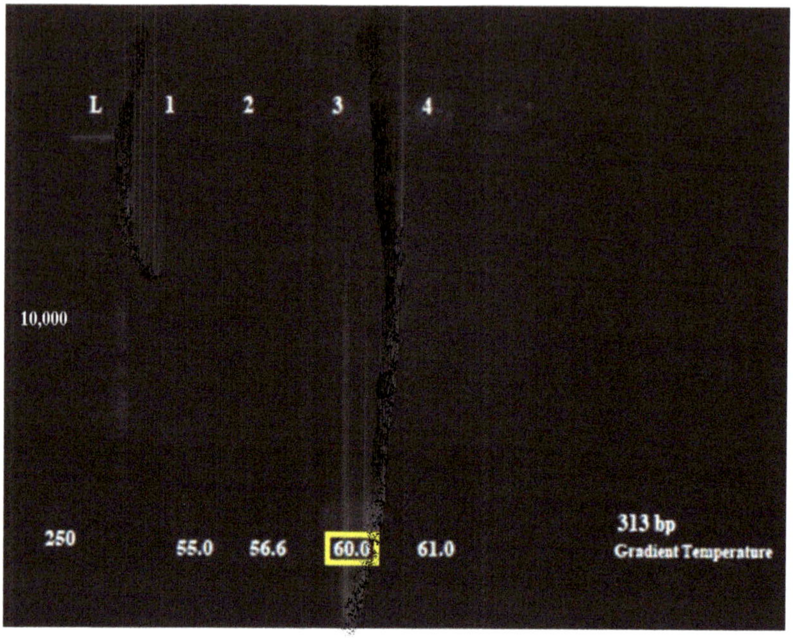

Figure 2. Electrophoresis gel (3% agarose) of the PCR products of the *mecA* gene for PCR optimization. The amplification optimized using MRSA isolates. Lane L = 1 kb ladder. The amplification of optimized isolate for annealing temperature 55.0 °C, 56.6 °C, 60.0 °C, and 61.0 °C. Lane 1 = amplification at annealing temperature 55.0 °C, Lane 2 = amplification at annealing temperature 56.6 °C, Lane 3 = amplification at annealing temperature 60.0 °C, and Lane 4 = amplification at annealing temperature 61.0 °C. A more apparent band was observed for the 60.0 °C reaction as shown in the yellow box.

2.4.3. Detection of *mecA* Gene by Multiplex PCR

The multiplex PCR was used to detect *mecA* gene-encoded ceftriaxone resistance directly from MRSA culture using the *mecA* gene and specific *S. aureus* 16S rRNA primers as an internal control for the *16S rRNA* gene, which is a conserved region in all prokaryotic bacteria. In the MRSA bacteria, the *mecA* gene should amplify at 313 bp and the *16S rRNA* gene at 528 bp. In comparison, the methicillin-susceptible *Staphylococcus Aureus* (MSSA) should only amplify the *16S rRNA* gene at 528 bp. An MRSA confirmation test was carried out using MRSA and MSSA isolates as the control. As shown in Figure 3, the MRSA isolates successfully amplified the *mecA* gene (313 bp) and *16S rRNA* gene (528 bp), while the MSSA isolates only amplified the *16S rRNA* gene (528 bp) as predicted.

Subsequently, different combinations of MEPL (1000 µg/mL and 2000 µg/mL) with ceftriaxone (1000 µg/mL) were tested against the MRSA isolate to investigate the influences of a different combinations of MEPL with ceftriaxone on the regulation of the *mecA* gene in the tested MRSA strain. As shown in Figure 4, the *mecA* gene was present in the MRSA isolate treated with MEPL and ceftriaxone at 1000 µg/mL; however, the combination of MEPL with ceftriaxone at 2000 µg/mL of MEPL and 1000 µg/mL of ceftriaxone successfully suppressed the presence of the *mecA* gene at 313 bp. In addition, as expected, the *mecA* gene was not expressed in the tested MSSA isolate.

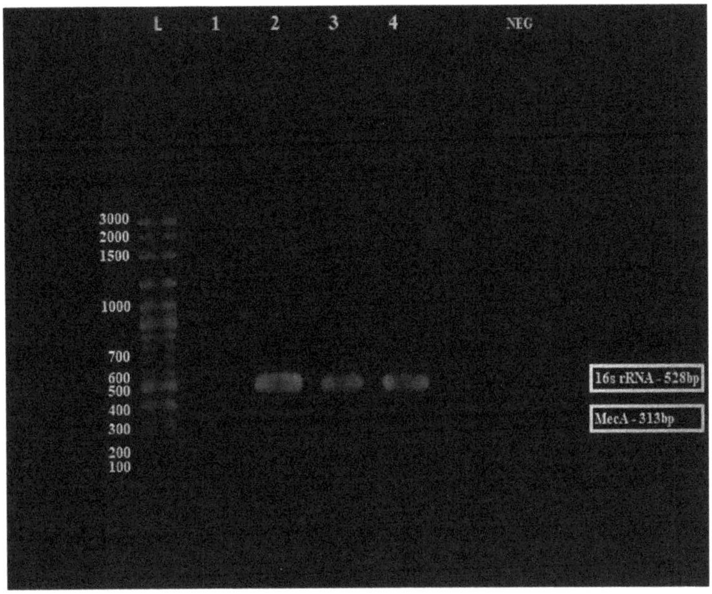

Figure 3. Electrophoresis gel (3% agarose) of the PCR products of the *mecA* gene for the discovery of the *mecA* gene from MRSA (Lane 2–3) and MSSA (Lane 4) isolates. Lane L = 100 bp ladder, Lane 1 = empty lane, Lane 2 and 3 = MRSA isolates, Lane 4 = MSSA control isolate, and Lane NEG = negative control. The yellow box: The *mecA* gene was expressed in MRSA strain while the *16S rRNA* gene was expressed in both MRSA and MSSA strains.

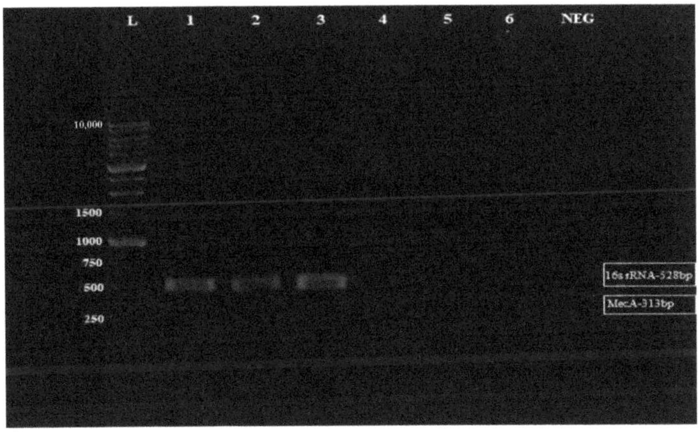

Figure 4. Electrophoresis gel (3% agarose) of the PCR products of the *mecA* gene and 16S rRNA for the detection of the *mecA* gene from MRSA (Lane 2–4) and MSSA (Lane 5) isolates. Lane L = 1 kb ladder, Lane 1 = Treated MRSA isolate (in a combination of 1000 µg/mL PLLME and 1000 µg/mL ceftriaxone), Lane 2 = Treated MRSA isolate (in a combination of 2000 µg/mL PLLME and 1000 µg/mL ceftriaxone), Lane 3 = untreated MRSA isolates, Lane 4 = blank, Lane 5 = MSSA isolate (control), Lane 6 = empty lane and Lane NEG = negative control. The yellow box: The *mecA* gene was not amplified in MRSA treated with the combination of 2000 µg/mL PLLME and 1000 µg/mL ceftriaxone.

Figure 5 shows the relative intensity (Table S3) of the amplified multiplex PCR products with the *mecA* gene and 16S rRNA bands formed on the electrophoresis gel using ImageJ software. The ImageJ software analysis of DNA bands can be used to quantify the *mecA* gene expression in the MRSA isolate. The ImageJ software analysis on the relative intensity of the *mecA* gene in MRSA provides quantitative data for the convenient evaluation of qualitative electrophoresis gel results. Therefore, with the aid of the ImageJ software, the quantification of the *mecA* gene band's relative intensity on the electrophoresis gel was further analyzed. The finding confirmed that the combination treatment of MEPL (2000 µg/mL) with ceftriaxone (1000 µg/mL) against MRSA isolates (Figure 4, Lane 3) displays a zero value for the *mecA* gene fragment (313 bp), which indicated the complete suppression of the *mecA* gene in MRSA.

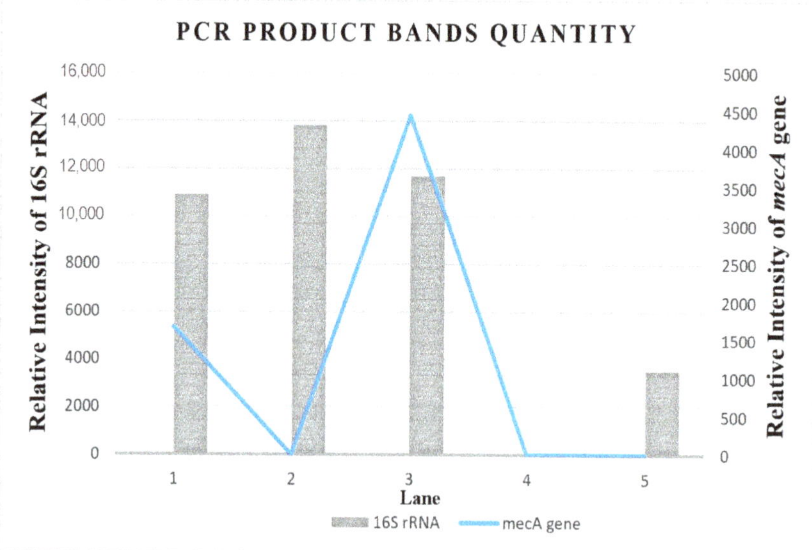

Figure 5. Relative band intensity by densitometry analysis of electrophoresis (3% agarose) gel of the PCR products of the *mecA* gene and 16S rRNA performed using ImageJ quantification software. Lane 1 = Treated MRSA isolate (in a combination of 1000 µg/mL MEPL and 1000 µg/mL ceftriaxone); Lane 2 = Treated MRSA isolate (in a combination of 2000 µg/mL MEPL and 1000 µg/mL ceftriaxone); Lane 3 = untreated MRSA isolates, Lane 4 = blank, Lane 5 = MSSA isolate (control).

2.4.4. Antimicrobial Compounds in MEPL

Ultra High-Performance Liquid Chromatography (UHPLC) analysis was performed to analyze and tentatively annotate the extracted metabolites in the MEPL with the aid of the chemical library of Metlin_AM_PCDL-N-170502.cdb. The UHPLC analysis of the MEPL showed the presence of several antimicrobial phytochemicals. Among these, beta-himachalene (1.9%), 5Z,8Z,11Z,14Z-octadecatetraenoic acid (8.3%), 9Z,12Z,15E-octadecatrienoicc acid (6.1%), and luteolin 7-rhamnosyl(1->6)galactoside (5.7%) were the antimicrobial compounds in MEPL extract. The chemical structures of the antimicrobial phytochemical compounds found in MEPL are presented in Figure 6.

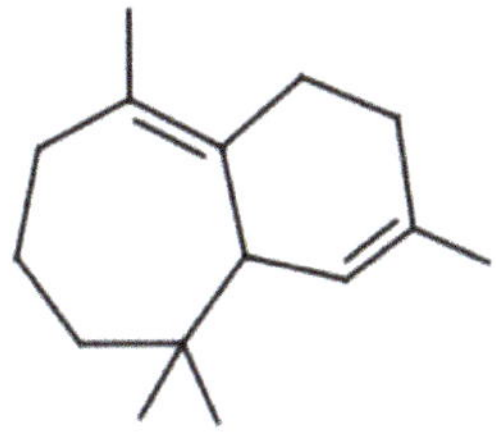

Beta-himachalene

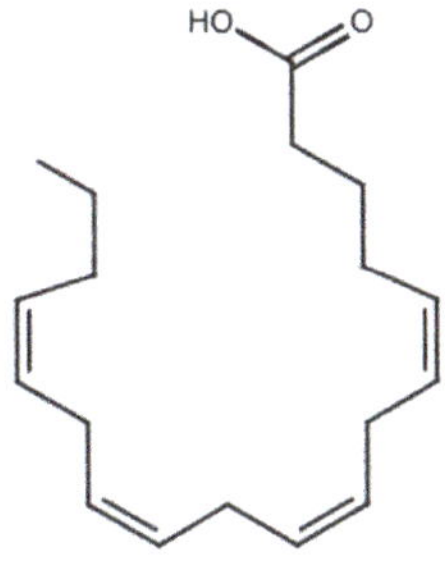

5Z,8Z,11Z,14Z-octadecatetraenoic acid

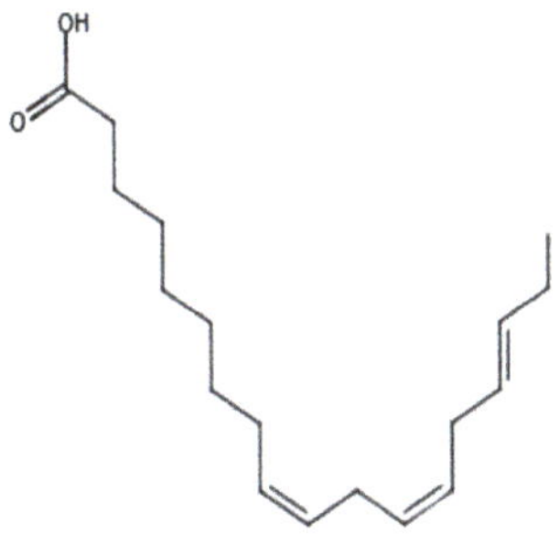

9Z,12Z,15E-octadecatrienoic acid

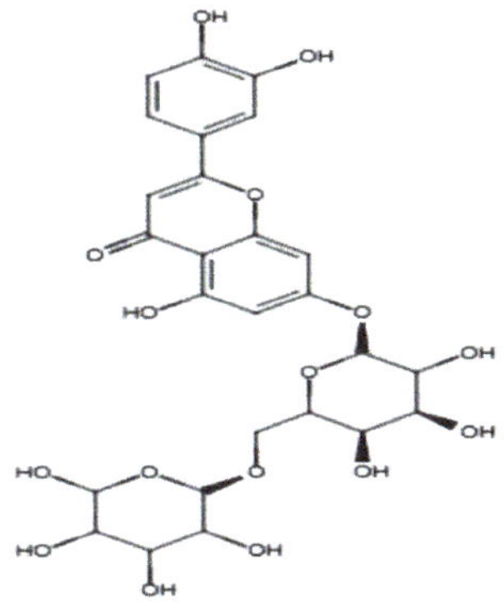

Luteolin 7-hamnosyl(1->6)galactoside

Figure 6. Antimicrobial phytochemical compounds found in the methanol extract of *Polyalthia longifolia* Leaf (MEPL) were detected using ultra high-performance liquid chromatography (UHPLC) equipped with the chemical library.

3. Discussion

Methicillin-resistant *Staphylococcus aureus* (MRSA) infection has become one of the most historic pathogenic bacterial infection associated with health issues in developing countries. Moreover, the crucial *mecA* gene contributes to methicillin resistance in MRSA strains, which encodes a novel penicillin-binding protein PBP2a. The global trend has represented a rise in MRSA infections with the high emergence of multidrug-resistant strains [12]. This bacterium has shown resistance to various antibiotics such as methicillin, penicillin, and amoxicillin, including ceftriaxone. Ceftriaxone is a third-generation cephalosporin and remains one of the most commonly used antibiotics for antimicrobial therapy due to its efficacy and low therapeutic index [13,14]. It is reported that ceftriaxone has a broad potency spectrum against Gram-positive and Gram-negative bacteria [15]. In addition, ceftriaxone is used frequently to treat MSSA infections [16,17]. It has been used as a first-line treatment against bacteremia alongside other antibiotic combinations [18]. The rise of microorganism resistance towards third-generation cephalosporins is a global burden and has led to antimicrobial treatment failure. The bacterial organism becomes inherently resistant to the increased use of antibiotics at a higher antibiotic dosage [19,20]. Besides, bacterial resistance to an antibiotic can also be attributed to random genetic mutation [21] or the uptake of plasmid DNA (horizontal gene transfer) from foreign cells [22]. Hence, MRSA has become the center of this public health concern due to its high virulence and resistance to a broad spectrum of antibiotics [23]. This widespread organism causes challenges to both the healthcare system and patients due to increased hospitalization costs and notable mortality/morbidity rates [24]. In addition to this complication, antibiotics often produce adverse effects, namely hypersensitivity, immune suppression, and allergic reactions [25–28]. The need to develop new antimicrobials as an alternative to synthetic antibiotics for MRSA treatment is achieved from various sources. Many developing countries commonly use medicinal plants in the treatment of multiple health complications. The application of medicinal plant extracts rich with pharmacological activity, such as *P. longifolia* and its associated phytochemicals, can significantly contribute to the treatment of infectious diseases. Hence, the current research was performed to evaluate the synergistic antibacterial activity of natural MEPL in combination with ceftriaxone against MRSA bacteria.

The synergistic antimicrobial action between medicinal plant extracts and conventional antibiotics has been extensively studied to overcome the antibiotic resistance problem [29–31]. Synergism takes place when two different molecules interact and strengthen their actions. On the other hand, any reduction in activity from the combination treatment is termed antagonism [32]. The synergistic properties of MEPL with ceftriaxone against MRSA were evaluated in this study. The results indicated positive synergism in the combination treatment of MEPL and ceftriaxone compared to ceftriaxone or MEPL alone against the MRSA bacteria. The MIC and MBC values of ceftriaxone and MEPL decreased in the combination treatment, indicating the synergistic antimicrobial activity of MEPL in combination with ceftriaxone. MEPL may promote synergistic antimicrobial properties by acting as synergistic activity enhancers in combination with ceftriaxone, enhancing the overall antibiotic effect. The advantages associated with the synergistic interactions are that synergism effect increases treatment efficiency, decreases undesirable side effects of the single drug, such as diarrhea, nausea, bloating, and indigestion, increases the bioavailability of free agents, and an adequate therapeutic effect is achieved with comparatively smaller doses when compared with individual synthetic antimicrobials [33]. Many researchers have reported that combination therapy, mainly plant extracts with synthetic antibiotics, exhibited a synergistic effect against *S. aureus* [34–37]. Interestingly, a recent study reported impaired cell division, extensive wrinkles, cell shrinkage, and the emergence of rougher cells with fibrous matrix and clustered cells, highlighting the synergistic effect of ethyl acetate *P. longifolia* in combination with ampicillin against MRSA cells [10,11]. Another study has also suggested that the membrane-disrupting activity of combination therapy between Trp-containing antimicrobial peptides (AMPs) with four classes of traditional chemical antibiotics, namely penicillin, ampicillin, and erythromycin, increases the access of small molecule antibiotics to the cell, which allows the synergistic activity to improve antimicrobial agents' effectiveness, increasing bacterial killing and prevent resistance development [38]. Moreover, AL-Ali et al. [39] reported the synergistic antimicrobial activity of various plant extracts in combination treatment against multi-drug resistance (MRSA) *S. aureus*. The combination of four plant extracts, namely *Mentha cervina*, *Mentha longifolia*, *Ocimum basilicum*, and *Origanum vulgare* showed good synergistic antibacterial activity against the multi-drug resistance (MDR) *S. aureus*. Besides, another independent study has reported the antimicrobial activities of the methanol, acetone, and 1,4-dioxan fractions of *P. longifolia* leaves [40]. The tested sample showed better antibacterial activity against Gram-positive bacterial and fungal strains than the Gram-negative bacterial strains studied.

Various secondary metabolites in the MEPL, as reported in the literature, such as flavonoids, alkaloids, and diterpenoids [41], can be responsible for the observed antimicrobial properties of the MEPL. Hence, screening of MEPL was performed to annotate the chemical profiles using UHPLC analysis equipped with the chemical library of Metlin_AM_PCDL-N-170502.cdb to identify the bioactive chemical constituents that could be responsible for the observed antimicrobial activity. UHPLC analysis led to the detection of the various chemical constituents, as shown in Figure 6. Moreover, the presence of himachalene and its derivatives [42], fatty acid octadecatetraenoic (9Z,12Z,15E-octadecatrienoicc acid and 5Z,8Z,11Z,14Z-octadecatetraenoic acid) [43,44], and luteolin and its derivatives [45] compounds were found in MEPL, which were previously reported to show good antimicrobial activity against various microbes including *S. aureus*, which might have contributed to the observed antimicrobial activity of the MEPL in this study. Besides, rutin was used to standardize the MEPL extract in this study since rutin enhanced the antibacterial activities, as reported in the literature [46]. As observed in this study, rutin also might contribute to the synergistic effect of the MEPL extract.

In addition, various reports in the literature reported the isolation of compounds from *P. longifolia* with antimicrobial and synergistic antibacterial activity. Interestingly, seven antimicrobial clerodane diterpenoids, namely 16(R and S)-hydroxy-cleroda-3,13(14)Z-dien-15,16-olide, 16-oxo-cleroda-3,13(14)E-dien-15-oic acid, methyl-16-oxo-cleroda-3,13(14)E-dien-15-oate, 2-oxokolavenic acid, 16(R and S) hydroxy-cleroda-3,13(14)Z-dien-15,16-olide-2-one, (4→2)abeo-16(R and S)-hydroxy-cleroda-2, 13(14)Z-dien-15, 16-olide-3-al, and 3β, 16α-dihydroxy-cleroda-4(18), 13(14)Z-dien-15,16-olide [47] were isolated from the methanol extract of *P. longifolia* leaves, which are widely reported for their antibacterial and antifungal properties [48]. Furthermore, diterpenoids induce bacterial membrane disruption [49], which may allow other compounds to enter cells to initiate antibacterial activity in a combination therapy mode. Therefore, the presence of diterpenoids [49] and flavonoids [50] in the MEPL, as reported in the literature, can be hypothesized to be synergistic and enhance the antibiotic function by disrupting the membrane of the MRSA and making it susceptible to ceftriaxone. In particular, the presence of clerodane diterpene 16α-hydroxycleroda-3, 13 (14) Z-dien-15, 16-olide (CD) has been reported to be synergistic against MRSA through the disruption of the cell membrane [51]. In addition, the combination of CD, a bioactive compound in MEPL, reduced the MIC of fluoroquinolones, such as norfloxacin, ciprofloxacin, and ofloxacin, against MRSA through significant inhibition of the efflux pump [52]. Efflux pumps have been cited as the main reason for the emergence of multidrug resistance bacteria towards various antibiotics among Gram-positive and Gram-negative bacteria [53]. It was reported that CD downregulates the expression of efflux pump genes, such as *norA*, *norB*, *norC*, *mdeA*, and *mepA*, which are the genes responsible for expelling antibiotics outside the *S. aureus* cells [54]. Therefore, it can be deduced that the bioactive compounds in the MEPL may play a similar role in inhibiting the efflux pump in *S. aureus* and synergistically reversing the resistance of MRSA towards ceftriaxone.

This study also attempted to assess whether the combination of MEPL with ceftriaxone influences the presence of the *mecA* gene by observing the presence of the *mecA gene* on the agarose gel upon treatment. In this study, MEPL from *P. longifolia* with ceftriaxone inhibits the manifestation of the resistant *mecA* gene in the studied strain. In the presence of β-lactam derivatives, the MRSA strains will not demonstrate growth inhibition and can retain their capacity to expand the zone of inhibition [55]. The methicillin-resistant *mecA* gene in MRSA isolates encodes PBP2a, a transpeptidase that inhibits the antibiotic's antimicrobial action. Another study has reported that the *mecA* gene can be a useful molecular marker for MRSA isolates [56]. In contrast, *S. aureus* isolates lacking *the mecA* gene can be considered as MSSA strains [57]. The *mecA*-positive strains differ in the expression levels to methicillin resistance, which may be complex and difficult to diagnose [58]. Therefore, molecular techniques, such as polymerase chain reaction (PCR), are suitable for detecting the methicillin resistance *mecA* gene. The multiplex PCR technique utilized in this study is a rapid tool and considered the "gold standard" for detecting the methicillin resistance *mecA* gene due to its efficacy and accuracy [59]. Optimization of the PCR protocol is routine and necessary for better sensitivity and specificity. Adequate DNA templates and optimum annealing temperature are crucial factors for successfully amplifying the *mecA* gene [60]. This was evidently supported by the current research results, where the positive control MRSA isolate amplification was improved with the appropriate annealing temperature and DNA template.

Besides, the influence of MEPL on the *mecA* gene in MRSA bacteria was proven by the finding of the checkerboard method conducted in this research to assess the synergistic action of MEPL and ceftriaxone. The checkerboard method results indicate the synergistic effect of ceftriaxone combined with MEPL against MRSA by enhancing the antimicrobial effect. The *mecA* gene analysis in the MRSA treated with ceftriaxone (1000 µg/mL) combined with MEPL (2000 µg/mL) by multiplex PCR examination showed the absence of the *mecA* gene band. This finding indicated that the gene-specific primers could not identify and bind to the region coding the *mecA* gene. This finding disclosed the effective influence of MEPL on inhibiting the presence of the *mecA* gene in MRSA bacteria. The combination of ceftriaxone and MEPL influenced the presence of the *mecA* gene in MRSA to make the local strain susceptible to ceftriaxone. Interestingly, several studies report on the influence of medicinal plant extracts on bacterial gene expression, namely *T. integrifolia, Eurycoma longifolia* Jack, and *Helmintostachys zeylanica* against *Salmonella typhimurium* strains via the Ames Test [56,61]. Alkaloids, such as β-carboline, have been a vital influence against bacterial DNA [57,58,62,63]. It was reported that β-carboline alkaloids, such as harman and harmine of *Passiflora* spp. (Passifloraceae), are responsible for DNA damage of *Saccharomyces cerevisiae* [59,64]. In another study, the mutagenic properties of the methanolic extract of *Byrsonima crassa* Niedenzu was reported due to amentoflavone. Plant extracts containing flavonoids, such as Quercitin, have also been implicated in mutagenesis [60,65]. Therefore, flavonoids [50] and alkaloids [61,66] in MEPL might be responsible for the observed suppression of the expression of the *mecA* gene, which warrants further detailed studies.

The present research studies the antimicrobial effects and modulation of *mecA* gene expression by MEPL combined with ceftriaxone against an MRSA strain for possible application as a natural product agent. The MEPL combination with ceftriaxone exhibited vigorous antimicrobial activity against the MRSA isolate. Moreover, MEPL showed a synergistic antibacterial effect with ceftriaxone against the tested MRSA strain and suppressed the presence of the resistant *mecA* gene. From the findings of this research, it was established that MEPL could reinstate the effectiveness of ceftriaxone against MRSA. Consequently, the findings of this research propose that the MEPL and ceftriaxone combination could develop novel natural remedies based on combination antibiotics therapy against MRSA infection. Furthermore, various in vitro and in vivo experiments, such as the genotoxic effect evaluated via plasmid relaxation assay, acute oral toxicity studies in animal models, and the *Allium cepa* assay [67], showed that MEPL was not toxic and safe in human applications. The in vivo acute oral toxicity study showed that MEPL was safe even at a single dose of 5000 mg/kg body weight in female albino Wistar rats. Besides, the literature also reports that MEPL exhibits various biologically beneficial effects. The antimicrobial activity of *P. longifolia* leaf extracts were also reported by Chanda and Nair [40] against 91 clinically significant pathogenic microbial strains. The polyphenol-rich MEPL exhibited good antioxidant and hepatoprotective activities against paracetamol-induced oxidative damage [67,68]. Besides, the MEPL also supported the X-ray irradiated mouse survival rate increases compared to 100% mortality in the untreated mice [69], and renoprotection against radiation-induced nephropathy by an anti-oxidative molecular mechanism [70]. These findings highlight that MEPL decreased oxidative stress and nephropathy in rats due to its anti-inflammatory activities. Moreover, MEPL also showed good cytotoxicity against HeLa cancer cells via inducing apoptotic cell death and miRNA regulation [71–74]. A recent study also showed that MEPL exhibited good antiaging activities in S. cerevisiae by modulating oxidative stress, enhancing GSH content, and increasing SOD and SIRT1 gene expression [75].

4. Materials and Methods

4.1. *Polyalthia longifolia Leaf*

Mature leaves of *P. longifolia* were collected (Voucher specimen number: USM/HERBARIUM/11306) from University Sains Malaysia (USM), Pulau Pinang, Malaysia main campus, in February 2020. The leaves were rinsed thoroughly with tap water and air dried under shade inside the laboratory for about 2 weeks until the leaves were dried entirely. Dried leaf parts were homogenized to a fine powder using a regular blender and stored in airtight bottles.

4.2. *Preparation of Polyalthia longifolia Leaf Extract*

Dried powder *P. longifolia* leaves then underwent methanol extraction using a cold percolation process on a rotary shaker [76]. A mass of 100 g of dried powder of *P. longifolia* leaves was added into a conical flask and soaked in 1000 mL of methanol. The flask was sealed with aluminum foil and kept on a rotary shaker (Ohaus, Parsippany, NJ, USA) at 190–220 rpm for 3 days. After 3 days, the content of the flask was filtered at different levels, initiated via 8 layers of muslin cloth followed by Whatman No. 1 filter paper to get the crude extract. The filtrates were then collected and concentrated in a rotary vacuum evaporator (Eyela, Bohemia, NY, USA) [77] at 120 rpm and 200 pi at 41 °C overnight to remove solvents from samples through the evaporation process. The concentrated extract was then collected in a glass Petri dish, kept in the oven (60 °C), and incubated to remove excessive methanol further from the sample. A constant weight of the completely solvent-free filtrates was obtained after incubation in the oven. The filtrates were then stored at 4 °C in air-tight bottles. The final product of the methanol extract of *Polyalthia longifolia* leaf (MEPL) was used to conduct the antibacterial study. The MEPL stock solution was dissolved and prepared in 5% DMSO at a final 10 mg/mL concentration. The rutin measure in MEPL extract was established on the peak area calculated from the calibration curve equation of commercially (Sigma-Aldrich, St. Louis, MO, USA) available rutin (5, 10, 100, 400, 600, 800, and 1000 μg/mL) compound (standard) (y = 275,885x, r^2 = 0.9977). The amount of rutin in the MEPL was found to be 8.96 μg (0.896%) in 1000 μg [69].

4.3. *Test Microorganism Collection and Maintenance*

The Gram-positive bacterium MRSA and MSSA were collected from the Penang General Hospital Microbiology Unit (GH), Penang, Malaysia. The MRSA and MSSA strains were aseptically removed with an inoculating loop and streaked in a zig-zag pattern onto the freshly prepared nutrient agar (NA) plate. The MRSA and MSSA strains on the NA plate were grown for 24 h at 37 °C. The stock culture was then stored at 4 °C. The stock culture was sub-cultured every 3 weeks to maintain viability.

Antimicrobial Susceptibility Test

Antimicrobial susceptibility testing (AST) of bacterial isolates is a collective and significant technique in most clinical laboratories. In this study, AST was conducted using the Kirby Bauer technique [78] based on the Clinical Laboratory Standard Institutions [79] guidelines on molten Mueller Hinton Agar (MHA). The steps involved in this assay are as follows.

4.4. *Culture Media and Inoculum Preparation*

The test organisms were grown on molten Mueller Hinton Agar (MHA) at 37 °C during the antibacterial susceptibility test. The molten MHA was prepared according to the manufacturer's instruction (Oxoid, Basingstoke, UK), autoclaved and poured onto sterile Petri dishes, and solidified at room temperature. An inoculum suspension (1.5×10^8 cells/mL) equal to 0.5 McFarland was prepared by inoculating 5 similar colonies with a wire loop in up to 5 mL of tryptone soya broth (TSB) and incubated at 37 °C for 8 h up until mild-to-moderate turbidity growths could be seen.

4.5. *Agar Disc Diffusion Assay of Ceftriaxone*

A sterile cotton swab was dipped into the prepared inoculum of the MRSA suspension, which was rotated resolutely against the tube's upper inside wall to rapidly removed excess fluid and then streaked through the entire surface of MHA plates. The plate was allowed to dry at room temperature with the inoculum, with the lid in place, for about 10 min. Standard antibiotic discs of ceftriaxone (8 μg/mL, 16 μg/mL, 32 μg/mL, and 64 μg/mL), also known as blank cartridges (Oxoid, Basingstoke, UK), were placed on the upper layer of the seeded agar plate and gently pressed on the disc's handle while making sure all the discs were completely attached to the medium. The plates were incubated for 24 h at

37 °C. The formation of a clear zone of inhibition of ≥21 mm in diameter was considered a significant susceptibility of the organism to the MEPL extract. The experiment was replicated three times, and the mean value is displayed in this study. By measuring the diameter of the zone of inhibition, the antimicrobial activity was determined and recorded in millimeters with the aid of sliding calipers, and the organisms present were classified as sensitive, resistant, or intermediate, referring to CLSI guidelines (Table 3). The 5% DMSO was used as a negative control.

Table 3. Diameter of inhibition zone interpretative criteria for *S. aureus*.

	Diameter of Zone of Inhibition (mm)			
	Potency	**Resistant**	**Intermediate**	**Sensitive**
Ceftriaxone	30 μg	≤13	14–20	≥21

4.6. *Agar Well Diffusion Method of Antibacterial Susceptibility Test for MEPL*

The agar well diffusion method evaluated the antimicrobial activity of the MEPL with certain modifications. MRSA was grown on nutrient broth (NB) and incubated at 37 °C for 24 h. A total of 1600 μL of overnight NB culture was added to 120 mL of molten MHA and mixed well; the mixture was then poured into a sterile Petri dish and set aside to allow the plate to dry at room temperature. A sterile 5 mm in diameter cork-borer was used on the set agar to create wells. Subsequently, 25 μL of diluted plant extract in a sequence of 8 mg/mL, 7 mg/mL, 6 mg/mL, 5 mg/mL, 4 mg/mL, 3 mg/mL, 2 mg/mL, and 1 mg/mL was applied to the wells, and the plates were then incubated at 37 °C overnight. The bacterial growth was assessed based on the diameter of the inhibition zone. The tests were performed in triplicate, and average values were recorded. A 5% DMSO solution was used as a negative control.

4.7. *Evaluation of the Minimum Inhibitory Concentration (MIC) and Minimum Bactericidal Concentration (MBC) of the MRSA Isolate against Ceftriaxone and MEPL*

4.7.1. Determination of the MIC of Ceftriaxone and MEPL against the MRSA Isolate

To determine the MIC concentration of ceftriaxone and MEPL against the MRSA isolate, the broth dilution method was used under the CLSI guideline. Two-fold serial dilutions of the ceftriaxone at a concentration of (8000—62.5 μg/mL) and the MEPL in the arrangement of 16,000–62.5 μg/mL in 5% DMSO were prepared in sterile capped universal bottles. Subsequently, 2 mL of overnight incubated (37 °C) MRSA suspension was added to 2 mL of each concentration of the antibiotic ceftriaxone and MEPL dilution followed by vortexing and 18 h incubation at 37 °C. The negative control comprised MHB and ceftriaxone antibiotic, while the positive control was MHB and MRSA suspension. Another 2 mL broth prepared in the universal bottle was inoculated MRSA and kept overnight in a refrigerator at 4 °C separately. The tube was used as a standard for determining complete inhibition. The MIC value was determined as the lowest concentration of ceftriaxone and MEPL inhibiting the MRSA by referring to turbidity. Besides, comparing to the standard tube incubated previously in the refrigerator was used to assess the inhibition of the organism's growth.

4.7.2. Determination of the MBC of Ceftriaxone and MEPL against the MRSA Isolate

Immediately after the MIC determination, the tubes with ceftriaxone and MEPL inhibiting the MRSA growth were used to determine the MBC. Subsequently, about 100 μL of the inoculum was added to a sterile NA media plate and incubated for 24 h in a 37 °C incubator to observe possible bacterial growth. The lowest concentration of ceftriaxone and MEPL in the subculture that showed no bacterial growth on the plate was considered the MBC [80].

4.8. Investigation of the Synergistic Properties of MEPL with Ceftriaxone

4.8.1. Preparation of MEPL and Ceftriaxone for Synergistic Study

Two-fold serial dilutions of the extracts (16,000–62.5 μg/mL) and ceftriaxone (8000–62.5 μg/mL) were prepared. A combination drug was prepared at a ratio of 1:1 of MEPL:ceftriaxone from the highest to lowest concentration to investigate of the synergistic properties of MEPL with Ceftriaxone.

4.8.2. Measurement of the Fractional Inhibitory Concentration (FIC) by Checkerboard Analysis

Ninety-six well microtiter plates were used to measure the FIC concentration for synergistic activity between MEPL and ceftriaxone [81,82]. The inoculum suspension was prepared in MHB. A total volume of 100 μL of two-fold dilution of EPL/ceftriaxone combination (1:1 ratio) was added to 900 μL of the inoculum suspension into each well of the microtiter plates, bringing the final total volume to 1 mL. The ceftriaxone was placed in columns in ascending concentrations starting at zero MIC and ending at two times the MIC. The MEPLs were similarly distributed among the rows. Accordingly, each well of the 96-well microtiter plate had a unique combination of different concentrations of the antibiotic and MEPL. Two control wells were preserved for each test batch. These included the test control (the well containing MEPL/antibiotic and the medium without inoculum) and organism control (the well containing the growth medium and the inoculum). The plate was incubated overnight at 37 °C. The MIC value was determined as the lowest concentration of ceftriaxone and MEPL inhibiting the MRSA by referring to turbidity.

Calculation of the Fractional Inhibitory Concentration (FIC) Index

The ΣFICs were computed with the formulae below [83]:

- FIC of plant extracts = MIC of MEPL in combination/MIC of MEPL alone
- FIC of antibiotic = MIC of antibiotic in combination/MIC of antibiotic alone

$$\text{FIC index} = \text{FIC of MEPL} + \text{FIC of antibiotic}$$

Synergy was defined as a FIC index ≤ 0.5.
The additive effect was defined as a FIC index > 0.5 but $\leq$4.0.
Antagonism was defined as a FIC index > 4.0.

4.9. Presence of the mecA Gene in MRSA Treated with Different Combinations of MEPL and Ceftriaxone

4.9.1. Concentration-Dependent Assay of Ceftriaxone and MEPL against MRSA and MSSA Isolates

The one-day-old cultures of MRSA and MSSA isolates were inoculated in 50 mL MH broth and incubated at 37 °C at 120 rpm agitation. The next day, the MRSA and MSSA isolates were treated with different combinations of plant extract dosages and antibiotics as 1000 μg/mL ceftriaxone with 1000 μg/mL or 2000 μg/mL MEPL. The cultures were incubated at 37 °C at a 120 rpm agitation rate for 24 h. The next day, the culture was pelleted at 0.12× *g* (120 rpm) speed for 10 min in a tabletop centrifuge. The pellets were then subjected to genomic DNA extraction.

4.9.2. Genomic DNA Extraction

The ready-to-use DNA mini kit from Stratec Molecular GmbH Berlin, Germany, was used to separate bacterial DNA from MRSA and MSSA strains. The genomic DNA of MRSA and MSSA strains was purified using the bacterial DNA purification mini kit (Stratec Molecular, Berlin, Germany) following the manufacturer's protocol, and stored at −20 °C.

4.9.3. DNA Quantification

The eluted genomic DNA was quantified by measuring UV absorption using a NanoDrop spectrophotometer (BioRad, Hercules, CA, USA). The integrity of each eluted DNA sample was evaluated by subjecting it to 0.8% (*w*/*v*) agarose gel electrophoresis analysis, and the DNA samples were kept at −20 °C for future analysis.

4.9.4. Multiplex Polymerase Chain Reaction (PCR)

The genomic DNA was further subjected to multiplex PCR amplification to detect targeted genes (*mecA* and *16S rRNA*). The multiplex PCR amplification was carried out using a Bio-Rad thermal cycler. PCR was carried out in 50 µL reaction mixtures, 25 µL Quick-Load 2X power Taq Master Mix applied with 1 µL reverse primer (1 µM) and 1 µL (1 µM) forward primer and genomic DNA (30 ng/µL). Sterile distilled water was added to bring the total volume to 50 µL. The negative control comprised just the Quick-Load 2X power Taq Master Mix, primers, and sterile water. The list of primers used in this study is listed in Table 4. The conditions of the gradient multiplex PCR (30 cycles) used in this study are as follows: denaturation at 95 °C for 30 s, annealing at 55–61 °C for 30 s, and eventually elongation at 72 °C for 30 s [84]. After the optimization of the annealing temperature, a conventional multiplex PCR was carried out using the same conditions of 95 °C for 30 s, followed by annealing at 60 °C for 30 s, and eventually elongation at 72 °C for 30 s. All PCR products were then assessed using 3% (*w*/*v*) gel electrophoresis.

Table 4. PCR primer sequences for the detection of MRSA.

Primers	Oligonucleotide Primer Sequences (5′ to 3′)	Amplicon Size (bp)
mecA 761R *mecA* 449F	CTT GTA CCC AAT TTT GAT CCA TTT G AAA CTA CGG TAA CAT TGA TCG CAA	313
16S rRNA 914R 16S rRNA 387F	AAC CTT GCG GTC GTA CTC CC CGA AAG CCT GAC GGA GCA AC	528

4.9.5. Agarose Gel Electrophoresis Analysis of the PCR Product

The agarose gel (3%, *w*/*v*) was prepared using 3 g gel powder dissolved in 100 mL of TBE buffer before microwave heating for up for 2 min. A 1 µL loading dye ratio to 3 µL PCR liquid (1:3) was used for all reactions. A volume of 4 µL of the sample and appropriate DNA ladder was loaded into each well and ran at a constant 65 V power supply for 40 min. Once the bromophenol blue stain hit more than two-thirds, the gel was examined by staining with ethidium bromide under a UV trans-illuminator (Appleton Woods, Birmingham, UK) to observe the specific band locations of the amplified DNA, which were recorded in an automatic gel documentation scanner. The intensity of the bands was quantified by using the ImageJ software.

4.10. LC–ESI–MS/MS Identification of Antimicrobial Compounds in MEPL

Identification of the antimicrobial compounds was carried out by using the Agilent 1200 series Ultra High-Performance Liquid Chromatography (UHPLC) system (Agilent Technologies, Santa Clara, CA, USA), coupled with an Agilent 6520 Accurate-Mass quadrupole-time of flight (Q-TOF) mass spectrometer with a dual electrospray ionization source (ESI). The UHPLC system, equipped with the chemical library of Metlin_AM_PCDL-N-170502.cdb, consisted of a vacuum solvent degassing unit, a capillary pump, and an automatic sample injector. The ESI operated in positive and negative modes with an m/z range from 100–3200. ESI conditions were as follows: fragmentor voltage 125 V; nebulizer pressure 45 psi; capillary voltage 3500 V; gas temperature 300 °C, gas flow 10 L/min, and skimmer 65 V. The chromatography was performed using Agilent Zorbax Eclipse XDB-C18, Narrow-Bore 2.1 × 150 mm, 3.5 microns (Agilent Technologies, Santa Clara, CA, USA). The auto-sampler compartment was maintained at 4 °C, and the mobile phase was 0.1% formic acid in water (A) and 0.1% formic acid in acetonitrile (B). The multi-step linear gradient was applied as follows: 5% solvent B for 5 min and the gradient keep isocratic at 100% solvent B from 20 min to 25 min. The initial condition was held for 5 min before the subsequent analysis. The injection volume was 1 μL, and the flow rate was 0.5 mL/min. The chromatographic separation was performed using C_{18} column (Agilent Eclipse XDB-C18 Narrow-bore, 150 mm × 2.1 mm, 3.5-micron) and the column temperature was 25 °C. The compounds in MEPL were identified via the Metlin database by using the spectra of chromatograms obtain from liquid chromatography mass spectrometric analysis, which determined the molecular mass of the compounds in the crude extract. The mass spectra of the compounds derived from UHPLC were run against the Metlin_AM_PCDL-N-170502.cdb library for the identification of homologous compounds via Agilent Mass Hunter software. The determination of the novelty of the identified compounds was performed on Scifinder software. Conversely, previously testified compounds were subjected to a literature search for biological activities, specifically for antimicrobial activity.

5. Conclusions

In conclusion, the results obtained in this study demonstrate the potential of MEPL to be a candidate for combination therapy against MRSA bacteria because it has a synergistic antibacterial effect with ceftriaxone in the tested strain. The killing effect of the combinatorial treatment is connected with the inhibition of the presence of the *mecA* gene in staphylococcal resistance to β-lactams antibiotics. The antimicrobial compound analysis the in MEPL extract showed the presence of several antimicrobial compounds known for their antibacterial activity. These discoveries provided a novel choice for clinicians to use natural MEPL in combination with antibiotics in MRSA infection treatment. Further study is also needed in an animal model to evaluate MEPL and ceftriaxone combination therapy in vivo efficacy.

Supplementary Materials: The following supporting information can be downloaded at: https://www.mdpi.com/article/10.3390/antibiotics12030477/s1. Table S1. Extracted MRSA genomic DNA concentration and purity. Table S2. The relative intensity of PCR products of the *mecA* gene for PCR optimization was obtained using ImageJ quantification software. Table S3: The relative intensity of PCR products of the *mecA* gene and 16S rRNA was obtained using ImageJ quantification software.

Author Contributions: Conceptualization, V.R., S.S. (Sreenivasan Sasidharan), S.S. (Sumaira Sahreen), Y.C., L.A.A.-K., M.P., M.A., N.A. and S.C.B.G.; methodology, V.R., S.S. (Sreenivasan Sasidharan), S.S. (Sumaira Sahreen), Y.C., L.A.A.-K., M.P., M.A., N.A. and S.C.B.G.; software, V.R.; validation, V.R. and S.S. (Sreenivasan Sasidharan); formal analysis, V.R.; investigation, V.R. and S.S. (Sreenivasan Sasidharan); resources, S.S. (Sreenivasan Sasidharan); data curation, V.R.; writing—original draft preparation, V.R.; writing—review and editing, S.S. (Sreenivasan Sasidharan); visualization, V.R.; supervision, S.S. (Sreenivasan Sasidharan); project administration, S.S. (Sreenivasan Sasidharan); funding acquisition, S.S. (Sreenivasan Sasidharan). All authors have read and agreed to the published version of the manuscript.

Funding: This work was partly funded by the Research University Grants (RUI; Grant No.: 1001/CIPPM/8012229) from the Universiti Sains Malaysia, Malaysia.

Institutional Review Board Statement: Not applicable.

Informed Consent Statement: Not applicable.

Data Availability Statement: Not applicable.

Acknowledgments: L.A.A. extends her appreciation to Princess Nourah bint Abdulrahman University Researchers Supporting Project Number PNURSP2023R82, Princess Nourah bint Abdulrahman University, Riyadh, Saudi Arabia.

Conflicts of Interest: The authors declare no conflict of interest.

References

1. Garoy, E.Y.; Gebreab, Y.B.; Achila, O.O.; Tekeste, D.G.; Kesete, R.; Ghirmay, R.; Kiflay, R.; Tesfu, T. Methicillin-Resistant, *Staphylococcus aureus* (MRSA): Prevalence and Antimicrobial Sensitivity Pattern among Patients-A Multicenter Study in Asmara, Eritrea. *Can. J. Infect. Dis. Med. Microbiol.* **2019**, *2019*, 8321834. [CrossRef]
2. Fair, R.J.; Tor, Y. Antibiotics and Bacterial Resistance in the 21st Century. *Perspect. Med. Chem.* **2014**, *6*, PMC-S14459. [CrossRef] [PubMed]
3. Kaur, D.C.; Chate, S.S. Study of antibiotic resistance pattern in methicillin resistant *Staphylococcus aureus* with special reference to newer antibiotic. *J. Glob. Infect. Dis.* **2015**, *7*, 78. [CrossRef]
4. Wielders, C.L.; Fluit, A.C.; Brisse, S.; Verhoef, J.; Schmitz, F.J. *mecA* gene is widely disseminated in *Staphylococcus aureus* population. *J. Clin. Microbiol.* **2002**, *40*, 3970–3975. [CrossRef]
5. Stapleton, P.D.; Taylor, P.W. Methicillin resistance in *Staphylococcus aureus*: Mechanisms and modulation. *Sci Prog* **2002**, *85 Pt 1*, 57–72. [CrossRef]
6. Katkar, K.V.; Suthar, A.C.; Chauhan, V.S. The chemistry, pharmacologic, and therapeutic applications of *Polyalthia longifolia*. *Pharmacogn. Rev.* **2010**, *4*, 62. [PubMed]
7. Parchman, M.L.; Munoz, A. Risk Factors for Methicillin-Resistant *Staphylococcal aureus* Skin and Soft Tissue Infections Presenting in Primary Care: A South Texas Ambulatory Research Network (STARNet) Study. *J. Am. Board Fam. Med.* **2009**, *22*, 375–379. [CrossRef] [PubMed]
8. Green, K.W.; Zelbst, P.J.; Meacham, J.; Bhadauria, V.S. Green supply chain management practices: Impact on performance. *Supply chain management Supply Chain Manag.* **2012**, *17*, 290–305. [CrossRef]
9. Dumyati, G.; Stevens, V.; Hannett, G.E.; Thompson, A.D.; Long, C.; MacCannell, D.; Limbago, B. Community-associated Clostridium difficile infections, Monroe County, New York, USA. *Emerg. Infect. Dis.* **2012**, *18*, 392. [CrossRef]
10. Kirubakari, B.; Chen, Y.; Kanwar, J.R.; Shin, L.N.; Sasidharan, S. Studies on In Vitro Interaction of *Ampicillin* and *Polyalthia longifolia* Leaf Ethyl Acetate Fraction (PLEAF) by Checkerboard Method Against Methicillin Resistant *Staphylococcus aureus* (MRSA). *Curr. Bioact. Compd.* **2020**, *16*, 1049–1062. [CrossRef]
11. Kirubakari, B.; Chen, Y.; Sasidharan, S. Synergistic effect of *Polyalthia longifolia* leaf and antibiotics against clinical isolates of methicillin-resistant *Staphylococcus aureus* (MRSA) by microscopic technique. *Antiinflamm. Antiallergy Agents Med. Chem.* **2020**, *19*, 323–334. [CrossRef] [PubMed]
12. Van Duin, D.; Paterson, D.L. Multidrug-resistant bacteria in the community: Trends and lessons learned. *Dis. Clin.* **2016**, *30*, 377–390.
13. Gube, A.A.; Gonfa, R.; Tadesse, T. Evaluation of Antibiotic Use in Medical Ward of Fitche District Hospital, North Showa Zone, Oromia Region, Ethiopia. *Adv. Pharmacoepidemiol. Drug Saf.* **2017**, *6*, 217.
14. Lee, H.; Jung, D.; Yeom, J.S.; Son, J.S.; Jung, S.I.; Kim, Y.S.; Kim, C.K.; Chang, H.H.; Kim, S.W.; Ki, H.K.; et al. Evaluation of ceftriaxone utilization at multicenter study. *Korean J. Intern. Med.* **2009**, *24*, 374–380. [CrossRef] [PubMed]
15. Richards, D.G.; Wolz, J.P.; Herman, L.M. Vocal mimicry of computer-generated sounds and vocal labeling of objects by a bottlenosed dolphin, *Tursiops truncatus*. *J. Comp. Psychol.* **1984**, *98*, 10. [CrossRef] [PubMed]
16. Patel, U.C.; Mckissic, E.L.; Kasper, D.; Lentino, J.R.; Pachucki, C.T.; Lee, T.; Lopansri, B.K. Outcomes of ceftriax-one use compared to standard of therapy in Methicillin Susceptible *Staphylococcal aureus* (MSSA) blood-stream infections. *Int. J. Clin. Pharm.* **2014**, *36*, 1282–1289. [CrossRef] [PubMed]
17. Kamfose, M.M.; Muriiti, F.G.; Knight, T.; Lasserson, D.; Hayward, G.J.A. Intravenous ceftriaxone versus multiple dosing regimens of intravenous anti-Staphylococcal antibiotics for Methicillin-Susceptible *Staphylococcus aureus* (MSSA): A Systematic Review. *Antibiotic* **2020**, *9*, 39. [CrossRef] [PubMed]
18. Lother, S.A.; Press, N. Once-Daily Treatments for Methicillin-Susceptible *Staphylococcus aureus* Bacteremia: Are They Good Enough? *Curr. Infect. Dis. Rep.* **2017**, *19*, 43. [CrossRef] [PubMed]
19. McGowan, J.E. Antimicrobial resistance in hospital organisms and its relation to antibiotic use. *Rev. Infect. Dis.* **1983**, *5*, 1033–1048. [CrossRef]
20. Bbosa, G.; Mwebaza, N.; Odda, J.; Kyegombe, D.; Ntale, M. Antibiotics/antibacterial drug use, their marketing and promotion during the post-antibiotic golden age and their role in emergence of bacterial resistance. *Health* **2014**, *6*, 410–425. [CrossRef]
21. Wooford, N.; Ellington, M.J. The emergence of antibiotic resistance by mutation. *Clin. Microbiol. Infect.* **2007**, *13*, 5–18. [CrossRef] [PubMed]
22. Von Wintersdorff, C.J.; Penders, J.; van Niekerk, J.M.; Mills, N.D.; Majumder, S.; van Alphen, L.B.; Savelkoul, P.H.; Wolffs, P.F. Dissemination of antimicrobial resistance in microbial ecosystems through horizontal gene transfer. *Front. Microbiol.* **2016**, *7*, 173. [CrossRef] [PubMed]
23. Holcomb, H.G.; Durbin, K.J.; Cho, M.; Choi, K.J.; Darling, N.D.; Angerio, A.D. Methicillin-resistant *Staphylococcus aureus* as a threat to public health: A cellular approach. Georgetown Undergraduate. *J. Health. Sci.* **2008**, *5*, 2008.
24. Al-Zoubi, M.S.; Al-Tayyar, I.A.; Hussein, E.; Jabali, A.A.; Khudairat, S. Antimicrobial susceptibility pattern of *Staphylococcus aureus* isolated from clinical specimens in Northern area of Jordan. *Iran. J. Microbiol.* **2015**, *7*, 265–272. [PubMed]
25. Davis, J. Inactivation of antibiotics and dissemination of resistance genes. *Science* **1994**, *264*, 375–382. [CrossRef] [PubMed]
26. Ahmad, I.; Mehmood, Z.; Mohammad, F. Screening of Some Indian Medicinal Plants for their Antimicrobial Properties. *J. Ethnopharmacol.* **1998**, *62*, 183–193. [CrossRef]

27. Loper, J.E.; Henkels, M.D.; Roberts, R.G.; Grove, G.G.; Willett, M.J.; Smith, T.J. Evaluation of Streptomycin, Oxytetracycline and Copper Resistance of *Erwinia amylavora* isolated from pear orchards in Washington State. *Plant Dis.* **1991**, *75*, 287–290. [CrossRef]
28. Service, R.F. Antibiotics That Resist Resistance. *Science* **1991**, *270*, 724–727. [CrossRef]
29. Liu, C.S.; Cham, T.M.; Yang, C.H.; Chang, H.W.; Chen, C.H.; Chuang, L.Y. Antibacterial properties of Chinese herbal medicines against nosocomial antibiotic resistant strains of *Pseudomonas aeruginosa* in Taiwan. *Am. J. Chin. Med.* **2007**, *35*, 1047–1060. [CrossRef]
30. Alexopoulos, A.; Kimbaris, A.C.; Plessas, S.; Mantzourani, I.; Theodoridou, I.; Stavropoulou, E.; Polissiou, M.G.; Bezirtzoglou, E. Antibacterial activities of essential oils from eight Greek aromatic plants against clinical isolates of *Staphylococcus aureus*. *Anaerobe* **2011**, *17*, 399–402. [CrossRef]
31. Toroglu, S. In-vitro antimicrobial activity and synergistic/antagonistic effect of interactions between antibiotics and some spice essential oils. *J. Environ. Biol.* **2011**, *32*, 23–29. [PubMed]
32. Chung, P.Y.; Navaratnam, P.; Chung, L.Y. Synergistic antimicrobial activity between pentacyclic triterpenoids and antibiotics against *Staphylococcus aureus* strains. *Ann. Clin. Microbiol. Antimicrob.* **2011**, *10*, 25. [CrossRef] [PubMed]
33. Inui, T.; Wang, Y.; Deng, S.; Smith, D.C.; Franzblau, S.G.; Paul, G.F. Counter-current chromatography-based analysis of synergy in an anti-tuberculosis ethnobotanical. *J. Chromatogr. A* **2008**, *1151*, 211–215. [CrossRef] [PubMed]
34. Adwan, G.; Mhanna, M. Synergistic effects of plant extracts and antibiotics on *Staphylococcus aureus* strains isolated from clinical specimen. *Asian Pac. J. Trop. Med.* **2009**, *2*, 46–51.
35. Ahmed, Z.; Khan, S.S.; Khan, M.; Tanveer, A.; Lone, Z.A. Synergistic effect of *Salvadora persica* extracts, tetracycline and penicillin against *Staphylococcus aureus*. *Afr. J. Basic Appl. Sci.* **2010**, *2*, 25–29.
36. Aboulmagd, E.; Al-Mohammed, H.I.; Al-Badry, S. Synergism and post-antibiotic effect of green tea ex-tract and imipenem against methicillin-resistant *Staphylococcus aureus*. *Microbiol. J.* **2011**, *1*, 89–96. [CrossRef]
37. Betoni, J.E.; Mantovani, R.P.; Barbosa, L.N.; Di Stasi, L.C.; Fernandes Junior, A. Synergism between plant ex-tract and antimicrobial drugs used on *Staphylococcus aureus* diseases. *Mem. Inst. Oswaldo Cruz* **2006**, *101*, 387–390. [CrossRef] [PubMed]
38. Shang, D.; Liu, Y.; Jiang, F.; Ji, F.; Wang, H.; Han, X. Synergistic Antibacterial Activity of Designed Trp-Containing Antibacterial Peptides in Combination with Antibiotics Against Multidrug-Resistant *Staphylococcus epidermidis*. *Front. Microbiol.* **2019**, *10*, 2719. [CrossRef]
39. AL-Ali, K.; Abdelrazik, M.; Hemeg, H.; Ozbak, H. Antibacterial activity of four herbal extracts against methicillin resistant bacteria isolates collected from Almadinah Hospitals. Saudi Arabia. *Int. J. Acad. Res.* **2014**, *2*, 27–34.
40. Chanda, S.; Nair, R. Antimicrobial activity of *Polyalthia longifolia* (sonn.) thw. var. pendula leaf extracts against 91 clinically important pathogenic microbial strains. *Chin. Med.* **2010**, *1*, 31–38. [CrossRef]
41. Phadnis, A.P.; Patwardhan, S.A.; Dhaneshwar, N.N. Clerodane diterpenoids from *Polyalthia longifolia*. *Phytochemistry* **1988**, *27*, 2899–2901. [CrossRef]
42. Chaudhary, A.; Sood, S.; Das, P.; Kaur, P.; Mahajan, I.; Gulati, A.; Singh, B. Synthesis of novel antimicrobial aryl himachalene derivatives from naturally occurring himachalenes. *EXCLI J.* **2014**, *13*, 1216.
43. Silva, A.; Silva, S.A.; Carpena, M.; Garcia-Oliveira, P.; Gullón, P.; Barroso, M.F.; Prieto, M.A.; Simal-Gandara, J. Macroalgae as a Source of Valuable Antimicrobial Compounds: Extraction and Applications. *Antibiotics* **2020**, *9*, 642. [CrossRef]
44. Rahman, M.M.; Ahmad, S.H.; Mohamed, M.T.; Ab Rahman, M.Z. Antimicrobial compounds from leaf extracts of *Jatropha curcas*, *Psidium guajava*, and *Andrographis paniculata*. *Sci. World J.* **2014**, *2014*, 635240. [CrossRef]
45. Qian, W.; Liu, M.; Fu, Y.; Zhang, J.; Liu, W.; Li, J.; Li, X.; Li, Y.; Wang, T. Antimicrobial mechanism of luteolin against *Staphylococcus aureus* and *Listeria monocytogenes* and its antibiofilm properties. *Microb. Pathog.* **2020**, *142*, 104056. [CrossRef] [PubMed]
46. Arima, H.; Ashida, H.; Danno, G. Rutin enhanced antibacterial activities of flavonoids against *Bacillus cereus* and *Salmonella enteritidis*. *Biosci. Biotechnol. Biochem.* **2002**, *66*, 1009–1014. [CrossRef]
47. Lee, T.H.; Wang, M.J.; Chen, P.Y.; Wu, T.Y.; Wen, W.C.; Tsai, F.Y.; Lee, C.K. Constituents of *Polyalthia longifolia* var. *pendula*. *J. Nat. Prod.* **2009**, *72*, 1960–1963. [CrossRef]
48. Murthy, M.M.; Subramanyam, M.; Bindu, M.H.; Annapurna, J. Antimicrobial activity of clerodane diterpenoids from *Polyalthia longifolia* seeds. *Fitoterapia* **2005**, *76*, 336–339. [CrossRef]
49. Sashidhara, K.V.; Singh, S.P.; Sarkar, J.; Sinha, S.J. Cytotoxic clerodane diterpenoids from the leaves of *Polyalthia longifolia*. *Nat. Prod. Res.* **2010**, *24*, 1687–1694. [CrossRef]
50. Nahari, D.S.; Prasetyawan, S.; Beltran, M.A.G.; Aulanni'am, A. Separation of flavonoids in the extract *Polyalthia longifolia* (Sonn.) Thw. leaves from Indonesia and the Philippines. *J. Phys. Conf. Ser.* **2019**, *1374*, 012001. [CrossRef]
51. Gupta, V.K.; Tiwari, N.; Gupta, P.; Verma, S.; Pal, A.; Srivastava, S.K.; Darokar, M.P. A clerodane diterpene from *Polyalthia longifolia* as a modifying agent of the resistance of methicillin resistant *Staphylococcus aureus*. *Phytomedicine* **2016**, *23*, 654–661. [CrossRef] [PubMed]
52. Gupta, V.K.; Verma, S.; Pal, A.; Srivastava, S.K.; Srivastava, P.K.; Darokar, M.P. In vivo efficacy and synergistic interaction of 16α-hydroxycleroda-3, 13 (14) Z-dien-15, 16-olide, a clerodane diterpene from *Polyalthia longifolia* against methicillin-resistant *Staphylococcus aureus*. *Appl. Microbiol. Biotechnol.* **2013**, *97*, 9121–9131. [CrossRef] [PubMed]
53. Mandal, S.M.; Roy, A.; Ghosh, A.K.; Hazra, T.K.; Basak, A.; Franco, O.L. Challenges and future prospects of antibiotic therapy: From peptides to phages utilization. *Front. Pharmacol.* **2014**, *5*, 105. [CrossRef]
54. Li, X.Z.; Nikaido, H. Efflux-mediated drug resistance in bacteria: An update. *Drugs* **2009**, *69*, 1555–1623. [CrossRef]

55. Tang, Y.W.; Kilic, A.; Yang, Q.; McAllister, S.K.; Li, H.; Miller, R.S.; McCormac, M.; Tracy, K.D.; Stratton, C.W.; Han, J.; et al. StaphPlex system for rapid and simultaneous identification of antibiotic resistance determinants and Panton-Valentine leukocidin detection of staphylococci from positive blood cultures. *J. Clin. Microbiol.* **2007**, *45*, 1867–1873. [CrossRef]
56. Moisan, H.; Pruneau, M.; Malouin, F. Binding of ceftaroline to penicillin-binding proteins of *Staphylococcus aureus* and *Streptococcus pneumoniae*. *J. Antimicrob. Chemother.* **2010**, *65*, 713–716. [CrossRef]
57. Gradelski, E.; Valera, L.; Aleksunes, L.; Bonner, D.; Fung-Tomc, J. Correlation between genotype and phenotypic categorization of Staphylococci based on methicillin susceptibility and resistance. *J. Clin. Microbiol.* **2001**, *39*, 2961–2963. [CrossRef] [PubMed]
58. Hartman, B.J.; Tomasz, A. Low-affinity penicillin-binding protein associated with beta-lactam resistance in *Staphylococcus aureus*. *J. Bacteriol.* **1984**, *158*, 513–516. [CrossRef]
59. Özel, G.; Aslan, V.; Erdem, G.B.; Çağatay, M.; Sencan, I.; Mert, A. Comparison of oxacillin, cefoxitin, ceftizoxime, and moxalactam disk diffusion methods for detection of methicillin susceptibility in staphylococci. *Mikrobiyoloji Bulteni* **2011**, *45*, 258–265.
60. Prasad, K.N.; Kumar, R.; Tiwari, D.P.; Mishra, K.K.; Ayyagari, A. Comparison of various conventional methods with a polymerase chain reaction assay for detecting methicillin-resistant & susceptible *Staphylococcus aureus* strains. *Indian J. Med. Res.* **2000**, *112*, 198–202.
61. Mohd-Fuat, A.R.; Kofi, E.A.; Allan, G.G. Mutagenic and cytotoxic properties of three herbal plants from Southeast Asia. *Trop. Biomed.* **2007**, *24*, 49–59.
62. Allen, J.R.F.; Holmstedt, B.R. The simple β-carboline alkaloids. *Phytochemistry* **1980**, *19*, 1573–1582. [CrossRef]
63. Patel, K.; Gadewar, M.; Tripathi, R.; Prasad, S.; Patel, D.K. A review on medicinal importance, pharmacological activity and bioanalytical aspects of beta-carboline alkaloid "Harmine". *Asian Pac. J. Trop. Biomed.* **2012**, *2*, 660–664. [CrossRef]
64. Boeira, J.M.; Viana, A.F.; Picada, J.N.; Henriques, J.A. Genotoxic and recombinogenic activities of the two beta-carboline alkaloids harman and harmine in *Saccharomyces cerevisiae*. *Mutat. Res.* **2002**, *500*, 39–48. [CrossRef] [PubMed]
65. Rietjens, I.M.; Boersma, M.G.; Van Der Woude, H.; Jeurissen, S.M.; Schutte, M.E.; Alink, G.M.J.; Mutagenesis, M.M.O. Flavonoids and alkenylbenzenes: Mechanisms of mutagenic action and carcinogenic risk. *Mutat. Res.* **2005**, *574*, 124–138. [CrossRef] [PubMed]
66. Faizi, S.; Khan, R.A.; Azher, S.; Khan, S.A.; Tauseef, S.; Ahmad, A. New antimicrobial alkaloids from the roots of *Polyalthia longifolia* var. *pendula*. *Planta Med.* **2003**, *69*, 350–355. [CrossRef]
67. Jothy, S.L.; Chen, Y.; Kanwar, J.R.; Sasidharan, S. Evaluation of the Genotoxic Potential againstH2O2-Radical-Mediated DNA Damage and Acute Oral Toxicity of Standardized Extract of *Polyalthia longifolia* Leaf. *Evid. Based Complement. Altern. Med.* **2013**, *2013*, 925380. [CrossRef]
68. Jothy, S.L.; Aziz, A.; Chen, Y.; Sasidharan, S. Antioxidant Activity and Hepatoprotective Potential of *Polyalthia longifolia* and *Cassia spectabilis* Leaves against Paracetamol-Induced Liver Injury. *Evid. Based Complement. Altern. Med.* **2012**, *2012*, 561284. [CrossRef]
69. Jothy, S.L.; Saito, T.; Kanwar, J.R.; Chen, Y.; Aziz, A.; Yin-Hui, L.; Sasidharan, S. Radioprotective activity of *Polyalthia longifolia* standardized extract against X-ray radiation injury in mice. *Phys. Med.* **2016**, *32*, 150–161. [CrossRef]
70. Mostafa, N.M.; Edmond, M.P.; El-Shazly, M.; Fahmy, H.A.; Sherif, N.H.; Singab, A.N.B. Phytoconstituents and renoprotective effect of *Polyalthia longifolia* leaves extract on radiation-induced nephritis in rats via TGF-β/smad pathway. *Nat. Prod. Res.* **2021**, *36*, 4187–4192. [CrossRef]
71. Vijayarathna, S. Fundamental Studies on the Mechanism of *Polyalthia longifolia* (Sonn.) Thwaites Polyphenols Action in HeLa Cells in Relation to microRNA Regulation. PhD Thesis, Universiti Sains Malaysia, Pinang, Malaysia, 2017.
72. Vijayarathna, S.; Chen, Y.; Kanwar, J.R.; Sasidharan, S. Standardized *Polyalthia longifolia* leaf extract (PLME) inhibits cell proliferation and promotes apoptosis: The anti-cancer study with various microscopy methods. *Biomed. Pharmacother.* **2017**, *91*, 366–377. [CrossRef] [PubMed]
73. Vijayarathna, S.; Oon, C.E.; Chen, Y.; Kanwar, J.R.; Sasidharan, S. *Polyalthia longifolia* Methanolic Leaf Extracts (PLME) induce apoptosis, cell cycle arrest and mitochondrial potential depolarization by possibly modulating the redox status in hela cells. *Biomed. Pharmacother.* **2017**, *89*, 499–514. [CrossRef] [PubMed]
74. Vijayarathna, S.; Shanmugapriya Khanwar, J.R.; Sasidharan, S. Standardized Polyalthia longifolia methanolic leaf extracts (PLME) inhibits HeLa cells through inducing microRNAs expression and apoptosis. In Proceedings of the International Conference on Traditional & Alternative Medicine, Kerala, India, 27 September 2017.
75. Hemagirri, M.; Sasidharan, S. In vitro antiaging activity of polyphenol rich *Polyalthia longifolia* (Annonaceae) leaf extract in *Saccharomyces cerevisiae* BY611 yeast cells. *J. Ethnopharmacol.* **2022**, *290*, 115110. [CrossRef]
76. Vaghasiya, Y.; Patel, H.; Chanda, S. Antibacterial activity of methanol extract of *Mangifera indica* against some human pathogens and its phytochemical study. *Afr. J. Biotechnol.* **2011**, *10*, 15788–15794. [CrossRef]
77. Manasa, M.; Vivek, M.N.; Yashoda, K.; Onkarappa, R.; Prashith, K.T.R. Antimicrobial activity of leaf and pericarp extracts of *Polyalthia longifolia* (Annonaceae). *J. Pharm. Sci.* **2004**, *393*, 221–225.
78. Bauer, A.W.; Kirby, W.M.; Sherris, J.C.; Turck, M. Antibiotic susceptibility testing by a standardized single disk method. *Am. J. Clin. Pathol.* **1966**, *45*, 493–496. [CrossRef]
79. CLSI Standard. *M100 Performance Standards for Antimicrobial Susceptibility Testing*, 30th ed.; Clinical and Laboratory Standards Institute (CLSI): Wayne, PA, USA, 2020.

80. Akinyemi, K.O.; Oladapo, O.; Okwara, C.E.; Ibe, C.C.; Fasure, K.A. Screening of crude extracts of six medicinal plants used in South-West Nigerian unorthodox medicine for anti-methicillin resistant *Staphylococcus aureus* activity. *BMC Complement Altern. Med.* **2005**, *5*, 6. [CrossRef]
81. Andrews, J.M. Determination of minimum inhibitory concentrations. *J. Antimicrob. Chemother.* **2001**, *48*, 5–16. [CrossRef] [PubMed]
82. Sampaio, F.C.; Pereira, M.d.S.V.; Dias, C.S.; Costa, V.C.; Conde, N.C.; Buzalaf, M.A. In vitro antimicrobial activity of *Caesalpinia ferrea* Martius fruits against oral pathogens. *J. Ethnopharmacol.* **2009**, *124*, 289–294. [CrossRef]
83. Agarwal, A.; Jain, N.; Jain, A. Synergistic effect of cefixime and cloxacillin combination against common bacterial pathogens causing community acquired pneumonia. *Indian J. Pharmacol.* **2007**, *39*, 251–252.
84. Zhao, W.H.; Hu, Z.Q.; Okubo, S.; Hara, Y.; Shimamura, T. Mechanism of Synergy between Epigallocatechin Gallate and b-Lactams against Methicillin-Resistant *Staphylococcus aureus*. *Antimicrob. Agents Chemother.* **2001**, *45*, 1737–1742. [CrossRef] [PubMed]

antibiotics MDPI

Article

Antimicrobial Polymeric Surfaces Using Embedded Silver Nanoparticles

Pooja Sharma [1,2,*], Luisa Fialho [3,4], Nuno Miguel Figueiredo [1], Ricardo Serra [1], Albano Cavaleiro [1,5] and Sandra Carvalho [1,2,5]

1 CEMMPRE, Mechanical Engineering Department, University of Coimbra, 3030-788 Coimbra, Portugal
2 CFUM-UP, Centro de Física das Universidades do Minho e do Porto, University of Minho, Campus of Azurém, 4800-058 Guimaraes, Portugal
3 i3S—Instituto de Investigação e Inovação em Saúde, Universidade do Porto, Rua Alfredo Allen, 208, 4200-135 Porto, Portugal
4 INEB—Instituto de Engenharia Biomédica, Universidade do Porto, Rua Alfredo Allen, 208, 4200-135 Porto, Portugal
5 IPN—LED&MAT, Instituto Pedro Nunes, Rua Pedro Nunes, 3030-199 Coimbra, Portugal
* Correspondence: b12473@fisica.uminho.pt

Abstract: Pathogens (disease-causing microorganisms) can survive up to a few days on surfaces and can propagate through surfaces in high percentages, and thus, these surfaces turn into a primary source of pathogen transmission. To prevent and mitigate pathogen transmission, antimicrobial surfaces seem to be a promising option that can be prepared by using resilient, mass-produced polymers with partly embedded antimicrobial nanoparticles (NPs) with controlled size. In the present study, a 6 nm thick Ag nanolayer was sputter deposited on polycarbonate (PC) substrate and then thermally annealed, in a first step at 120 °C (temperature below Tg) for two hours, for promoting NP diffusion and growth, and in a second step at 180 °C (temperature above Tg) for 22 h, for promoting thermal embedding of the NPs into the polymer surface. The variation in the height of NPs on the polymer surface with thermal annealing confirms the embedding of NPs. It was shown that the incorporation of silver nanoparticles (Ag NPs) had a great impact on the antibacterial capacity, as the Ag NP-embedded polymer surface presented an inhibition effect on the growth of Gram-positive and Gram-negative bacteria. The tested surface-engineering process of incorporating antimicrobial Ag NPs in a polymer surface is both cost-effective and highly scalable.

Keywords: polycarbonate; silver nanoparticles; thermal embedding; glass transition temperature; antimicrobial activity

Citation: Sharma, P.; Fialho, L.; Figueiredo, N.M.; Serra, R.; Cavaleiro, A.; Carvalho, S. Antimicrobial Polymeric Surfaces Using Embedded Silver Nanoparticles. *Antibiotics* **2023**, *12*, 207. https://doi.org/10.3390/antibiotics12020207

Academic Editor: Helena P. Felgueiras

Received: 31 December 2022
Revised: 15 January 2023
Accepted: 16 January 2023
Published: 18 January 2023

1. Introduction

Infectious diseases (IDs) are a leading cause of death worldwide and affect so many people globally per year. Pathogens (bacteria, fungi and viruses), including virulent and drug-resistant strains, have consistently been found in high percentages on surfaces all over the world [1,2]. Most pathogens are able to survive on surfaces for up to several months that can act as primary sources of pathogen transmission if no disinfection is performed [3–5]. This transmission of pathogens through inanimate surfaces and equipment is a crucial challenging health problem in intensive care units all over the world [5]. Another common example is the transmission of impetigo (a very contagious skin disease that affects mostly children) through contaminated toys [6].

To prevent this surface transmission of diseases, novel multifunctional antiviral/antimicrobial surfaces need to be developed. Silver nanoparticles (Ag NPs) have been shown to exhibit broad-spectrum antiviral/antimicrobial properties, including against virulent and drug-resistant strains [7]. Although they have been associated with health hazards [7], when released in small concentrations, Ag is non-toxic to humans, leading to an increased interest in their application in medical devices [8,9], and also being highlighted as an anti-inflammatory agent [7]. Ag NPs have also been investigated as a potential anticancer agent [10,11]. In vivo assays demonstrate the effectiveness of Ag NPs on antimicrobial activity, epithelization, and collagen deposition [12]. Ag antimicrobial activity mechanism may occur in various forms such as ionic, reactive oxygen species (ROS), or by NPs internalization, depending on the embedding matrix [7,13,14].

Further extension of applications of biocide metals can be made possible via polymer/metal composites. A large percentage of the steady growth of the polymer market is due to their interesting properties, such as stiffness/flexibility, opacity/transparency, durability, barrier behavior for gases and liquids, chemical and heat resistance, and low production cost [15]. This set of properties has unlocked their extensive use in several industries and applications, including in medical devices, packaging products, and delivery systems for solid and liquid pharmaceuticals [15]. Thus, the development and deployment of antimicrobial polymers could be a highly desirable strategy to avoid surface transmission of pathogens. Such polymers could be prepared either by embedding a biocide agent into the bulk polymer or by stabilizing bioactive NPs over the polymeric surfaces [16,17].

Polycarbonate (PC) is a transparent thermoplastic with carbonate functional groups, which can be melted and forced into a mold with high pressure to give it the desired shape. It has remarkable mechanical and optical properties, being resistant to impact and fracture [18]. Moreover, it may show eco-friendly processing and recyclability. Due to its desirable characteristics, PC is extensively used worldwide in products such as water bottles, monitor screens, aircraft interiors, screens, cover sheets, automotive light covers, etc. [18–20]. Therefore, the development of a post-processing, affordable, scalable method for making PC surfaces antimicrobial might have a great positive impact on human health and safety.

The direct deposition of NPs with biocide activity over the surface of materials is an efficient way to provide it with antiviral/antimicrobial functionality. Physical, chemical, and biological methods can be used to synthesize such NPs [10,11,13]. However, the deposited NPs usually show very low adhesion and, hence, can easily agglomerate and/or be lost to the environment by simple mechanical action. A promising method to stabilize NPs on a surface consists of the thermal embedding of NPs over an amorphous substrate [21]. Per surface science, the immersion on a rigid metal NP of a soft material can occur when the surface tension of the NP/air interface is greater than the sum of the surface tensions at the matrix/air and NP/matrix interfaces [22–24]. Considering our system, the surface tension of Ag (1200 mJ/m^2) is significantly higher than that of PC (30–40 mJ/m^2) [25,26]. However, in the case of a solid-state polymer such as PC, the NP embedding requires a change in Gibbs free energy to promote the work of adhesion. This energy can be provided by external heating (thermal annealing) at a temperature near the thermal softening temperature (Tg) of the amorphous material, leading to an increase in polymer chain mobility at the polymer/air interface and facilitating the indentation [20,27].

In this study, PC polymeric surfaces were coated with Ag NPs and subsequently thermally annealed below and above the substrate's Tg. The samples were characterized concerning surface morphology, NP adhesion, reflectivity and antibacterial properties.

2. Materials and Methods

2.1. Ag Nanoparticles Deposition

An Ag nanolayer with a nominal thickness of ~6 nm was deposited using a pure (99.99%) Ag planar target (50.8 mm in diameter by 3 mm in thickness) in a pure (99.999%) Ar atmosphere. Ar flux in the inlet was set to 30 sccm, corresponding to a constant deposition pressure of 0.3 Pa. The target-to-substrate distance was maintained at 12 cm, and the rotation speed of the substrate holder was set to 30 rpm. The substrate holder was kept at a floating potential. PC substrates of 10 × 10 × 5 mm were used. Preceding the depositions, the PC substrates were cleaned in an ultrasonic bath, in two steps using different solvents: ethanol, and distilled water, each one for 10 min. A DC power supply (Advanced Energy

Pinnacle Plus, CO, USA) was used with a constant power of 100 W during deposition. The deposition chamber was pumped with rotary (Pfeiffer Vacuum, Germany DUO 20 M, pumping speed 20 m^3/h) and diffusion (BOC Edwards Diffstak, England 160/700, pumping speed 2736 m^3/h) pumps to a base pressure below 5×10^{-4} Pa.

2.2. Thermal Annealing Treatments

In thermogravimetric (TG) measurements, a Netzsch TG 209 F1 Libra was used, where a nitrogen flow of 20 mL/min was used, with a heating ramp at 10 K/min from 30 to 600 °C. Regarding the conditions defined in the differential scanning calorimetry (DSC) measurements, a nitrogen flow of 40 mL/min and a heating rate of 20 K/min between −95 and 300 °C was implemented. The equipment used was the Netzsch DSC 204 F1 Phoenix. For data processing of results, the Proteus 8.0® program was used. DSC was performed on the PC substrate to determine its glass transition temperature (Tg). Based on this observation, thermal annealing of the deposited substrates was performed in a two-stage sequence: (i) in the first step, the substrates were annealed in the oven for two hours at 120 °C (a temperature below the substrate Tg) to promote thermal dewetting of the Ag nanolayer and NP formation; (ii) in a second step, these samples were subsequently heated for 22 h at 180 °C (a temperature above the substrate Tg) to promote NP growth and NP thermal embedding.

2.3. General Characterizations

In this work, in order to test the adhesion of the coatings/NPs to the polymeric substrates, 1 cm viton balls were adapted into a scratch tester stylus to perform scratch tests using a progressive load setup up to 15 N. The vertical load rate was set to 100 N/min, whereas the horizontal velocity was set to 30 mm/min. Experiments were performed in a CSEM Revetest automatic scratch tester.

The surface morphology of samples was analyzed by atomic force microscopy (AFM) on a Bruker diInnova using a silicon tip with 6 nm of tip radius. The scan area for all films was 5×5 μm^2. Grain size, shape, and height distributions were obtained for each sample using the open-source software Gwyddion 2.61. For the reflectivity measurement, the device used was a portable spectrophotometer of the brand ColorEye® XTH, model Greta Macbeth with SCI (specular component included). The reflectivity values are given as a function of the wavelength of the light, corresponding in this case to a range between 360 and 750 nm, with an increment of 10 nm between the points.

2.4. Antibacterial Assessment

The antibacterial activity of the films was tested using two bacteria, a Gram-positive *Staphylococcus aureus* (ATCC 6538) and a Gram-negative *Escherichia coli* (CECT 434). The zone of inhibition (ZoI) or halo test, adapted from the Kirby–Bauer method, was performed to evaluate the antibacterial activity. Firstly, the samples were esterized by exposure to UV light for 1 h. The inoculum of each bacterium was prepared by inoculation of a single colony in 30 mL of Tryptic Soy Broth (TSB, Frilabo, Milheirós, Portugal) and incubated at 37 °C and 120 rpm overnight. The optical density (OD) of the inoculum was measured at 620 nm and properly diluted in culture media to 1×108 CFU/mL. An aliquot of cellular suspension (100 μL) was spread in Tryptic Soy Agar (TSA, Merck, Darmstadt, Germany) Petri dishes. After medium solidification, the samples were placed separately on the top of the agar plate, placed on the side with the film in contact with the agar, and incubated for 24 h at 37 °C. After the incubation period, the halo (transparent medium zone, which translates that there is no bacteria growth) formed around the samples was photographed to record the results (images captured by Image LabTM software 3.0). All the experiments were repeated with at least three independent assays. After the halo test, the samples were removed and washed three times in distilled water, followed by sequential dehydration in graded ethanol solutions (70, 95, and 100% (*v*/*v*) for 10, 10, and 20 min, respectively), and stored in a desiccator. Then, the cultured samples were coated with an Au/Pd thin film by sputtering using the SPI Module Sputter Coater equipment and analyzed by SEM (Quanta 400 FEG ESEM/EDAX Genesis X4M, Thermo Fisher Scientific, Waltham, MA, USA).

3. Results and Discussion

3.1. DSC and TG

Figure 1 shows the TG/DTG analysis of the PC substrate. No change in mass is observed at around 100–200 °C, indicating that the PC is free from any absorbed solvent or moisture. The major mass loss of around 80% occurred from 400 to 550 °C, which is related to the decomposition of the polymer matrix. The PC starts to degrade from approximately 440 °C and degraded up to 86.3% in mass at 496.9 °C. As the temperature increases further, the mass loss increases up to ~35.3% at 534.2 °C. These temperatures are close to those reported in the literature for the degradation temperature of PC [28].

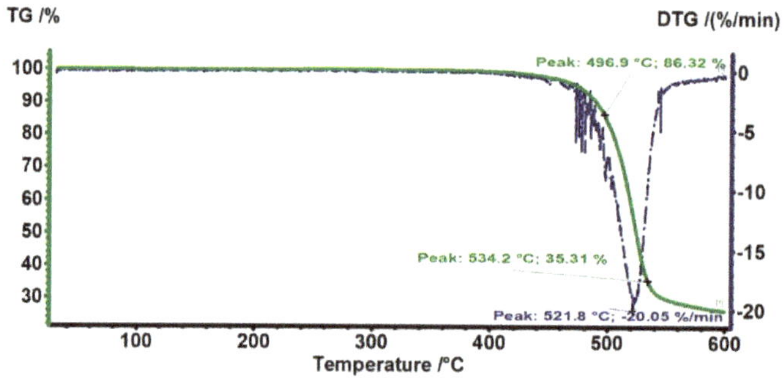

Figure 1. TG thermograms of PC, where TG % is the mass percentage of the polymer sample remaining after heating the polymer to a certain temperature.

Figure 2 shows the TG/DTG analysis of the PC substrate, made during 24 h at 180 °C. The results indicate that there is no significant mass loss within this time period. It can be seen that there is only about a 3% percent mass loss, which states the stability of the PC polymer during thermal annealing at 180 °C for 24 h.

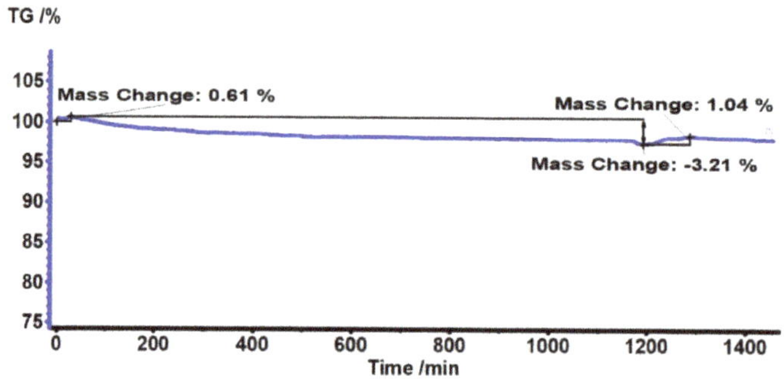

Figure 2. TG thermograms of PC substrate heated at 180 °C for 24 h, where TG % is the mass percentage of the polymer sample remaining after heating the polymer to a certain temperature.

The mechanical and physical properties of an amorphous or semi-crystalline polymer are dependent on the behavior of the temperature, which is defined by the glass transition temperature. When this Tg is reached, the transition of the amorphous region from a rigid state to a more flexible and ductile state takes place. When the temperature is below Tg, the molecular chains of amorphous materials are not mobile. On the other hand, at a temperature above the Tg, the amount of energy supplied to the material allows these chains to acquire mobility, exhibiting a "rubbery" behavior [29].

DSC characterization was performed to determine the glass transition temperature (Tg) of PC polymers, as shown in Figure 3. The DSC curve of the PC in the heating cycle exhibits a glass transition temperature of ≈150.4 °C. No amorphous polymer can exhibit a melting transition, as melting is a first-order transition occurring only for crystalline polymers [27].

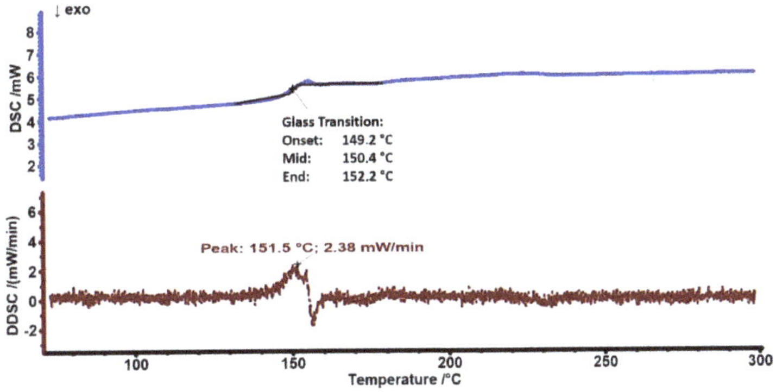

Figure 3. DSC and DDSC thermograms of PC substrate.

3.2. Surface Morphology

The surface morphology of samples was examined by AFM, as shown in Figure 4. Calculated surface analysis parameters, such as NP lateral size and height distribution curves, are shown in Figure 5. Figure 4a,b are shown the two-dimensional AFM images of the pure and Ag-coated PC substrates, respectively. Despite the nominal thickness of the Ag coating being around 6 nm, such an amount was clearly below the percolation limit for this system, originating in the presence of discontinuous, well-separated small islands on the surface of the substrate. For the as-deposited sample, the average NP diameter was 45 nm, and the average NP height was 14 nm (Figure 5a). After the thermal annealing treatment at 120 °C, the nanoparticles increased in both lateral size and height, developing bimodal size/height distribution curves centered at around 85/30 nm and 170/90 nm (Figure 5b), respectively. This can be explained by the thermally activated surface diffusion and coalescence of bigger nanoparticles along with the nucleation and growth of newer nanoparticles from smaller islands and/or individual atoms. Interestingly, after an additional thermal annealing step at 180 °C (temperature above the substrate Tg, determined by DSC to be 150 °C) both bimodal size/height distribution curves were centered at lower values, close to 78/16 nm and 114/16 nm (Figure 5c). These results, in particular the decreasing NP heights, demonstrate the successful thermal embedding of Ag nanoparticles into the PC surface. If there was no surface embedding process taking place, the combined 120 °C + 180 °C thermal annealing process would obviously generate bigger nanoparticles than the single 120 °C thermal annealing process (due to more energy being given to the system), which is not the case. On a different perspective, when annealing above the substrate Tg, both the NP growth and surface embedding process can take place over time, although most of the metal NP diffusion and growth occurs during the first 15–30 min of annealing, after which the surface embedding process becomes dominant [30]. Thus, even considering that the thermal embedding process was dominant during the 24 h annealing treatment above Tg, some extra energy was still given to the nanoparticles that could translate into some NP growth. We can justify the fact that bigger NP sizes were observed in sample PC/Ag/120 °C than in sample PC/Ag/180 °C by the diffusion and growth of NPs that naturally takes place, even at room temperature, for the case of un-stabilized Ag nanostructures over surfaces [31] (i.e., between the thermal annealing treatment and the SEM analysis, diffusion and growth occurred for the 120 °C condition but not for the 120 + 180 °C condition where the NPs were stabilized). Thus, the surface-embedded Ag NPs have the advantage of being stable over time, in different environments, and even when subject to mechanical contact action (as will be demonstrated ahead in Section 3.4).

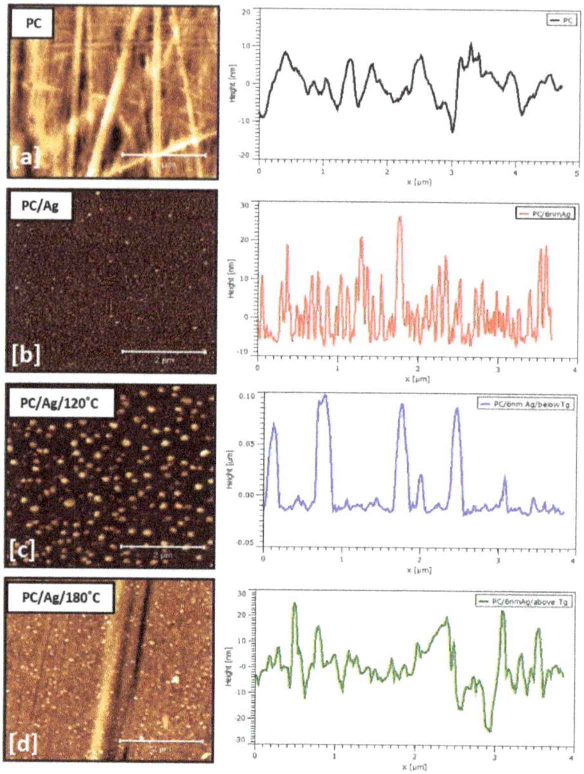

Figure 4. AFM surface micrograph and line profile of sample: (**a**) PC; (**b**) PC/Ag; (**c**) PC/Ag/120 °C; (**d**) PC/Ag/180 °C.

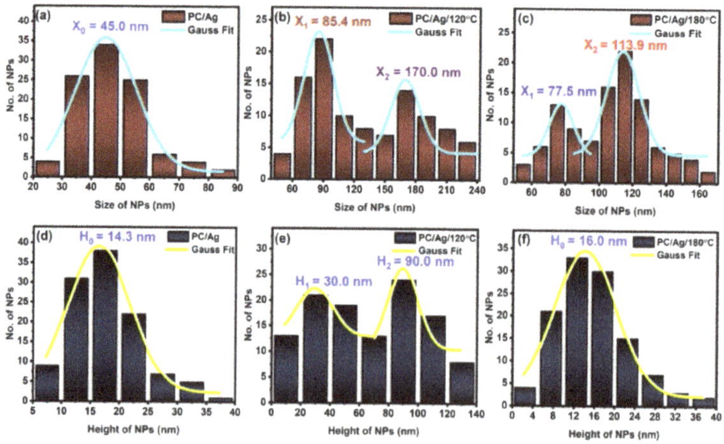

Figure 5. Ag NPs' size distribution curves for sample: (**a**) PC/Ag; (**b**) PC/Ag/120 °C; (**c**) PC/Ag/180 °C and Ag NPs' height distribution curves for sample (**d**) PC/Ag; (**e**) PC/Ag/120 °C; (**f**) PC/Ag/180 °C.

3.3. Reflectivity

The optical properties of a nanocomposite containing metal NPs are an important, non-destructive way of accessing the information on the NPs' morphology and environment.

Figure 6 shows the reflectivity plot of pristine and coated PC substrates. All the coated samples showed the localized surface plasmon resonance (LSPR) scattering bands typical of Ag nanoparticles of sizes above 30–40 nm [32]. In PC/6 nm Ag sample (as-deposited film), the spectrum shows a broad LSPR band centered around 524 nm. Such broad red-shifted LSPR bands are characteristic of an ensemble of ellipsoidal nanoparticles having low aspect ratios and/or wide distribution of sizes/shapes [33]. From the optical results alone, we can conclude that the nanolayer of 6 nm is mostly discontinuous, as was attested before by AFM analysis. After annealing at 120 °C for 2 h, a narrowing and a blue-shift of the LSPR scattering band to ~490 nm due to NP shape renormalization (a tendency of the particles to become more spherical) is observed [34], confirming previous AFM observations on the general increase in NPs' aspect ratio. Further annealing above Tg causes a pronounced narrowing and red-shift of the LSPR scattering band to 548 nm. The narrowing of the LSPR peak can be understood as a narrowing of the NP size and shape distribution curves [34,35] caused by NP diffusion and recrystallization. On the other hand, the pronounced red-shift of the LSPR peak is a confirmation of the NP thermal embedding process. As the NPs penetrate the surface of the polymer, their average surrounding refractive index increases, which causes the buildup of polarization charges that weakens the total restoring force on the dielectric side of the interface, resulting in a lower energy (red-shifted) LSPR scattering peak [35]. Therefore, these optical results are well consistent with the idea of thermally embedded NPs in the PC/Ag/180 °C case.

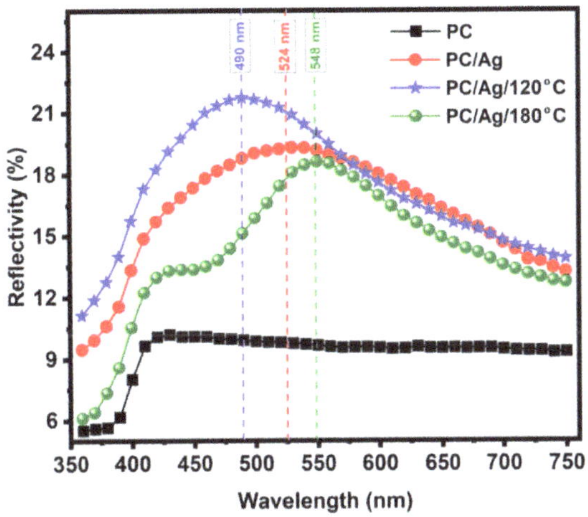

Figure 6. Reflectivity spectra of the pure and coated substrates, before and after each annealing step.

3.4. Adhesion Test

The adhesion of coatings to substrates is an important physical characteristic for evaluating the performance and reliability of coated components. Several techniques allow to check the adhesion of the coating-substrate: the tape test, indentation test, scratch test, laser fragmentation, etc. [36]. In order to test the adhesion of NPs/nanolayer to the PC surface, 1 cm viton balls were adapted into a scratch tester stylus in order to perform scratch tests using a progressive load setup up to 15 N. Such setup was created to mimic the contact-slide of a human finger over a surface. Visual pictures of the tested zones are shown in Figure 7 for the as-deposited, 120 °C-annealed and 180 °C-annealed PC/Ag samples.

As expected, the as-deposited sample was easily scratched due to the low adhesion of the 6 nm Ag nanolayer/nanoparticles to the substrate (white zone circumscribed by a yellow circle in Figure 7a). The 120 °C-annealed sample, containing Ag in the form of bigger NPs, showed better adhesion results but still showed some degree of NP delamination from the polymer surface (Figure 7b). For the 180 °C annealed sample (PC/Ag/180 °C), containing thermally embedded Ag nanoparticles, no delamination zone was observed (Figure 7c), demonstrating the highly robust nature and stability of the multifunctional nanocomposite surfaces obtained by thermal embedding of Ag NPs into PC.

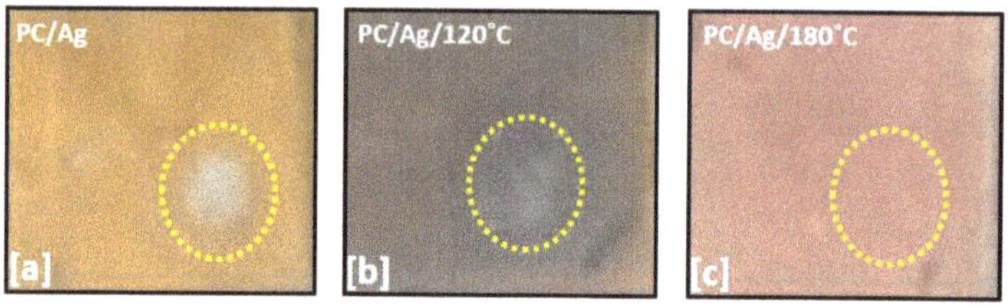

Figure 7. Pictures of the scratch-tested areas of the sample: (**a**) PC/Ag; (**b**)PC/Ag/120 °C; (**c**) PC/Ag/180 °C. Yellow dotted circles were added around the tested zones to facilitate interpretation.

3.5. Antibacterial Activity

Taking into consideration the best result from the adhesion test (thermally annealed sample), the antibacterial activity of sample PC/Ag/180 °C was evaluated, using the uncoated polymeric surface as control, by the zone of inhibition assays (halo test) and SEM analysis after the halo test against *E. coli* and *S. aureus* (Figure 8).

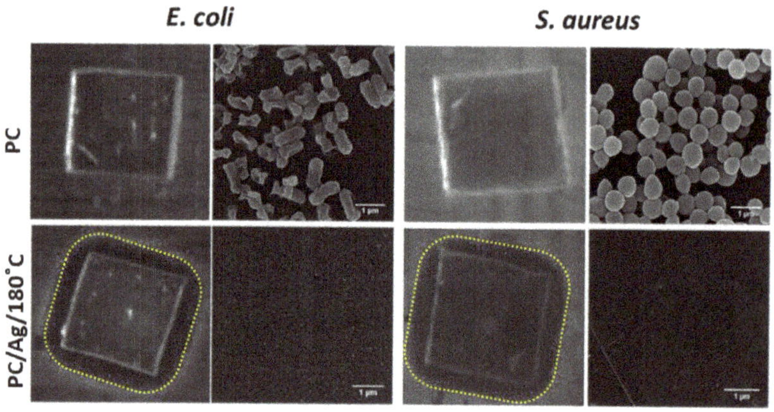

Figure 8. Antibacterial activity against *E. coli* and *S. aureus* of uncoated polymeric surface and polymeric surfaces coated Ag NPs followed by thermal annealing treatment at 180 °C (PC/Ag/180 °C), evaluated by the halo test and SEM micrographs of samples after the halo test (scale bar: 1 μm).

As expected, the uncoated polymeric surface did not show any effect on both *E. coli* and *S. aureus* bacteria. The sample with thermally embedded Ag NPs (PC/Ag/180 °C sample) presented an inhibition effect on the growth of both bacteria, as an inhibition halo (transparent biological medium with no bacteria growth) was observed around the samples. The SEM micrographs of the samples demonstrated a vast bacteria colonization on the uncoated polymeric surface (PC samples) and a total absence of bacteria on the polymeric surface with thermally embedded Ag NPs (PC/Ag/180 °C sample).

4. Conclusions

Robust, mechanically stable nanocomposite PC/Ag surfaces were successfully obtained by sputter depositing an Ag nanolayer over PC, followed by thermal annealing at a temperature above the glass transition point of the substrate. With AFM analysis, it was evident that there was a growth of NPs taking place after annealing at 120 °C, confirming the thermally activated diffusion and growth process. Similarly, the height variation of NPs with thermal annealing at 180 °C (above Tg) confirmed the embedding of the NPs on the PC surface. Reflectivity results were consistent with the idea of NP embedding and shape renormalization at this temperature as well. The adhesion test demonstrated that the thermal embedding process resulted in better adhesion of NPs to the surface.

The Ag NP-embedded polymer surfaces (PC/Ag/180 °C) showed strong antimicrobial activity on *E. coli* and *S. aureus* bacteria, making them antimicrobial surfaces.

Overall, we have demonstrated a cost-effective approach for creating robust antimicrobial surfaces on mass-produced polymers.

Author Contributions: Conceptualization, P.S.; data curation, P.S., L.F. and R.S.; formal analysis, P.S.; funding acquisition, A.C. and S.C.; investigation, P.S.; methodology, P.S., L.F., N.M.F. and R.S.; project administration, A.C. and S.C.; software, P.S.; supervision, A.C. and S.C.; validation, N.M.F., A.C. and S.C.; writing—original draft, P.S.; writing—review and editing, P.S., L.F., N.M.F., R.S., A.C. and S.C. All authors have read and agreed to the published version of the manuscript.

Funding: This research was funded by CEMMPRE's research and development unit project (UIDB/00285/2020) and Science DiabetICC Footwear with reference POCI-01-0247-FEDER-039784, supported by Portugal 2020. This research was supported by the COMPETE 2020 Portuguese and European Union initiative through Project POCI-01-0247-FEDER-072607, R&D and the production of logos for the automotive industry. This work was supported by the Portuguese Foundation for Science and Technology (FCT) in the framework of Strategic Funding (co-financed via UIDB/00285/2020) and LA/P/0112/2020.

Institutional Review Board Statement: Not applicable.

Data Availability Statement: Not applicable.

Acknowledgments: Authors would like to acknowledge funding from the Portuguese Government and European Union.

Conflicts of Interest: The authors declare no conflict of interest.

References

1. World Health Organization. Coronavirus Disease (COVID-19) Outbreak. 2021. Available online: https://www.who.int (accessed on 24 December 2022).
2. Perlman, S. Another Decade, Another Coronavirus. *N. Engl. J. Med.* **2020**, *382*, 760–762. [CrossRef]
3. Al Safar, M.; Amoodi, H.; Al-Satti, M.; Alsaidi, A. The precautions efficacy taken among health care workers while performing tracheostomies on COVID-19 patients: Systematic Review. *Int. J. Otolaryngol. Head Neck Surg.* **2020**, *11*, 283–291. [CrossRef]
4. Setti, L.; Passarini, F.; De Gennaro, G.; Barbieri, P.; Perrone, M.G.; Borelli, M.; Palmisani, J.; Di Gilio, A.; Piscitelli, P.; Miani, A. Airborne transmission route of COVID-19: Why 2 meters/6 feet of inter-personal distance could not be enough. *Int. J. Environ. Res. Public Health* **2020**, *17*, 2932. [CrossRef]
5. Russotto, V.; Cortegiani, A.; Raineri, S.M.; Giarratano, A. Bacterial contamination of inanimate surfaces and equipment in the intensive care unit. *J. Intensive Care* **2015**, *3*, 54. [CrossRef]
6. Gerhardts, A.; Henze, S.V.; Bockmühl, D.; Höfer, D. Fabric-skin models to assess infection transfer for impetigo contagiosa in a kindergarten scenario. *Eur. J. Clin. Microbiol. Infect. Dis.* **2015**, *34*, 1153–1160. [CrossRef]
7. Flores-López, L.Z.; Espinoza-Gómez, H.; Somanathan, R. Silver nanoparticles: Electron transfer, reactive oxygen species, oxidative stress, beneficial and toxicological effects. *Mini Rev. J. Appl. Toxicol.* **2019**, *39*, 16–26. [CrossRef] [PubMed]
8. Ferreira, T.P.; Nepomuceno, N.C.; Medeiros, E.L.; Medeiros, E.S.; Sampaio, F.C.; Oliveira, J.E.; Oliveira, M.P.; Galvao, L.S.; Bulhoes, E.O.; Santos, A.S. Antimicrobial coatings based on poly (dimethyl siloxane) and silver nanoparticles by solution blow spraying. *Prog. Org. Coat.* **2019**, *133*, 19–26. [CrossRef]
9. Jo, Y.K.; Seo, J.H.; Choi, B.H.; Kim, B.J.; Shin, H.H.; Hwang, B.H.; Cha, H.J. Surface-independent antibacterial coating using silver nanoparticle-generating engineered mussel glue. *ACS Appl. Mater. Interfaces* **2014**, *6*, 20242–20253. [CrossRef] [PubMed]
10. Baran, A.; Baran, M.F.; Keskin, C.; Kandemir, S.I.; Valiyeva, M.; Mehraliyeva, S.; Khalilov, R.; Eftekhari, A. Ecofriendly/rapid synthesis of silver nanoparticles using extract of waste parts of artichoke (cynara scolymus l.) and evaluation of their cytotoxic and antibacterial activities. *J. Nanomater.* **2021**, *2021*, 110. [CrossRef]
11. Baran, A.; Keskin, C.; Baran, M.F.; Huseynova, I.; Khalilov, R.; Eftekhari, A.; Irtegun-Kandemir, S.; Kavak, D.E. Ecofriendly synthesis of silver nanoparticles using ananas comosus fruit peels: Anticancer and antimicrobial activities. *Bioinorg. Chem. Appl.* **2021**, *2021*, 2058149. [CrossRef]

12. El-Aassar, M.R.; Ibrahim, O.M.; Fouda, M.M.G.; El-Beheri, N.G.; Agwa, M.M. Wound healing of nanofiber comprising polygalacturonic/hyaluronic acid embedded silver nanoparticles: In-vitro and in-vivo studies. *Carbohydr. Polym.* **2020**, *238*, 117484. [CrossRef] [PubMed]
13. Carvalho, I.; Lima, M.J.; Nobre, D.; Marques, S.M.; Castro, D.; Leite, T.R.; Henriques, M.; Duarte, F.; Ramalho, A.; Carvalho, S. Silver oxide coatings deposited on leathers to prevent diabetic foot infections. *Surf. Coat. Technol.* **2022**, *442*, 128338. [CrossRef]
14. Kruk, T.; Szczepanowicz, K.; Kręgiel, D.; Szyk-Warszyńska, L.; Warszyński, P. Nanostructured multilayer polyelectrolyte films with silver nanoparticles as antibacterial coatings. *Colloids Surf. B Biointerfaces* **2016**, *137*, 158–166. [CrossRef]
15. Sastri, V.S. *Plastic in Medical Devices: Properties, Requirements and Applications*; Elsevier: Burlington, MA, USA, 2010.
16. Kovacs, G.J.; Vincett, P.S. Formation and thermodynamic stability of a novel class of useful materials: Close-packed monolayers of submicron monodisperse spheres just below a polymer surface. *J. Colloid Interface Sci.* **1982**, *90*, 335–351. [CrossRef]
17. Kovacs, G.J.; Vincett, P.S. Subsurface particulate film formation in softenable substrates: Present status and possible new applications. *Thin Solid Film.* **1983**, *100*, 341–353. [CrossRef]
18. Legrand, D.G.; Bendler, J.T. *Handbook of Polycarbonate Science and Technology*; Marcel Dekker, Inc.: New York, NY, USA, 2000; Volume 3, 374p, ISBN 0-8247-9915-1, 1482273691, 9781482273694.
19. Kausar, A. A review of filled and pristine polycarbonate blends and their applications. *J. Plast. Film Sheeting* **2018**, *34*, 60–97. [CrossRef]
20. Bonde, H.C.; Fojan, P.; Popok, V.N. Controllable embedding of size-selected copper nanoparticles into polymer films. *Plasma Process. Polym.* **2020**, *17*, 1900237. [CrossRef]
21. Erichsen, J.; Kanzow, J.; Schürmann, U.; Dolgner, K.; Günther-Schade, K.; Strunskus, T.; Zaporojtchenko, V.; Faupel, F. Investigation of the surface glass transition temperature by embedding of noble metal nanoclusters into monodisperse polystyrenes. *Macromolecules* **2004**, *37*, 1831–1838. [CrossRef]
22. Kovacs, G.J.; Vincett, P.S. Subsurface particle monolayer and film formation in softenable substrates: Techniques and thermodynamic criteria. *Thin Solid Film.* **1984**, *111*, 65–81. [CrossRef]
23. Rudoy, V.M.; Dement'eva, O.V.; Yaminskii, I.V.; Sukhov, V.M.; Kartseva, M.E.; Ogarev, V.A. Metal nanoparticles on polymer surfaces: 1. A new method of determining glass transition temperature of the surface layer. *Colloid J.* **2002**, *64*, 746–754. [CrossRef]
24. Muhammad, H.; Juluri, R.R.; Fojan, P.; Popok, V. Polymer films with size-selected silver nanoparticles as plasmon resonance-based transducers for protein sensing. *Biointerface Res. Appl. Chem.* **2016**, *6*, 1564–1568.
25. Zhang, W.; Zhang, Y.H.; Ji, J.H.; Zhao, J.; Yan, Q.; Chu, P.K. Antimicrobial properties of copper plasma-modified polyethylene. *Polymer* **2006**, *47*, 7441–7445. [CrossRef]
26. Kenawy, E.R.; Worley, S.D.; Broughton, R. The chemistry and applications of antimicrobial polymers: A state-of-the-art review. *Biomacromolecules* **2007**, *8*, 1359–1384. [CrossRef]
27. Teichroeb, J.H.; Forrest, J.A. Direct imaging of nanoparticle embedding to probe viscoelasticity of polymer surfaces. *Phys. Rev. Lett.* **2003**, *91*, 016104. [CrossRef] [PubMed]
28. Larosa, C.; Patra, N.; Salerno, M.; Mikac, L.; Merijs Meri, R.; Ivanda, M. Preparation and characterization of polycarbonate/multiwalled carbon nanotube nanocomposites. *Beilstein J. Nanotechnol.* **2017**, *8*, 2026–2031. [CrossRef] [PubMed]
29. Daniels, P.H.; Cabrera, A. Plasticizer compatibility testing: Dynamic mechanical analysis and glass transition temperatures. *J. Vinyl Addit. Technol.* **2015**, *21*, 7–11. [CrossRef]
30. Ruffino, F.; Torrisi, V.; Marletta, G.; Grimaldi, M.G. Effects of the embedding kinetics on the surface nano-morphology of nano-grained Au and Ag films on PS and PMMA layers annealed above the glass transition temperature. *Appl. Phys. A* **2012**, *107*, 669–683. [CrossRef]
31. Prakash, J.; Pivin, J.C.; Swart, H.C. Noble metal nanoparticles embedding into polymeric materials: From fundamentals to applications. *Adv. Colloid Interface Sci.* **2015**, *226*, 187–202. [CrossRef]
32. Kreibig, U.; Vollmer, M. Theoretical Considerations. In *Optical Properties of Metal Clusters*; Springer: Berlin/Heidelberg, Germany, 1995; pp. 13–201.
33. Figueiredo, N.M.; Cavaleiro, A. Dielectric properties of shape-distributed ellipsoidal particle systems. *Plasmonics* **2020**, *15*, 379–397. [CrossRef]
34. Sharma, P.; Singhal, R.; Vishnoi, R.; Agarwal, D.C.; Banerjee, M.K.; Chand, S.; Kanjilal, D.; Avasthi, D.K. Effect of Ag ion implantation on SPR of Cu-C60 nanocomposite thin film. *Plasmonics* **2018**, *13*, 669–679. [CrossRef]
35. Figueiredo, N.M.; Vaz, F.; Cunha, L.; Cavaleiro, A. Au-WO3 nanocomposite coatings for localized surface plasmon resonance sensing. *Materials* **2020**, *13*, 246. [CrossRef] [PubMed]
36. Ladani, L.; Harvey, E.; Choudhury, S.F.; Taylor, C.R. Effect of varying test parameters on elastic–plastic properties extracted by nanoindentation tests. *Exp. Mech.* **2013**, *53*, 1299–1309. [CrossRef]

antibiotics

MDPI

Article

Microbial Community Structure among Honey Samples of Different Pollen Origin

Elisavet Stavropoulou [1,2,3,*], Nikolaos Remmas [2], Chrysoula (Chrysa) Voidarou [4], Georgia Vrioni [1], Theodoros Konstantinidis [5], Spyridon Ntougias [2,*] and Athanasios Tsakris [1]

1 Department of Microbiology, Medical School, National Kapodistrian University of Athens, 11527 Athens, Greece
2 Department of Environmental Engineering, Democritus University of Thrace, Vas. Sofias 12, 67132 Xanthi, Greece
3 Centre Hospitalier Universitaire Vaudois (CHUV), 1101 Lausanne, Switzerland
4 Department of Agriculture, School of Agriculture, University of Ioannina, 47100 Arta, Greece
5 Laboratory of Hygiene and Environmental Protection, Department of Medicine, Democritus University of Thrace, Dragana, 68100 Alexandroupolis, Greece
* Correspondence: elisavet.stavropoulou@chuv.ch or elisabeth.stavropoulou@gmail.com (E.S.); sntougia@env.duth.gr (S.N.); Tel.: +30-2541079313 (S.N.)

Abstract: Honey's antibacterial activity has been recently linked to the inhibitory effects of honey microbiota against a range of foodborne and human pathogens. In the current study, the microbial community structure of honey samples exerting pronounced antimicrobial activity was examined. The honey samples were obtained from different geographical locations in Greece and had diverse pollen origin (fir, cotton, fir–oak, and *Arbutus unedo* honeys). Identification of honey microbiota was performed by high-throughput amplicon sequencing analysis, detecting 335 distinct taxa in the analyzed samples. Regarding ecological indices, the fir and cotton honeys possessed greater diversity than the fir–oak and *Arbutus unedo* ones. *Lactobacillus kunkeei* (basionym of *Apilactobacillus kun-keei*) was the predominant taxon in the fir honey examined. *Lactobacillus* spp. appeared to be favored in honey from fir-originated pollen and nectar since lactobacilli were more pronounced in fir compared to fir–oak honey. *Pseudomonas*, *Streptococcus*, *Lysobacter* and *Meiothermus* were the predominant taxa in cotton honey, whereas *Lonsdalea*, the causing agent of acute oak decline, and *Zymobacter*, an osmotolerant facultative anaerobic fermenter, were the dominant taxa in fir–oak honey. Moreover, methylotrophic bacteria represented 1.3–3% of the total relative abundance, independently of the geographical and pollen origin, indicating that methylotrophy plays an important role in honeybee ecology and functionality. A total of 14 taxa were identified in all examined honey samples, including bacilli/anoxybacilli, paracocci, lysobacters, pseudomonads, and sphingomonads. It is concluded that microbial constituents of the honey samples examined were native gut microbiota of melliferous bees and microbiota of their flowering plants, including both beneficial bacteria, such as potential probiotic strains, and animal and plant pathogens, e.g., *Staphylococcus* spp. and *Lonsdalea* spp. Further experimentation will elucidate aspects of potential application of microbial bioindicators in identifying the authenticity of honey and honeybee-derived products.

Keywords: fir and fir–oak honey; *Arbutus unedo* honey; methylotrophs; *Apilactobacillus kun-keei*; *Lonsdalea*; *Zymobacter*

Citation: Stavropoulou, E.; Remmas, N.; Voidarou, C.; Vrioni, G.; Konstantinidis, T.; Ntougias, S.; Tsakris, A. Microbial Community Structure among Honey Samples of Different Pollen Origin. *Antibiotics* **2023**, *12*, 101. https://doi.org/10.3390/antibiotics12010101

Academic Editor: Helena P. Felgueiras

Received: 18 December 2022
Revised: 2 January 2023
Accepted: 3 January 2023
Published: 6 January 2023

1. Introduction

The honey market constitutes an important agricultural sector, with the global honey production being estimated at approximately 6.6 billion USD in 2015 [1]. Moreover, honeybees are essential for agriculture, ecology, and the maintenance of life, as the pollination assures the reproduction of plants; thus, beekeeping is actually an issue of concern all around the world [2]. In the European Union (EU), the second largest producer of honey in the globe after China, and EU Member States such as Spain, Romania, Greece, Poland, France, and Italy [3], are the main producers not only of honey, but also of propolis, bee pollen, bee bread, bee wax, and royal jelly (COM(2019) 635 final).

According to latest data from the European Commission, 19 million beehives owned by 615,000 beekeepers produced 275,000 tons of honey in 2020, albeit covering only 60% of the demand in the EU [4], whilst the remaining 40% was covered primarily by Ukraine and secondarily by China [5]. The regulatory framework regarding honey production, properties, and labeling is specified in the Council Directive (EU) 2001/110/EC relating to honey.

In the recent decades, a series of factors, including the extensive application of pesticides, the penetration of invasive species, and deforestation, have added excessive pressure on the global bee population [6]. However, honey is a valuable natural food product, as well as an ingredient used in various personal care products, as well as in traditional medicine. Honey, a hypersaturated sugar solution of high viscosity and osmotic pressure, consists mainly of fructose and glucose, which account for 54.3–87.5% of its content, while minor concentrations of other monosaccharides and oligosaccharides are detected [7]. Moreover, limited amounts of minerals, pollen, amino acids, proteins, organic acids, and their esters are present in honey [7]. Any detected variation can be attributed to the geographical region and flowering, climate conditions, production practices, and storage [7,8]. Moreover, honey has a pH near 4 [9,10] and low water activity, which lies between 0.45 and 0.6 [9,11]. The presence of flavonoids and carotenoids, as well as of minerals, such as Fe, Zn, Cu, and Mn, defines the color of honey [10,12].

In addition to its dietary value, honey is of high medicinal importance since its antimicrobial and wound-healing properties have been noticed since ancient times [13]. Honey's antibacterial activity, which is linked to enzymatic generation of hydrogen peroxide, wound moisturization ability, and high viscosity, acting as a barrier against infections, are key mode-of-action parameters regarding its medicinal properties [14]. In addition to the generation of hydrogen peroxide via the action of glucose oxidase present in the nectar, the antimicrobial activity and healing properties of honey can be attributed to the antibacterial properties of phenolic acids and flavonoids [14,15]. Moreover, the extremely low water activity and the high osmolarity, as well as the acidic pH, of honey further strengthen the inhibitory effects against bacterial pathogens [16]. On the other hand, spatial and temporal variation in bee pollen and nectar influence the antimicrobial properties of honey [17].

Recently, the antimicrobial activity of honey has also been linked to the inhibitory effects of honey microbiota against foodborne and human pathogens. Pajor et al. [18] investigated the inhibitory effect of bacterial strains isolated from honey against pathogens, reporting that *Bacillus* spp. exhibited antimicrobial activity against *Listeria monocytogenes* ATCC 7644, whereas similar inhibitory activity was also induced by these bacilli against certain *Staphylococcus aureus* and *S. epidermidis* strains. In addition, Voidarou et al. [19] evaluated the therapeutic properties of oregano honey against gastric ulcers and gastritis caused by *Helicobacter pylori*, pointing out that diethyl ether extracts of the honey and the honey itself reduced urease activity exhibited by the specific pathogen. Moreover, Masoura and Gkatzionis [20] examined possible antimicrobial effects of thyme and Manuka honeys against methicillin-resistant *Staphylococcus aureus* (MRSA) strains. The low pH and the high H_2O_2 concentration were the key factors influencing the antibacterial activity of monofloral thyme honey. Similarly, Jia et al. [21] also reported the antagonistic action of *Bacillus* sp. A2 from honey against the yeast *Candida albicans* and the bacterial species *Escherichia coli* and *Staphylococcus aureus*. However, employment of high-throughput techniques to uncover microbial diversity in honey is limited. Specifically, Wen et al. [22] performed pyrosequencing to analyze microbial diversity during ripening of vitex honey, reporting the dominance of *Bacillus* spp. and *Lactococcus* spp. and yeasts of the genus *Metschnikowia*.

Thus, the aim of the present study was to identify, for the first time, the predominant microbial communities in *Apis mellifera* honey samples of various pollen origin (fir, cotton, fir–oak, and arbutus) exhibiting high antimicrobial activity against common foodborne and human pathogens, and to comparatively evaluate their microbial community structure and ecological indices, through the application of high-throughput sequencing techniques.

2. Results and Discussion

Four honey samples from bees fed with pollen and nectar of various melliferous plant species were examined in terms of their microbial community structure. Regarding diversity indices, the fir and cotton honeys exhibited significantly greater diversity than the fir–oak and *Arbutus unedo* honeys (*a* < *0.05*, in Duncan's multiple tests for Chao1, Fisher, Shannon, and Simpson diversity indices) (Figure 1). No statistically significant differences were identified between fir and cotton honey, or between fir–oak and *Arbutus unedo* honey regarding all ecological indices estimated, except for the Shannon index, where *Arbutus unedo* honey showed the least score (Figure 1).

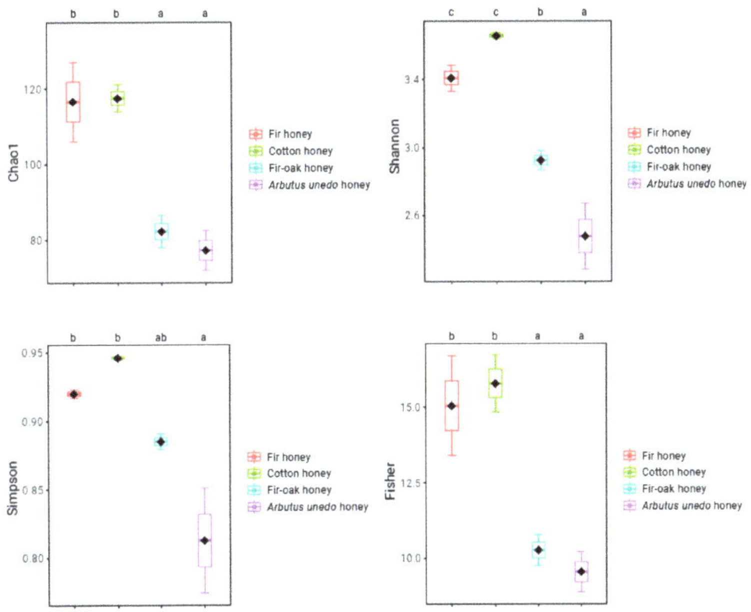

Figure 1. Diversity indices of the examined honey samples. Bars represent ± standard errors of means. Analysis of variance (ANOVA) using Duncan's multiple post hoc tests at a significance level of 5% (*a* < *0.05*) were carried out to identify statistically significant differences. No letter in common at the top of the subgraphs is indicative of statistically significant differences.

The examined fir honey was dominated by lactic acid bacteria of the genus *Lactobacillus* (19.80 ± 0.59% of the total relative abundance), followed by members of the genera *Bradyrhizobium* (11.93 ± 0.20%) and *Pseudomonas* (11.42 ± 1.68%) (Figure 2). Specifically, *L. kunkeei* (current taxonomic name *Apilactobacillus kun-keei*) represented 19.34 ± 0.54% of the total relative abundance at species level in fir honey (Table 1).

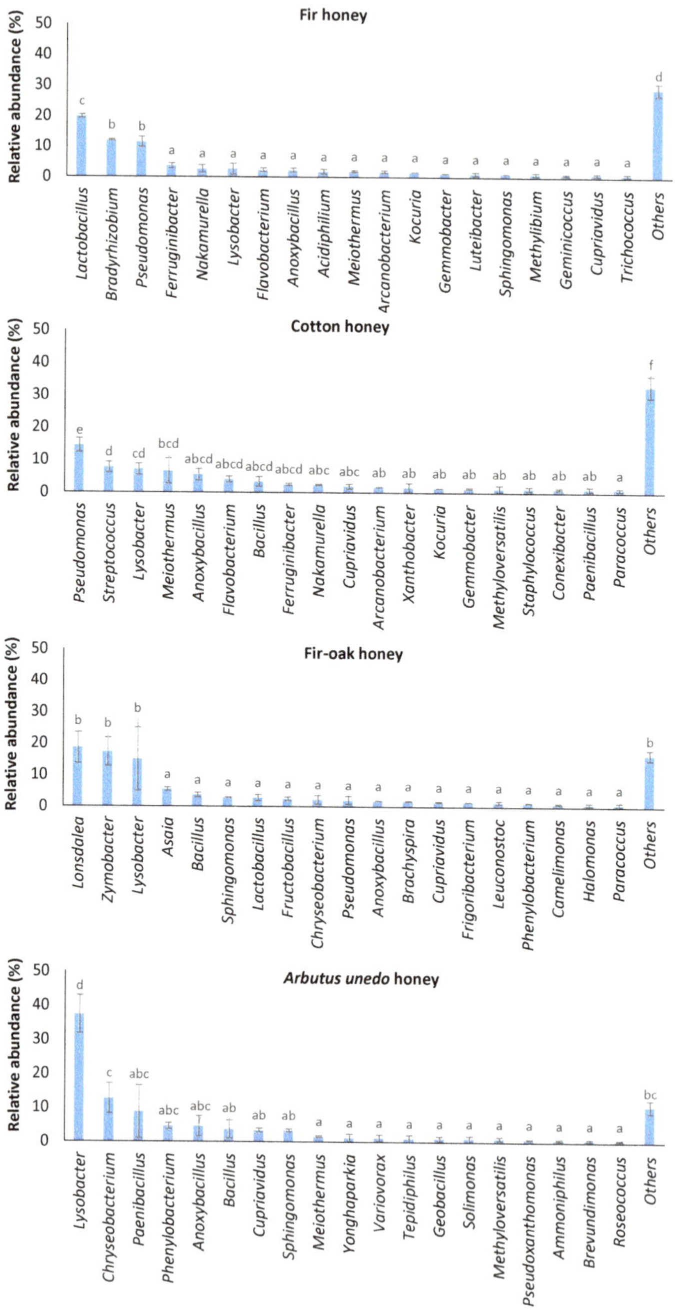

Figure 2. Major bacterial taxa identified in honey samples of different pollen origin. Bars represent ± standard errors of means. Analysis of variance (ANOVA) using Duncan's multiple post hoc tests at a significance level of 5% (*a* < *0.05*) were carried out to identify statistically significant differences. No letter in common at the top of the bars is indicative of statistically significant differences.

Table 1. Relative abundance of *Lactobacillus* species detected in fir honey.

Taxon	Relative Abundance (%)
Lactobacillus kunkeei	19.34 ± 0.54
Lactobacillus johnsonii	0.31 ± 0.03
Lactobacillus alvei	Marginally detected
Lactobacillus sakei	Marginally detected
Lactobacillus mellis	Marginally detected
Lactobacillus melliventris	Marginally detected

Lactobacilli appeared to be favored in honey produced from fir-originated pollen and nectar, with *Lactobacillus* population being more pronounced in fir honey compared to fir–oak honey, indicating a proliferation of this taxon under feeding of honeybees with fir (Figure 3). Interestingly, *Bradyrhizobium* has been recently reported to exert antagonistic activity against various pathogens [23]. Recently, co-existence of lactic acid bacteria and *Bradyrhizobium* resulted in enhanced nodulation with positive impact on plant growth, a fact that may be indicative of co-evolution of these microbiota in certain honeybee hosts [24].

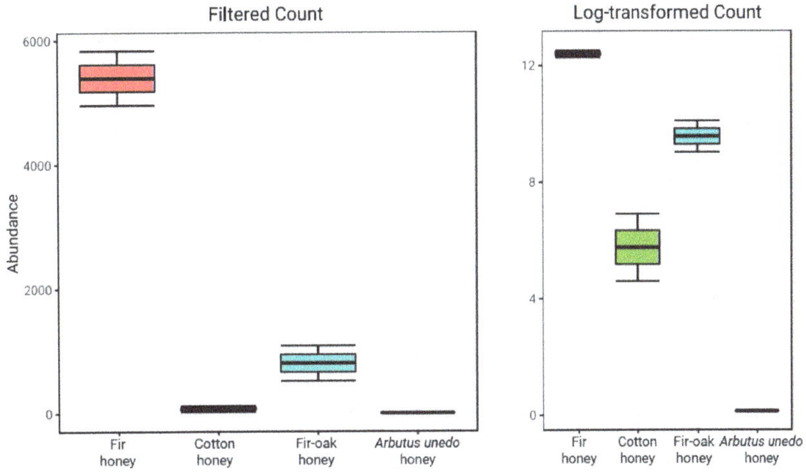

Figure 3. Abundance of *Lactobacillus* genus in the examined honey samples. Bars represent ± standard errors of means.

The predominant taxa in cotton honey were *Pseudomonas*, *Streptococcus*, *Lysobacter*, and *Meiothermus*, representing 14.35 ± 2.10%, 7.66 ± 1.69%, 7.07 ± 1.75%, and 6.53 ± 3.93% of the total relative abundance, respectively (Figure 2). *Pseudomonas* is an inhabitant of bee pollen, contributing to the decomposition of pollen walls [25]. As a result, this taxon is part of honeybee microbiome, which is commonly detected in honey [25–27]. Notably, streptococci/lactococci, lactobacilli, and enterococci are common microbial constituents of honeybee-collected pollen [28]. Moreover, *Lysobacter* was recently detected as a minor component of intestine honeybee microbiota [29]. Interestingly, *Meiothermus* was reported to be a beneficial bacterium of phytophagous insects, i.e., *Phasmotaenia lanyuhensis* [30].

The major bacterial taxa in fir–oak honey were *Lonsdalea*, causing acute oak decline [31], and *Zymobacter*, a facultative anaerobic fermenter [32], which covered 18.53 ± 4.96% and 17.22 ± 4.48% of the total reads, respectively (Figure 2). Recent findings revealed that "*insects visiting drippy blight diseased red oak trees are contaminated with the pathogenic bacterium Lonsdalea quercina*" [33]; therefore, this bacterium appears to be transmitted from infected oak trees to honeybees and subsequently to fir/oak-originated honey.

The presence of *Zymobacter* may result in a reduced *Lactobacillus* population in these honeybees, since this rarely appearing sugar-tolerant bacterium of *Halomonadaceae* [34] may be favored as a specialized alternative fermenter [35]. Moreover, the acetic acid bacterium *Asaia* is considered as an emerging symbiont of *Apis mellifera* [36]. *Kocuria* spp., which were among the major taxa identified in fir-related honeys (fir and fir–oak honeys; Figure 2), have been identified as the most abundant species in honeybees obtained from beehives in Turkey, which, however, were entomopathogens of honybees [37].

Lysobacter (37.31 ± 5.58% of the total relative abundance), *Chryseobacterium* (12.66 ± 4.44%), and *Paenibacillus* (8.81 ± 7.73%) were the predominant microbiota in *Arbutus unedo* honey (Figure 2). Although *Chryseobacterium* is rarely detected in honey [26], this taxon belongs to the native microbiota of *Arbutus unedo* (strawberry tree), since this bacterium was detected in all strawberry tree specimens examined by Martins et al. [38]. Apart from *Lysobacter*, which is a representative of honeybee gut microbiota [29] and a plant-associated microbe with plant-protective properties [39], various *Paenibacillus* spp. have been also detected in the microbiome of honeybee intestine [40]. Although certain members of the genus *Paenibacillus* are entomopathogenic to honeybees [41], various *Bacillus* spp. can be beneficial as biological control agents delivered from honeybees to plants [42].

Among other major taxa (Figure 2), *Sphingomonas* spp., which are also considered as common constituents of honeybee gut microbiota [43,44], were detected in all honey samples examined. Moreover, *Methylobacterium* has been previously detected in *Ceratina* bees [45].

Performance of correlation network analysis in honeys examined showed the strong relationships among certain members of *Actinobacteria* (*Kocuria*, *Nakamurella*, *Tessaracoccus*, *Acidothermus*, *Nocardia*, *Arcanobacterium*, and *Propionicicella*), *Bacteroidota* representatives (*Flavobacterium* and *Ferruginibacter*), the *Rhodobacteraceae* genera *Amaricoccus* and *Gemmobacter*, and the methylotrophic bacterium *Methylotenera*. Such microbiota may be involved in the decomposition of complex compounds present in flowering plants and honey, such as phenolics, flavonoids, lignin, and cellulose [46]. A second distinct cluster was formed solely by the major microbiota of the fir–oak honey (*Lonsdalea*, *Zymobacter*, and lactic acid bacteria such as *Fructobacillus* and *Leuconostoc*), indicating the interaction of *Lonsdalea* with key fermentative taxa in this honey sample.

Lactobacilli and especially *L. kunkeei* (basionym of *Apilactobacillus kun-keei*) strains in the gut of melliferous bees have been reported to induce resistance against deltamethrin, a pesticide that is considered as the major threat for pollinators [47]. *Lactobacillus* spp. are members of the native gut microbiome of honeybees, which have been found to enhance antiviral properties, even during tetracycline treatment that increases sensitivity to viral infections [48]. Moreover, honey with a high *Lactobacillus* population was reported to exhibit the greatest antioxidant activity among other honey samples examined by Wu et al. [49]. *L. kunkeei* (*A. kun-keei*) in the gut of *Apis mellifera* has been reported to exert antimicrobial activity and act as probiotic against the honeybee pathogens *Paenibacillus larvae* and *Melissococcccus plutonius* [50,51]. Indeed, *L. kunkeei* is favored by a long evolutionary symbiotic relationship in the gut of honeybees [52]. Moreover, Goh et al. [53] reported that *Lactobacillus* strains, which were isolated from stingless bees and their products, exhibited antimicrobial properties against *Listeria monocytogenes*. Kaškonienė et al. [54] also reported that lactococci and lactobacilli during solid-state lactic acid fermentation of bee pollen exerted antibacterial activity against *Micrococcus luteus*, *Staphylococcus aureus*, and *Escherichia coli*. Similarly, lactic acid bacteria from the intestine tract of honeybees exhibited inhibitory effects against *Bacillus cereus*, *Escherichia coli*, *Pseudomonas aeruginosa*, *Salmonella typhimurium*, and *Staphylococcus aureus* [55].

Comparative analysis of microbial communities in the examined honey samples revealed that their major taxa were either indigenous microbiota of honeybees, mostly beneficial and to a lesser extent potential foodborne and human pathogens, and/or inhabitants of bee pollen and flowering plants, including both plant protective and phytopathogenic microorganisms. In addition, honey originated from cotton (an industrial plant) included a higher proportion of potential foodborne and human pathogens compared to honey originated from bees fed with pollen of wild plants, a fact that may be attributed to the higher abundance of antimicrobial compounds in forest plants [56,57].

A total of 14 taxa were identified in all honey samples (Table 2), independently from their geographic distribution and pollen origin, indicating that these genera may serve as authenticity bioindicators of honey and honeybee-derived products. All of these are either aerobes, capable of growing in low-oxygen conditions (lysobacters and sphingomonads) [58,59] or presenting weak anaerobic growth (*Microbacterium*), as well as being facultative anaerobes (e.g., bacilli/anoxybacilli, paracocci, pseudomonads, staphylococci, and phenylobacteria) or common fermenters of insect gut (*Propionibacterium* spp.). Therefore, restricted oxygen levels in the honeybee gut and/or honey seem to play a role in shaping the microbial community structure in honey.

Table 2. List of bacterial taxa identified in all honey samples examined in order to serve as potential authenticity bioindicators of honey.

Taxon [1]	Fir Honey	Cotton Honey	Fir-Oak Honey	*A. unedo* Honey
Pseudomonas	11.42 ± 1.68 (b)	14.35 ± 2.10 (b)	2.02 ± 1.23 (a)	0.32 ± 0.03 (a)
Lysobacter	2.67 ± 1.79 (a)	7.07 ± 1.75 (a)	14.93 ± 10.05 (ab)	37.31 ± 5.58 (b)
Anoxybacillus	2.24 ± 0.72 (a)	5.47 ± 1.80 (a)	1.78 ± 0.07 (a)	4.60 ± 2.97 (a)
Bacillus	0.30 ± 0.30 (a)	3.35 ± 1.47 (a)	3.61 ± 0.76 (a)	3.69 ± 2.66 (a)
Meiothermus	1.91 ± 0.29 (a)	6.53 ± 3.93 (a)	0.61 ± 0.07 (a)	1.65 ± 0.26 (a)
Sphingomonas	1.14 ± 0.03 (a)	0.63 ± 0.49 (a)	2.86 ± 0.05 (b)	3.33 ± 0.43 (b)
Cupriavidus	0.97 ± 0.42 (a)	1.94 ± 0.72 (ab)	1.52 ± 0.18 (ab)	3.41 ± 0.59 (b)
Phenylobacterium	0.20 ± 0.06 (a)	0.58 ± 0.23 (a)	1.21 ± 0.03 (a)	4.61 ± 0.87 (b)
Paracoccus	0.44 ± 0.05 (a)	1.04 ± 0.43 (a)	0.79 ± 0.54 (a)	0.48 ± 0.23 (a)
Staphylococcus	0.35 ± 0.28 (a)	1.23 ± 0.64 (a)	0.23 ± 0.09 (a)	0.57 ± 0.32 (a)
Methylibium	1.02 ± 0.48 (a)	0.22 ± 0.06 (a)	0.69 ± 0.12 (a)	0.29 ± 0.03 (a)
Propionibacterium	0.65 ± 0.26 (a)	0.61 ± 0.21 (a)	0.27 ± 0.17 (a)	0.17 ± 0.01 (a)
Methylobacterium	0.50 ± 0.12 (b)	0.67 ± 0.02 (b)	0.07 ± 0.05 (a)	0.05 ± 0.04 (a)
Microbacterium	0.31 ± 0.03 (a)	0.26 ± 0.10 (a)	0.23 ± 0.02 (a)	0.18 ± 0.04 (a)

[1] No letter in common within the same row is indicative of statistically significant differences.

In all honey samples examined, methylotrophs represented an important fraction of microbial population. In particular, methylotrophs covered 1.3–3.0% of the total relative abundance (Table 3), independently of the geographical and pollen origin. Methanol has been reported to be among the most abundant volatile compounds of honey [60,61]. Interestingly, plant-colonizing methylotrophs of the genus *Methylobacterium* were capable of delivering insecticidal proteins and have been associated with the plant protective properties of the common biocontrol agent *Bacillus thuringiensis* [62]. Moreover, volatiles such as methanol promote the growth of the phyllosphere microbiota [63], serving as an energy source for methylotrophic epiphytes [64]. Thus, both methanol concentration and co-evolution of honeybees with their plant hosts and phyllosphere microbiota appear to shape methylotrophic communities in honey.

Table 3. Relative abundance of methylotrophic bacteria identified in the examined honey samples.

Taxon	Fir Honey	Cotton Honey	Fir-Oak Honey	*Arbutus unedo* Honey
Methylibium	1.02 ± 0.48	0.22 ± 0.06	0.69 ± 0.12	0.29 ± 0.03
Methylobacterium	0.50 ± 0.12	0.67 ± 0.02	0.07 ± 0.05	0.05 ± 0.04
Methylocapsa	0.18 ± 0.18	n.d.	n.d.	n.d.
Methylopila	0.02 ± 0.02	n.d.	0.16 ± 0.16	n.d.
Methylosinus	0.25 ± 0.16	n.d.	0.12 ± 0.08	n.d.
Methylotenera	0.40 ± 0.01	0.85 ± 0.26	n.d.	n.d.
Methyloversatilis	0.60 ± 0.60	1.26 ± 0.99	0.61 ± 0.53	1.00 ± 0.58
Total relative abundance [1]	2.97 ± 1.20 (a)	3.00 ± 1.18 (a)	1.65 ± 0.44 (a)	1.34 ± 0.59 (a)

[1] No letter in common within the same row is indicative of statistically significant differences; n.d., not detected.

In conclusion, native microbiota of the honeybee microbiome, such as fermentative bacteria, either with probiotic properties (e.g., lactobacilli) or potential foodborne and human pathogens (e.g., staphylococci and streptococci), as well as common inhabitants of bee pollen (e.g., pollen wall decomposers, such as *Pseudomonas* spp.) and flowering plants (e.g., beneficial microorganisms of plants and phytophagous insects, such as *Lysobacter* and *Meiothermus* spp., respectively), including plant pathogens of honeybee hosts that are transmitted from flowering plants to honey via honeybees (e.g., *Lonsdalea*, the causing agent of acute oak decline), constitute the microbial community structure in natural honeys.

3. Materials and Methods

3.1. Collection of Honey Samples

Honey samples from various locations in Greece were obtained aseptically and further examined in terms of their antimicrobial properties. Four honey samples from beehives situated in different geographical areas of Greece, i.e., from the Prefecture of Epirus (the one honey sample was produced from honeybees fed with fir and oak and the other with *Arbutus unedo* pollen and nectar), the Regional unit of Phthiotis, Mendenitsa region (from fir pollen and nectar) and the Regional unit of Karditsa, Palama region (from cotton pollen and nectar), were collected and further studied, due to their high antimicrobial activity (as examined in Stavropoulou et al. [65,66]). These honey samples were subsequently subjected to DNA extraction and phylogenetic analysis of microbial communities through Illumina sequencing.

3.2. DNA Extraction from Honey Samples of Different Pollen Origin and Performance of Amplicon Sequencing

The collected honey samples exhibiting high antimicrobial activity against foodborne and human pathogens were subjected to DNA extraction. Genomic DNA was extracted from honey through the use of NucleoSpin Tissue Kit (Macherey-Nagel, Düren, Germany), by following the instructions of the manufacturer. Aliquots of 60 μL of 10 mg/mL lysozyme and 60 μL of 10 mg/mL lysostaphin, as well as 6 μL of 60 U/μL lyticase (enzymes supplied by Sigma-Aldrich, Germany), were added per treated sample to facilitate the lysis of Gram(+) bacteria and yeast strains, respectively. The V4–V5 region of the 16S rRNA gene was amplified using the primers 515F (5′–GTG YCA GCM GCC GCG GTA A–3′) and 909R (5′–CCC CGY CAA TTC MTT TRA GT–3′). No amplification was achieved using primer set ITS1F (5′–CTT GGT CAT TTA GAG GAA GTA A–3′) and ITS4R (5′–TCC TCC GCT TAT TGA TAT GC–3′) for fungi, including yeasts. Amplification of partial 16S rRNA gene was conducted through a thermal scheme comprising of 3 min denaturation at 94 °C, succeeded by 28 cycles of 30 s denaturation at 94 °C, 40 s primer annealing at 53 °C, and 1 min DNA elongation at 72 °C, before a 5 min DNA thermal extension at 72 °C to complete the thermocycling reaction. The amplification reaction for the ITS region was performed by using 2 min denaturation at 94 °C, succeeded by 35 cycles of 30 s denaturation at 95 °C, 30 s primer annealing at 55 °C, and 1 min DNA elongation at 72 °C, before a 10 min DNA thermal extension at 72 °C to complete the thermocycling reaction. Illumina sequencing reactions were carried out at Mr DNA' (USA) in MiSeq equipment (Illumina, Inc., San Diego, CA, USA) after amplicon cleanup through the use of DNA purification beads.

3.3. Bioinformatic Analysis and Analysis of Variance (ANOVA)

Bacterial amplicons were proceeded to demultiplexing and trimming, and partial 16S rRNA gene sequences of abnormal size and low-quality were removed. The assembled reads were further subjected to denoising and chimera discard using USEARCH v.11 [67,68]. Bacterial sequences were clustered and ZOTUs (zero-radius OTU of denoised sequences) were generated, following National Center for Biotechnology Information (NCBI) taxonomy. Ecological indices, i.e., Chao1, Shannon, Simpson, and Fisher diversity indices, were sequentially calculated using the MicrobiomeAnalyst online suite of omics tools [69]. Relationships among bacterial taxa were identified in the MicrobiomeAnalyst platform via multiple correlation network matrix analyses based of the SparCC score at 0.97 correlation coefficient. The sequenced amplicons were deposited in the Sequence Read Archive (SRA) of the NCBI platform under the BioProject accession number PRJNA913297.

Analysis of variance (ANOVA) was conducted using Past v.4.10 [70] to identify statistically significant differences among the relative abundances and diversity indices of the bacterial communities identified in these Greek honey samples exhibiting high antimicrobial activity against foodborne and human pathogens.

Author Contributions: Conceptualization, E.S., A.T. and S.N.; methodology, E.S. and N.R.; formal analysis, E.S., N.R., C.V. and G.V.; resources, C.V. and T.K.; data curation, E.S., N.R., S.N. and A.T.; writing—original draft preparation, E.S., N.R. and S.N.; writing—review and editing, all; supervision, S.N. and A.T. All authors have read and agreed to the published version of the manuscript.

Funding: This research received no external funding.

Institutional Review Board Statement: Not applicable.

Informed Consent Statement: Not applicable.

Data Availability Statement: All data of this study are included in this article.

Conflicts of Interest: The authors declare no conflict of interest.

References

1. Zhou, X.; Taylor, M.P.; Salouros, H.; Prasad, S. Authenticity and geographic origin of global honeys determined using carbon isotope ratios and trace elements. *Sci. Rep.* **2018**, *8*, 14639. [CrossRef] [PubMed]
2. Potts, S.G.; Biesmeijer, J.C.; Kremen, C.; Neumann, P.; Schweiger, O.; Kunin, W.E. Global pollinator declines: Trends, impacts and drivers. *Trends Ecol. Evol.* **2010**, *25*, 345–353. [CrossRef]
3. EU Overview. Detailed Information on Honey Production in the European Union. Agriculture and Rural Development, European Commission. 2022. Available online: https://agriculture.ec.europa.eu/farming/animal-products/honey_en (accessed on 9 December 2022).
4. EU News Article. Beekeeping Sector: Results of the Pilot Study on Honey Bee Selection. Directorate-General for Agriculture and Rural Development, European Commission. 15 March 2022. Available online: https://agriculture.ec.europa.eu/news/beekeeping-sector-results-pilot-study-honey-bee-selection-2022-03-15_en (accessed on 9 December 2022).
5. Eurostat. EU Imported €405.9 Million Worth in Honey in 2021. Products Eurostat News, Eurostat. 2022. Available online: https://ec.europa.eu/eurostat/web/products-eurostat-news/-/edn-20220819-2 (accessed on 9 December 2022).
6. Alberoni, D.; Gaggìa, F.; Baffoni, L.; di Gioia, D. Beneficial microorganisms for honey bees: Problems and progresses. *Appl. Microbiol. Biotechnol.* **2016**, *100*, 9469–9482. [CrossRef]
7. Da Silva, P.M.; Gauche, C.; Gonzaga, L.V.; Costa, A.C.O.; Fett, R. Honey: Chemical composition, stability and authenticity. *Food Chem.* **2016**, *196*, 309–323. [CrossRef] [PubMed]
8. Bosancic, B.; Zabic, M.; Mihajlovic, D.; Samardzic, J.; Mirjanic, G. Comparative study of toxic heavy metal residues and other properties of honey from different environmental production systems. *Environ. Sci. Pollut. Res.* **2020**, *27*, 38200–38211. [CrossRef] [PubMed]
9. Iurlina, M.O.; Fritz, R. Characterization of microorganisms in Argentinean honeys from different sources. *Int. J. Food Microbiol.* **2005**, *105*, 297–304. [CrossRef] [PubMed]
10. Fangio, M.F.; Iurlina, M.O.; Fritz, R. Characterisation of Argentinean honeys and evaluation of its inhibitory action on *Escherichia coli* growth. *Int. J. Food Sci. Technol.* **2010**, *45*, 520–529. [CrossRef]
11. Bayram, N.E.; Kara, H.H.; Can, A.M.; Bozkurt, F.; Akman, P.K.; Vardar, S.U.; Çebi, N.; Yılmaz, M.T.; Sağdıç, O.; Dertli, E. Characterization of physicochemical and antioxidant properties of Bayburt honey from the North-east part of Turkey. *J. Apic. Res.* **2021**, *60*, 46–56. [CrossRef]
12. Pătruică, S.; Alexa, E.; Obiștioiu, D.; Cocan, I.; Radulov, I.; Berbecea, A.; Lazăr, R.N.; Simiz, E.; Vicar, N.M.; Hulea, A.; et al. Chemical composition, antioxidant and antimicrobial activity of some types of honey from Banat region, Romania. *Molecules* **2022**, *27*, 4179. [CrossRef]
13. Mandal, M.D.; Mandal, S. Honey: Its medicinal property and antibacterial activity. *Asian Pac. J. Trop. Biomed.* **2011**, *1*, 154–160. [CrossRef]
14. Yupanqui Mieles, J.; Vyas, C.; Aslan, E.; Humphreys, G.; Diver, C.; Bartolo, P. Honey: An advanced antimicrobial and wound healing biomaterial for tissue engineering applications. *Pharmaceutics* **2022**, *14*, 1663. [CrossRef]
15. Cheung, Y.; Meenu, M.; Yu, X.; Xu, B. Phenolic acids and flavonoids profiles of commercial honey from different floral sources and geographic sources. *Int. J. Food Prop.* **2019**, *22*, 290–308. [CrossRef]
16. Mundo, M.A.; Padilla-Zakour, O.I.; Worobo, R.W. Growth inhibition of foodborne pathogens and food spoilage organisms by select raw honeys. *Int. J. Food Microbiol.* **2004**, *97*, 1–8. [CrossRef] [PubMed]

17. Stevenson, P.C.; Nicolson, S.W.; Wright, G.A. Plant secondary metabolites in nectar: Impacts on pollinators and ecological functions. *Funct. Ecol.* **2017**, *31*, 65–75. [CrossRef]
18. Pajor, M.; Worobo, R.W.; Milewski, S.; Szweda, P. The antimicrobial potential of bacteria isolated from honey samples produced in the Apiaries located in Pomeranian Voivodeship in Northern Poland. *Int. J. Environ. Res. Public Health* **2018**, *15*, 2002. [CrossRef]
19. Voidarou, C.; Antoniadou, M.; Rozos, G.; Alexopoulos, A.; Giorgi, E.; Tzora, A.; Giorgi, E.; Tzora, A.; Skoufos, I.; Varzakas, T.; et al. An in vitro study of different types of Greek honey as potential natural antimicrobials against dental caries and other oral pathogenic microorganisms. Case study simulation of oral cavity conditions. *Appl. Sci.* **2021**, *11*, 6318. [CrossRef]
20. Masoura, M.; Gkatzionis, K. The antimicrobial mechanism of Greek thyme honeys against methicillin-resistant *Staphylococcus aureus* clinical isolates: A case study of comparison with Manuka honey. *Int. J. Food Sci. Technol.* **2022**, *57*, 7076–7084. [CrossRef]
21. Jia, L.; Kosgey, J.C.; Wang, J.; Yang, J.; Nyamao, R.M.; Zhao, Y.; Teng, X.; Gao, L.; Wabo, M.T.C.; Vasilyeva, N.V.; et al. Antimicrobial and mechanism of antagonistic activity of *Bacillus* sp. A2 against pathogenic fungus and bacteria: The implication on honey's regulatory mechanism on host's microbiota. *Food Sci. Nutr.* **2020**, *8*, 4857–4867. [CrossRef]
22. Wen, Y.; Wang, L.; Jin, Y.; Zhang, J.; Su, L.; Zhang, X.; Zhou, J.; Li, Y. The microbial community dynamics during the vitex honey ripening process in the honeycomb. *Front. Microbiol.* **2017**, *8*, 1649. [CrossRef]
23. Yang, S.; Xu, W.; Tan, C.; Li, M.; Li, D.; Zhang, C.; Feng, L.; Chen, Q.; Jiang, J.; Li, Y.; et al. Heat stress weakens the skin barrier function in sturgeon by decreasing mucus secretion and disrupting the mucosal microbiota. *Front. Microbiol.* **2022**, *13*, 860079. [CrossRef]
24. Kale, N.; Ashwini, M.; Jahagirdar, S.; Shirnalli, G. Potentials of lactic acid bacteria in enhancing nodulation of *Bradyrhizobium daqingense* and yield in soybean. *Legume Res.* **2022**, *45*, 507–513. [CrossRef]
25. Disayathanoowat, T.; Li, H.; Supapimon, N.; Suwannarach, N.; Lumyong, S.; Chantawannakul, P.; Guo, J. Different dynamics of bacterial and fungal communities in hive-stored bee bread and their possible roles: A case study from two commercial honey bees in China. *Microorganisms* **2020**, *8*, 264. [CrossRef] [PubMed]
26. Loncaric, I.; Ruppitsch, W.; Licek, E.; Moosbeckhofer, R.; Busse, H.J.; Rosengarten, R. Characterization of selected Gram-negative non-fermenting bacteria isolated from honey bees (*Apis mellifera carnica*). *Apidologie* **2011**, *42*, 312–325. [CrossRef]
27. Anjum, S.I.; Aldakheel, F.; Shah, A.H.; Khan, S.; Ullah, A.; Hussain, R.; Khan, H.; Ansari, M.J.; Mahmoud, A.H.; Mohammed, O.B. Honey bee gut an unexpected niche of human pathogen. *J. King Saud Univ. Sci.* **2021**, *33*, 101247. [CrossRef]
28. Belhadj, H.; Harzallah, D.; Khennouf, S.; Dahamna, S.; Ghadbane, M. Population dynamics of lactic acid bacteria during laboratory fermentation of honeybee-collected pollen. *Der. Pharm. Lett.* **2016**, *8*, 297–300.
29. Kim, E.; Seo, J.; Yang, S.H.; Kim, I.S.; Koo, Y. Intestine bacterial microbiota of Asian hornet (*Vespa velutina nigrithorax*) and honey bee. *Korean J. Environ. Agric.* **2018**, *37*, 135–140. [CrossRef]
30. Li, Y.-H.; Huang, Y.-F.; Chen, T.-H.; Wu, S.-S.; Tang, H.-C.; Hsiao, C.-Y.; Huang, L.-C.; Chang, J.-C.; Chiu, K.-P.; Nai, Y.-S. Comparison of gut microbiota of healthy and diseased walking sticks, *Phasmotaenia lanyuhensis*. *Arch. Insect Biochem. Physiol.* **2020**, *105*, e21749. [CrossRef]
31. Brady, C.L.; Cleenwerck, I.; Denman, S.; Venter, S.N.; Rodríguez-Palenzuela, P.; Coutinho, T.A.; de Vos, P. Proposal to reclassify *Brenneria quercina* (Hildebrand and Schroth 1967) Hauben *et al.* 1999 into a new genus, *Lonsdalea* gen. nov., as *Lonsdalea quercina* comb. nov., descriptions of *Lonsdalea quercina* subsp. *quercina* comb. nov., *Lonsdalea quercina* subsp. *iberica* subsp. nov. and *Lonsdalea quercina* subsp. *britannica* subsp. nov., emendation of the description of the genus *Brenneria*, reclassification of *Dickeya dieffenbachiae* as *Dickeya dadantii* subsp. *dieffenbachiae* comb. nov., and emendation of the description of *Dickeya dadantii*. *Int. J. Syst. Evol. Microbiol.* **2012**, *62*, 1592–1602.
32. Okamoto, T.; Taguchi, H.; Nakamura, K.; Ikenaga, H.; Kuraishi, H.; Yamasato, K. *Zymobacter palmae* gen. nov., sp. nov., a new ethanol-fermenting peritrichous bacterium isolated from palm sap. *Arch. Microbiol.* **1993**, *160*, 333–337. [CrossRef]
33. Sitz, R.A.; Aquino, V.M.; Tisserat, N.A.; Cranshaw, W.S.; Stewart, J.E. Insects visiting drippy blight diseased red oak trees are contaminated with the pathogenic bacterium *Lonsdalea quercina*. *Plant Dis.* **2019**, *103*, 1940–1946. [CrossRef]
34. Ntougias, S.; Zervakis, G.I.; Fasseas, C. *Halotalea alkalilenta* gen. nov., sp. nov., a novel osmotolerant and alkalitolerant bacterium from alkaline olive mill wastes, and emended description of the family *Halomonadaceae* Franzmann *et al.* 1989, emend. Dobson and Franzmann 1996. *Int. J. Syst. Evol. Microbiol.* **2007**, *57*, 1975–1983. [CrossRef] [PubMed]
35. Suenami, S.; Konishi Nobu, M.; Miyazaki, R. Community analysis of gut microbiota in hornets, the largest eusocial wasps, *Vespa mandarinia* and *V. simillima*. *Sci. Rep.* **2019**, *9*, 9830. [CrossRef] [PubMed]
36. Crotti, E.; Rizzi, A.; Chouaia, B.; Ricci, I.; Favia, G.; Alma, A.; Sacchi, L.; Bourtzis, K.; Mandrioli, M.; Cherif, A.; et al. Acetic acid bacteria, newly emerging symbionts of insects. *Appl. Environ. Microbiol.* **2010**, *76*, 6963–6970. [CrossRef]
37. Boğ, E.Ş.; Ertürk, Ö.; Yaman, M. Pathogenicity of aerobic bacteria isolated from honeybees (*Apis mellifera*) in Ordu Province. *Turk. J. Vet. Anim. Sci.* **2020**, *44*, 714–719. [CrossRef]
38. Martins, J.; Ares, A.; Costa, J.; Canhoto, J. A baseline of *Arbutus unedo* L. microbiome for future research: In vitro versus ex vitro. *Sci. Hortic.* **2022**, *292*, 110657. [CrossRef]
39. Puopolo, G.; Tomada, S.; Pertot, I. The impact of the omics era on the knowledge and use of *Lysobacter* species to control phytopathogenic micro-organisms. *J. Appl. Microbiol.* **2018**, *124*, 15–27. [CrossRef]
40. Yun, J.H.; Lee, J.Y.; Kim, P.S.; Jung, M.J.; Bae, J.W. *Paenibacillus apis* sp. nov. and *Paenibacillus intestini* sp. nov., isolated from the intestine of the honey bee *Apis mellifera*. *Int. J. Syst. Evol. Microbiol.* **2017**, *67*, 1918–1924. [CrossRef]

41. Ebeling, J.; Knispel, H.; Hertlein, G.; Fünfhaus, A.; Genersch, E. Biology of *Paenibacillus larvae*, a deadly pathogen of honey bee larvae. *Appl. Microbiol. Biotechnol.* **2016**, *100*, 7387–7395. [CrossRef]
42. Dedej, S.; Delaplane, K.S.; Scherm, H. Effectiveness of honey bees in delivering the biocontrol agent *Bacillus subtilis* to blueberry flowers to suppress mummy berry disease. *Biol. Control* **2004**, *31*, 422–427. [CrossRef]
43. Kačániová, M.; Terentjeva, M.; Žiarovská, J.; Kowalczewski, P.Ł. In vitro antagonistic effect of gut bacteriota isolated from indigenous honey bees and essential oils against *Paenibacillus larvae*. *Int. J. Mol. Sci.* **2020**, *21*, 6736. [CrossRef]
44. Shehabeldine, A.M.; Hashem, A.H.; Hasaballah, A.I. Antagonistic effect of gut microbiota of the Egyptian honeybees, *Apis mellifera* L. against the etiological agent of Stonebrood disease. *Int. J. Trop. Insect Sci.* **2022**, *42*, 1357–1366. [CrossRef]
45. Graystock, P.; Rehan, S.M.; McFrederick, Q.S. Hunting for healthy microbiomes: Determining the core microbiomes of *Ceratina*, *Megalopta*, and *Apis* bees and how they associate with microbes in bee collected pollen. *Conserv. Genet.* **2017**, *18*, 701–711. [CrossRef]
46. Cianciosi, D.; Forbes-Hernández, T.Y.; Afrin, S.; Gasparrini, M.; Reboredo-Rodriguez, P.; Manna, P.P.; Zhang, J.; Lamas, L.B.; Flórez, S.M.; Toyos, P.A.; et al. Phenolic compounds in honey and their associated health benefits: A review. *Molecules* **2018**, *23*, 2322. [CrossRef] [PubMed]
47. Dong, Z.X.; Tang, Q.H.; Li, W.L.; Wang, Z.W.; Li, X.J.; Fu, C.M.; Li, D.; Qian, K.; Tian, W.-L.; Guo, J. Honeybee (*Apis mellifera*) resistance to deltamethrin exposure by modulating the gut microbiota and improving immunity. *Environ. Pollut.* **2022**, *314*, 120340. [CrossRef]
48. Deng, Y.; Yang, S.; Zhao, H.; Luo, J.; Yang, W.; Hou, C. Antibiotics-induced changes in intestinal bacteria result in the sensitivity of honey bee to virus. *Environ. Pollut.* **2022**, *314*, 120278. [CrossRef]
49. Wu, J.; Han, B.; Zhao, S.; Zhong, Y.; Han, W.; Gao, J.; Wang, S. Bioactive characterization of multifloral honeys from *Apis cerana cerana*, *Apis dorsata*, and *Lepidotrigona flavibasis*. *Food Res. Int.* **2022**, *161*, 111808. [CrossRef]
50. Al-Ghamdi, A.; Al-Abbadi, A.A.; Khan, K.A.; Ghramh, H.A.; Ahmed, A.M.; Ansari, M.J. In vitro antagonistic potential of gut bacteria isolated from indigenous honey bee race of Saudi Arabia against *Paenibacillus larvae*. *J. Apic. Res.* **2020**, *59*, 825–833. [CrossRef]
51. Iorizzo, M.; Ganassi, S.; Albanese, G.; Letizia, F.; Testa, B.; Tedino, C.; Petrarca, S.; Mutinelli, F.; Mazzeo, A.; de Cristofaro, A. Antimicrobial activity from putative probiotic lactic acid bacteria for the biological control of American and European foulbrood diseases. *Vet. Sci.* **2022**, *9*, 236. [CrossRef]
52. Vásquez, A.; Forsgren, E.; Fries, I.; Paxton, R.J.; Flaberg, E.; Szekely, L.; Olofsson, T.C. Symbionts as major modulators of insect health: Lactic acid bacteria and honeybees. *PLoS ONE* **2012**, *7*, e33188. [CrossRef]
53. Goh, L.P.W.; Molujin, A.M.; Muthu, K.; Abdulla, R.; Sabullah, M.K.; Mohd Faik, A.A.; Gansau, J.A.; Jawan, R. Isolation and characterization of lactic acid bacteria from Sabah (North Borneo) stingless bees for probiotic and food applications. *Int. J. Food Prop.* **2021**, *24*, 564–578. [CrossRef]
54. Kaškonienė, V.; Adaškevičiūtė, V.; Kaškonas, P.; Mickienė, R.; Maruška, A. Antimicrobial and antioxidant activities of natural and fermented bee pollen. *Food Biosci.* **2020**, *34*, 100532. [CrossRef]
55. Jaafar, N.A.I.; Mohamad, S.A.S.; Razak, W.R.W.A. Antimicrobial activity and antibiotic resistance of lactic acid bacteria isolated from Malaysian stingless bee's gut. *Malays. J. Microbiol.* **2019**, *15*, 233–241.
56. Bağci, E.; Diğrak, M. Antimicrobial activity of essential oils of some *Abies* (Fir) species from Turkey. *Flavour Fragr. J.* **1996**, *11*, 251–256. [CrossRef]
57. Chandra, H.; Bishnoi, P.; Yadav, A.; Patni, B.; Mishra, A.P.; Nautiyal, A.R. Antimicrobial resistance and the alternative resources with special emphasis on plant-based antimicrobials—A review. *Plants* **2017**, *6*, 16. [CrossRef]
58. Reichenbach, H. The Genus *Lysobacter*. In *The Prokaryotes*; Dworkin, M., Falkow, S., Rosenberg, E., Schleifer, K.H., Stackebrandt, E., Eds.; Springer: New York, NY, USA, 2006.
59. Shao, Y.; Arias-Cordero, E.; Guo, H.; Bartram, S.; Boland, W. In vivo Pyro-SIP assessing active gut microbiota of the cotton leafworm, *Spodoptera littoralis*. *PLoS ONE* **2014**, *9*, e85948. [CrossRef] [PubMed]
60. Agila, A.; Barringer, S. Application of selected ion flow tube mass spectrometry coupled with chemometrics to study the effect of location and botanical origin on volatile profile of unifloral American honeys. *J. Food Sci.* **2012**, *77*, C1103–C1108. [CrossRef] [PubMed]
61. Ozcan-Sinir, G.; Copur, O.U.; Barringer, S.A. Botanical and geographical origin of Turkish honeys by selected-ion flow-tube mass spectrometry and chemometrics. *J. Sci. Food Agric.* **2020**, *100*, 2198–2207. [CrossRef]
62. Choi, Y.J.; Gringorten, J.L.; Bélanger, L.; Morel, L.; Bourque, D.; Masson, L.; Groleau, D.; Míguez, C.B. Production of an insecticidal crystal protein from *Bacillus thuringiensis* by the methylotroph *Methylobacterium extorquens*. *Appl. Environ. Microbiol.* **2008**, *74*, 5178–5182. [CrossRef]
63. Galbally, I.E.; Kirstine, W. The production of methanol by flowering plants and the global cycle of methanol. *J. Atmos. Chem.* **2002**, *43*, 195–229. [CrossRef]
64. Sharma, P.; Kumar, T.; Yadav, M.; Gill, S.S.; Chauhan, N.S. Plant-microbe interactions for the sustainable agriculture and food security. *Plant Gene* **2021**, *28*, 100325. [CrossRef]
65. Stavropoulou, E.; Ieronymaki, E.; Dimitroulia, E.; Constantinidis, T.C.; Vrioni, G.; Tsatsanis, C.; Tsakris, A. Anti-Inflammatory and antibacterial effects and mode of action of Greek arbutus, chestnut, and fir honey in mouse models of inflammation and sepsis. *Microorganisms* **2022**, *10*, 2374. [CrossRef] [PubMed]

66. Stavropoulou, E.; Voidarou, C.; Rozos, G.; Vaou, N.; Bardanis, M.; Konstantinidis, T.; Vrioni, G.; Tsakris, A. Antimicrobial evaluation of various honey types against carbapenemase-producing Gram-negative clinical isolates. *Antibiotics* **2022**, *11*, 422. [CrossRef] [PubMed]
67. Edgar, R.C. UNOISE2: Improved error-correction for Illumina 16S and ITS amplicon sequencing. *bioRxiv* **2016**. [CrossRef]
68. Edgar, R.C. Accuracy of microbial community diversity estimated by closed-and open-reference OTUs. *Peer J.* **2017**, *5*, e3889. [CrossRef]
69. Dhariwal, A.; Chong, J.; Habib, S.; King, I.L.; Agellon, L.B.; Xia, J. MicrobiomeAnalyst: A web-based tool for comprehensive statistical, visual and meta-analysis of microbiome data. *Nucleic Acids Res.* **2017**, *45*, W180–W188. [CrossRef] [PubMed]
70. Hammer, Ø.; Harper, D.A.; Ryan, P.D. PAST: Paleontological statistics software package for education and data analysis. *Palaeontol. Electron.* **2001**, *4*, 9.

antibiotics

MDPI

Article

Citizen Contribution for Searching for Alternative Antimicrobial Activity Substances in Soil

Rosa Fernández-Fernández [1,†], Beatriz Robredo [2,*,†], Enrique Navajas [2] and Carmen Torres [1,*]

1 Area of Biochemistry and Molecular Biology, Faculty of Science and Technology, University of La Rioja, 26006 Logroño, Spain
2 Area of Didactic of Experimental Sciences, Faculty of Science and Technology, University of La Rioja, 26006 Logroño, Spain
* Correspondence: beatriz.robredo@unirioja.es (B.R.); carmen.torres@unirioja.es (C.T.)
† These authors contributed equally to this study.

Abstract: Antimicrobial resistance (AMR) is problematic worldwide, and due to the loss of efficiency of many antibiotics, the pressure to discover alternative antimicrobial molecules has increased. Soil harbors a great biodiversity and biomass of microorganisms, and many antibiotics are produced by soil microbiota. Therefore, soil is a promising reservoir to find new antimicrobial agents. In this respect, novel pedagogical strategies regarding the AMR global crisis have recently been developed in different countries worldwide. Highlighted is the service-learning project "MicroMundo" integrated in a global Citizen Science project called "Tiny Earth". Hence, the present work aimed at determining the antimicrobial activity of soil bacteria, the biodiversity of the selected isolates as putative antimicrobial producers, and their antibiotic resistance profile. Moreover, through the MicroMundo project, we tried to illustrate the relevant link between science and education and the benefits of implementing service-learning methodologies to raise awareness of the AMR problem and to contribute to the search for new alternatives. A total of 16 teachers, 25 university students and 300 secondary school students participated in the search for antimicrobial activity on a collection of 2600 isolates obtained from a total of 130 soil samples analysed. In total, 132 isolates (5% of total tested) were selected as potential antimicrobial producers when two indicator bacteria were used (Escherichia coli and Staphylococcus epidermidis); the most frequent genus among these isolates was Bacillus, followed by Pseudomonas, Paenibacillus and Serratia. The antimicrobial activity (AA) of the 132 potential antimicrobial producers was studied in a second step against 15 indicator bacteria (of six genera and thirteen species, including relevant pathogens). Of the 132 potentially producing bacteria, 32 were selected for further characterization. In this respect, 18 isolates showed low AA, 12 isolates were considered as medium producers, and 2 highly antimicrobial-producing isolates were found (Brevibacillus laterosporus X7262 and Staphylococcus hominis X7276) showing AA against 80% of the 15 indicators tested. Moreover, 48% of the antimicrobial-producing bacteria were susceptible to all antibiotics tested. Due to citizen science, antimicrobial-producing bacteria of great interest have been isolated, managing to raise awareness about the problem of AMR.

Keywords: antimicrobial-producing bacteria; soil; antimicrobial resistance; MicroMundo

Citation: Fernández-Fernández, R.; Robredo, B.; Navajas, E.; Torres, C. Citizen Contribution for Searching for Alternative Antimicrobial Activity Substances in Soil. *Antibiotics* **2023**, *12*, 57. https://doi.org/10.3390/antibiotics12010057

Academic Editor: Helena P. Felgueiras

Received: 1 December 2022
Revised: 22 December 2022
Accepted: 23 December 2022
Published: 29 December 2022

1. Introduction

Antimicrobial resistance (AMR) has risen an awareness alarm, being one of the most urgent challenges for current medicine and society due to the emergence of multi-drug resistant (MDR) pathogens [1,2]. Concretely, the science community is especially concerned about the antimicrobial resistance associated with the ESKAPE pathogens (*Enterococcus faecium*, *Staphylococcus aureus*, *Klebsiella pneumoniae*, *Acinetobacter baumannii*, *Pseudomonas aeruginosa*, and *Enterobacter* species) [3].

The loss of efficacy of many antibiotics increases the pressure to identify new effective approaches [4]. In the last decade, intensive studies have looked at the potential of natural antibacterial molecules as next-generation therapeutics against pathogens [4,5]. Concretely, the ribosomally synthesized peptides of bacterial origin, also named as bacteriocins, are one of the most promising bioactive compounds with antimicrobial properties against other bacteria [5]. The ability to synthetize bioactive peptides is one of the oldest defensive mechanisms of microorganisms, and many microorganisms produce at least one bacteriocin [6]. These not essential secondary metabolites increase the bacterial chances of adaptation in a hostile environment, and they have been proposed as a good alternative to combat pathogens and MDR bacteria [5,7].

Soil harbors a great biodiversity and biomass of microorganisms [8,9]. Soil bacteria live in a crowded and highly competitive environment with limited resources and constantly changing conditions. There are many antibiotics produced by microorganisms isolated from soil, from penicillin, the first reported, to some new ones such as malacidins and teixobactin [10]. Actinomycetes, which are the most common bacteria in soil, produce 60% of antibiotics in clinical use [11]. Thus, the potential to find new antimicrobial compounds in this immense reservoir of microorganisms is enormous, and scientists are beginning to realize how little is known regarding soil microorganisms.

To address these problems, novel pedagogical strategies on the AMR global crisis have recently been developed in different countries worldwide. Among such strategies is "MicroMundo" [12], integrated in a global Citizen Science project on AMR called "Tiny Earth" (TE; https://tinyearth.wisc.edu/, accessed on 1 December 2022) originally implemented in 2012 in the United States, with "Small World Initiative" designation (SWI; http://www.smallworldinitiative.org/, accessed on 1 December 2022).

MicroMundo was developed in a service-learning environment, in which different educational levels are integrated [12]. The program seeks to raise awareness of the problem of AMR among students by participating in a creative research project that combines soil sample collection and laboratory work to discover new antimicrobial agents [13].

Hence, the present work aimed at determining the antimicrobial activity of bacteria of the soil, the biodiversity of the selected isolates as putative producers, and their antimicrobial resistance profile. Moreover, we tried to illustrate the relevant link between science and education and the benefits of implementing service-learning methodologies to raise awareness of AMR and to contribute to the search for new alternatives.

2. Material and Methods

2.1. MicroMundo: Service-Learning Methodology

During the 2020–2022 school years, the service-learning project called MicroMundo was developed in La Rioja region (Spain) through two scales of practical training, involving university and secondary education.

The initial step of the MicroMundo project was carried out at the university (University of La Rioja), taught by a qualified professor (SWIPI: Partner Instructor) to master students (n = 25) or biology secondary school teachers (n = 12) (SWITAs: technical assistants). The SWIPI trained the teaching abilities of SWITAs through periodic sessions, and at the end, they discussed the logistics of SWI actions to analyze the experimental results and to design data-recording sheets.

The second step was performed in secondary schools (La Rioja region) under the general supervision of SWIPI, using expendable material and equipment from the university. In this process, the 37 SWITAs (master students and secondary school teachers) and 299 students (SWISs) of 12 secondary schools were involved. The SWITAs trained the program to the SWISs through periodic sessions. They made up a total of 130 groups, each of whom analyzed a soil sample.

Practical work was divided into five 2 h sessions following the methodology previously explained [14]. Firstly, soil samples were diluted and grown on Tryptic Soy Agar (TSA) (Condalab, Madrid, Spain) plates for colony selection (20 isolates per sample). An antimicrobial activity test was performed with all 20 isolates obtained on each soil sample, using *Staphylococcus epidermidis* C2663 (Gram-positive bacteria) and *Escherichia coli* C408

(Gram-negative bacteria) as indicator microorganisms (Table S1). Indicator bacteria were inoculated in saline solution and spread as a lawn onto TSA plates. Then, bacteria to be tested for antimicrobial activity production were transferred with a sterile toothpick. Plates were evaluated by students after 24 h, and isolates with putative inhibition halloes were selected in this initial screening and transferred to the university for verification and further characterization (second screening).

2.2. Second Screening of Antimicrobial Activity by the Spot-on-Lawn Method

The selected strains with potential antimicrobial activity obtained in the first screening at school level were further analyzed and characterized in a second screening process at the university. For that purpose, these strains were tested by the *spot-on-lawn* method against 15 indicator strains of 6 different genera and 13 species, including relevant pathogens (Supplementary Table S1). Bacteria were grown in brain heart infusion (BHI) agar (Condalab, Madrid, Spain) for 24 h at 37 °C. To prepare test plates, a suspension of the indicator strain was prepared in a tube of 3 mL of sterile saline solution to obtain a turbidity of 0.5 MacFarland, and it was spread with a sterile swab in 0.3% yeast extract-supplemented solid Tryptic Soy Agar (TSA) plate (Condalab, Madrid, Spain). Then, each fresh solid culture of bacterium to be tested for antimicrobial activity was transferred with a sterile toothpick to the agar plates seeded with each of the 15 indicators tested. Plates were incubated at 37 °C for 24 h to evaluate the halos of inhibition. Isolates were considered antimicrobial producers (AP) when they showed a clear and sharp inhibition zone against at least one of the 15 indicator isolates. Depending on the antimicrobial activity (hallo of inhibition), it was expressed as + (<3 mm), ++ (3 < x < 10 mm), or +++ (>10 mm).

2.3. Bacterial Identification

Isolates with potential antimicrobial capacity obtained in the first screening (n = 132) were identified by matrix-assisted laser desorption/ionization time of flight (MALDI-TOF) mass spectrometry, using the standard protein extraction protocol recommended by the commercial device of Bruker Daltonics, Bremen, Germany. The antimicrobial-producing isolates of interest that could not be identified by MALDI-TOF were identified by amplification and sequencing of the 16S rDNA gene, using the following primer sequences (F-GTGCCAGCAGCCGCGGTAA, R-AGACCCGGGAACGTATTCAC) and PCR conditions: 94 °C, 2 min; 28 cycles (94 °C, 30 s; 45 °C, 1 min; 72 °C, 1 min) and final 72 °C, 7 min [15].

2.4. Antibiotic Susceptibility Testing of Antimicrobial-Producing Strains

The antibiotic susceptibility profile was determined in the antimicrobial-producing isolates verified in the second screening, by the disk diffusion test in Mueller Hinton (MH) agar (Condalab, Madrid, Spain). The antibiotics used for that purpose differed depending on the bacterial genera and are indicated in Supplementary Table S2. The EUCAST guidelines [16] were used for the antibiotic susceptibility testing, and *Staphylococcus* spp. breakpoints were used for Gram-positive bacteria, except for *Bacillus* spp., for which EUCAST indicates a specific breakpoint for some antimicrobial agents [16]. In the case of Gram-negative bacteria, the breakpoints of enterobacterales and *Pseudomonas* were used, depending on the genera.

2.5. Diversity of Antimicrobial-Producing Bacteria and Statistics

The 130 soil samples were grouped based on the geographic coordinates of the collection site. In this way, 5 geographical zones were distinguished: La Rioja East, La Rioja Central, La Rioja West, Logroño and Outside (Supplementary Figure S1). Statistics comparison between the antimicrobial-producing bacteria detected by the first screening and the non-antimicrobial-producing isolates included in the 5 established clusters was performed using the Fisher test, and significant differences were considered for $p < 0.05$. Moreover, diversity and statistical analyses were carried out considering each antimicrobial-producing isolate as an independent item. Isolates that could not be identified were excluded for diversity analysis; thus, a collection of 104 isolates was included. Renyi profile was performed to compare alpha diversity between the 5 clusters. The Renyi profile was used to compare the species diversity among zones with vegan 2.6–2 package from R (4.2.1). The averages of the Renyi profile values were calculated using rarefied samples. Significant differences in these values were assessed with an ANOVA test. Tukey's post hoc test was used to identify differences between zones. All analyses were carried out with R, version 4.1.2.

3. Results

A total of 130 soil samples were analyzed at school level, and 2600 isolates were obtained (20 isolates/sample) and tested in a first screening for antimicrobial activity; 132 isolates showed potential inhibitory capacity in the first screening test performed in the school and using only two indicator bacteria (*S. epidermidis* and *E. coli*). Statistical differences were not observed between antimicrobial-producing and non-antimicrobial-producing isolates based on their geographical location.

Identification by MALDI-TOF at genus level of these 132 putative antimicrobial-producing isolates detected in soil samples revealed 19 genera and 48 species with the following microbial diversity (number of isolates): *Acinetobacter* (1), *Arthrobacter* (2), *Bacillus* (40), *Bradybacterium* (1), *Brevibacillus* (1), *Enterobacter* (3), *Escherichia* (2), *Klebsiella* (1), *Microbacterium* (2), *Micrococcus* (1), *Paenibacillus* (12), *Pseudomonas* (27), *Staphylococcus* (2), *Serratia* (6), *Stenotrophomonas* (1), *Streptomyces* (1), *Olivibacter* (1), *Variovarax* (1) and *Viridibacillus* (1). Moreover, 26 out of the 132 isolates (19.7%) could not be identified by MALDI-TOF mass spectrometry (Table 1). Diversity at genera level is represented in Figure 1, where the number of isolates of *Bacillus, Paenibacillus, Pseudomonas, Serratia* as well as those unidentified are shown. Genera with low numbers of representants were considered together in a group called "Others" (n = 19 isolates). A high prevalence of the genus *Bacillus* followed by *Pseudomonas, Paenibacillus* and *Serratia* (all isolates belong to the species *S. plymuthica*) is of note.

Table 1. Genus and species identification of the 132 potential antimicrobial-producing isolates obtained in the first screening and those 32 isolates verified in the second screening.

Genus	Species	Number of Isolates First Screening	Number of Isolates Second Screening
Acinetobacter	*Acinetobacter radioresistens*	1	
Arthrobacter		2	
	Arthrobacter citreus	1	1
	Arthrobacter ilicis	1	
Bacillus		40	
	Bacillus marisflavi	1	
	Bacillus atrophaeus	2	2
	Bacillus cereus	6	1
	Bacillus cibi	1	
	Bacillus megaterium	4	1
	Bacillus mycoides	4	1
	Bacillus pumilus	7	5
	Bacillus safensis	2	2
	Bacillus simplex	3	
	Bacillus thuringiensis	2	
	Bacillus weihenstefanensis	3	
	Bacillus spp.	5	2

Table 1. *Cont.*

Genus	Species	Number of Isolates First Screening	Number of Isolates Second Screening
Bradybacterium	*Bradybacterium* spp.	1	1
Brevibacillus	*Brevibacillus laterosoporus*	1	1
Enterobacter	*Enterobacter cloacae*	3	
Escherichia	*Escherichia coli*	2	
Klebsiella	*Klebsiella aerogenes*	1	1
Microbacterium	*Microbacterium arborescensens*	2	1
Micrococcus	*Micrococcus luteus*	1	
Paenibacillus		12	
	Paenibacillus amylolyticus	6	
	Paenibacillus apiarus	2	2
	Paenibacillus gluconolyticus	1	
	Paenibacillus lautus	1	
	Paenibacillus polymyxa	1	1
	Paenibacillus xylanilyticus	1	
Pseudomonas		27	
	Pseudomonas brasicacearum	2	
	Pseudomonas brenneri	1	
	Pseudomonas caricapapayae	1	1
	Pseudomonas chlororaphis	3	1
	Pseudomonas kilonensis	5	3
	Pseudomonas koreensis	2	
	Pseudomonas mandelii	1	
	Pseudomonas mosselii	1	
	Pseudomonas putida	1	
	Pseudomonas savastanoi	1	
	Pseudomonas thivervalensis	2	
	Pseudomonas umsongensis	1	
	Pseudomonas spp.	6	2
Staphylococcus	*Staphylococcus hominis*	2	1
Serratia	*Serratia plymuthica*	6	
Stenotrophomonas	*Stenotrophomonas rhizophila*	1	
Streptomyces	*Streptomyces avidinii*	1	1
Olivibacter	*Olivibacter soli*	1	1
Variovorax	*Variovorax paradoxus*	1	
Viridibacillus	*Viridibacillus arenosi*	1	
Not identified		26	
Total		132	32

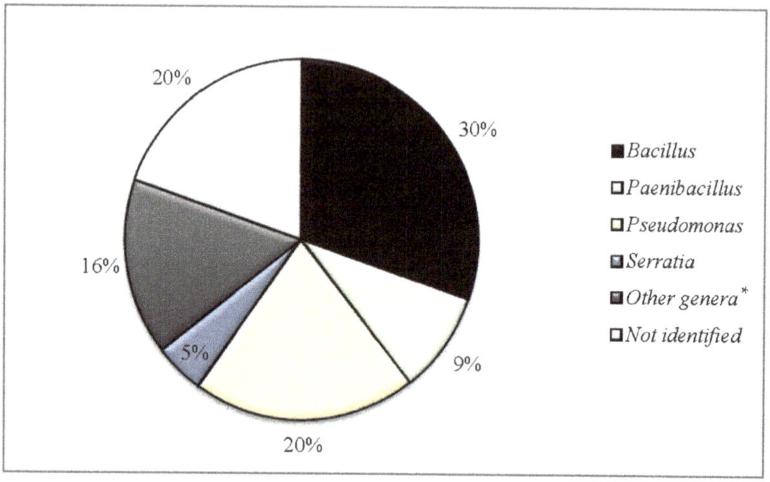

Figure 1. Diversity at genus level of the antimicrobial-producing bacteria isolated from soil samples in the first initial screening. * Other genera: include isolates of genera with less than six representants; see Table 1.

Renyi profiles allowed us to differentiate clusters according to the diversity of potential producers (Figure S1). The community with an overlapping diversity profile was considered the most diverse. Thus, the decreasing ranking of clusters was as follows: La Rioja Central, La Rioja West and Logroño (same alpha values), La Rioja East and Outside (Figure S2). This revealed that antimicrobial-producing isolates of La Rioja Central had the best distribution of richness (number of different genera/species) and abundance (number of isolates of each genus/species detected) (Table S3 and Figure S2).

3.1. *Verification of Antimicrobial Activity of Antimicrobial-Producing Bacteria in a Second Screening Process against 15 Indicator Bacteria*

The 132 isolates obtained in the first screening with potential antimicrobial activity production were tested at the university in a second screening by the *spot-on-lawn* method against 15 indicator bacteria. Then, 32 out of the 132 tested isolates were finally selected for presenting clear antimicrobial activity after several repetitions against at least one indicator bacteria (Table 2). Identification of 29 out of 32 isolates was obtained by MALDITOF, and the remaining 3 (X7264, X7265 and X7266) were identified by amplification and sequencing of the 16S rDNA gene. Twenty-three of the 32 antimicrobial-producing isolates were Gram-positive (15 species and 6 genera), and nine were Gram-negative (6 species and 3 genera), and they were further characterized. Most of antimicrobial antimicrobial-producing bacteria were of the genera *Bacillus* (43.8%) and *Pseudomonas* (21.9%) (Table 2).

The most susceptible indicators detected in this study with antimicrobial-producing isolates were the following: *S. epidermidis*, *M. luteus*, and methicillin-resistant and -susceptible *S. aureus* (MRSA and MSSA, respectively) (Tables 2 and 3). Seven Gram-positive producing isolates showed antimicrobial activity against the Gram-negative indicators used (*E. coli* and *P. aeruginosa*). It is to note the high inhibition produced by both Gram-positive and Gram-negative antimicrobial-producing bacteria against the indicators of the genus *Staphylococcus*, including methicillin-resistant staphylococci. In addition, three Gram-positive antimicrobial-producing isolates (*B. mycoides* X7258; *B. laterosporus* X7262 and *S. hominis* X7276) inhibited the indicator *L. monocytogenes*, a relevant pathogen in the food industry. Finally, excluding the species *E. cecorum* (with 48% of inhibition), *Enterococcus* was the most resistant indicator genus, only inhibited by Gram-positive antimicrobial-producing isolates, as expected (Tables 2 and 3).

Table 2. Activity profile of the 32 antimicrobial-producing isolates against the 15 indicator bacteria tested.

	Antimicrobial Activity [a] on the Indicator Bacteria [b]															
Producing Isolate	*E. coli*	*P. aeruginosa*	MRSA	MSSA	MRSP	MSSP	*S. delphini*	*S. sciuri*	*S. epidermidis*	*E. faecalis*	*E. faecium* [c]	*E. cecorum*	*E. gallinarum*	*L. monocytogenes*	*M. luteus*	No (%)
A. citreus X7246				+					+						+	3 (20)
B. cereus X7247			+			+			+			+			+	5 (33)
B. safensis X7248			+			+			+			+			+	5 (33)
B. safensis X7249	+		+			+			++			+			+	6 (40)
B. pumilus X7250	+		+	+		+	+		+			+			+	8 (53)
B. atrophaeus X7251			+			+			+			+			+	5 (33)
B. atrophaeus X7252	+		+	+		+			+			+			+	7 (47)
Bacillus spp. X7253				+					+						+	3 (20)
Bacillus spp. X7256.			+	+		+	+	+	+			+			+	8 (53)
B. pumilus X7254			+			+	+		+			+			+	6 (40)
B. megaterium X7255				+					+						+	3 (20)
B. pumilus X7257		+							+						+	3 (20)
B. mycoides X7258			+	+		+	+	+	+					+	+	8 (53)
B. pumilus X7259			+	+	+	+	+		+			+			+	8 (53)
B. pumilus X7260			+	+	+	+	+		+						+	7 (47)
Bradybacterium spp. X7261				+					+							2 (13)
B. laterosoporus X7262	+		+	+	+	+	+	+	+	+		+		+	+	12 (80)
M. arborescensens X7263			+	+	+	+	+		+						+	7 (47)
P. apiarus X7264	+		+				+		+						+	5 (33)
P. apiarus X7267			+						+						+	3 (20)
P. polymyxa X7268	+		+	+		+	+		+			+			+	8 (53)
S. hominis X7276			+	+		+	+	+	+	+	+	+	+	+	+	12 (80)
S. avidinii X7277			+	+					+						+	4 (27)
Number of inhibitions (%)	6 (26)	1 (4)	18 (78)	15 (65)	4 (17)	15 (65)	11 (48)	4 (17)	23 (100)	2 (9)	1 (4)	12 (52)	1 (4)	3 (13)	22 (92)	
O. soli X7265				+					+						+	3 (20)
K. aerogenes X7266	+			+		+										3 (20)
P. kilonensis X7269			+	+	+	+	+	+	+++						+	8 (53)
P. kilonensis X7270			+				+		+++						+	4 (27)
P. kilonensis X7271			+	+					+						+	4 (27)
Pseudomonas spp. X7272			+	+					+							3 (20)
Pseudomonas spp. X7273			+	+					+							3 (20)
P. chlororaphis X7274	+	+	+	+	+		+		+						+	8 (53)
P. caricapapayae X7275	+	+		+					+						+	5 (33)
Number of inhibitions (%)	3 (33)	2 (22)	6 (67)	8 (89)	2 (22)	2 (22)	3 (33)	1 (11)	8 (89)	0 (0)	0 (0)	0 (0)	0 (0)	0 (0)	6 (67)	

[a] Depending on the antimicrobial activity, it was expressed as + (<3 mm), ++ (3 < x < 10 mm), and +++ (>10 mm). Only positive results are indicated. [b] Abbreviations: MRSA, methicillin-resistant *S. aureus*; MSSA, methicillin-susceptible *S. aureus*; MRSP; methicillin-resistant *S. pseudintermedius*; MSSP, methicillin-susceptible *S. pseudintermedius*. [c] Vancomycin resistant strain.

Table 3. Antimicrobial activity profile of the 32 antimicrobial-producing isolates obtained in the second screening against 15 indicator bacteria (number of isolates).

	Number of Positive Results of the Antimicrobial-Producing Isolates against the Following Indicator Bacteria:						
Producing Isolates	*E. coli* (1)	*P. aeruginosa* (1)	MR-*Staphylococcus* [a] (2)	MS-*Staphylococcus* [a] (5)	*Enterococcus* (4)	*L. monocytogenes* (1)	*M. luteus* (1)
A. citreus (1)				2			1
Bacillus spp. (14)	3	1	13	41	9	1	14
Bradybacterium spp. (1)				2			
B. laterosoporus (1)	1		2	5	2	1	1
M. arborescensens (1)			2	4			1
Paenibacillus spp. (3)	2		3	7	1		3
S. hominis (1)			1	5	4	1	1
S. avidinii (1)			1	2			1
O. soli (1)				2			1
K. aerogenes (1)	1			2			
Pseudomonas spp. (7)	2	2	8	18			5

[a] Abbreviations: MR, methicillin resistant; MS, methicillin susceptible.

Three levels were differentiated regarding production based on the percentages of indicator bacteria inhibited by the antimicrobial-producing isolates: low (<35%), medium (from 35 to 70%) and high (>70%). In this respect, 18 isolates showed low antimicrobial activity, 12 isolates were considered as medium producers, and 2 isolates were found to be high producers with antimicrobial activity against 80% of the indicators tested (*Brevibacillus laterosporus* X7262 and *Staphylococcus hominis* X7276) (Table 2). Neither of the two highly producing isolates were active against *P. aeruginosa*, but *B. laterosporus* X7262 inhibited *E. coli*, the other Gram-negative indicator strain. On the other hand, both highly producing strains showed antimicrobial activity against methicillin-resistant and -susceptible (MR and MS) staphylococci (from 90% to 100% of inhibition), *L. monocytogenes* and *M. luteus*. The *S. hominis* X7276 strain revealed antimicrobial activity against all *Enterococcus* isolates used as indicators (100%), and *B. laterosporus* X7262 inhibited 50% of them (Figure 2).

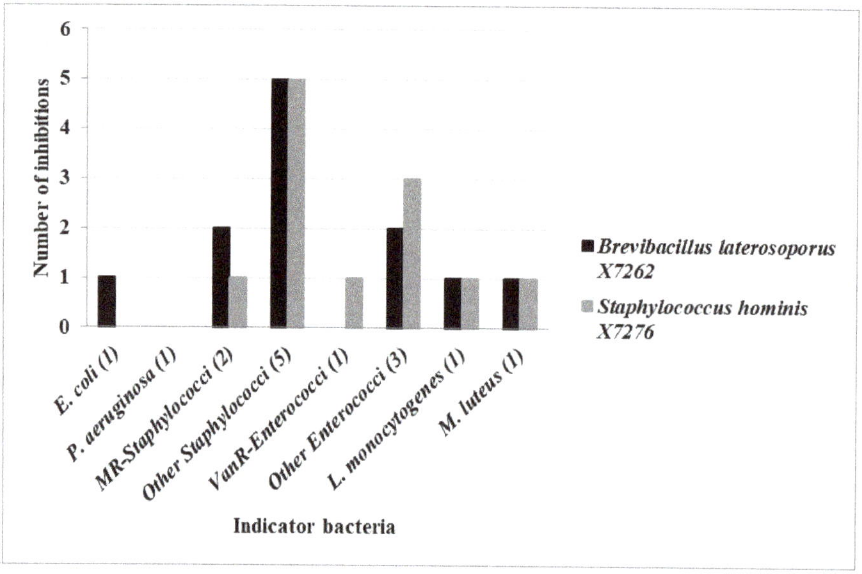

Figure 2. Number of indicator bacteria inhibited by the two highly producing isolates (*Brevibacillus laterosporus* X7262 and *Staphylococcus hominis* X7276).

3.2. Antibiotic Resistance Phenotype of the Antimicrobial-Producing Isolates

In total, 48% of the 32 antimicrobial-producing isolates showed susceptibility to all the antibiotics tested. Considering Gram-positive isolates, resistance was mostly detected for cefoxitin (22%) and penicillin and tobramycin (17%). Meropenem/imipenem and ciprofloxacin resistance were also found among *Bacillus* isolates. Four Gram-positive producing isolates (17%) were multidrug resistant (MDR). With respect to Gram-negative, 33% of them were susceptible to all the antibiotics tested (all belonging to *Pseudomonas* genus). Four *Pseudomonas* spp. showed resistance to ticarcillin, and two of them were also resistant to aztreonam. Moreover, two Gram-negative isolates were resistant to ampicillin and cefoxitin, one of them being MDR (*Olivibacter* X7265) (Table 4).

Table 4. Antimicrobial resistance phenotype of the selected 32 antimicrobial-producing isolates obtained in the second screening.

Type of Bacteria	Number of Isolates	Genus	Species [a]	Antimicrobial Resistance Phenotype [b]
Gram-positive	1	*Arthrobacter*	*A. citreus*	Susceptible
	7	*Bacillus* spp.	*B. pumilus* ^{2}, *B. safensis*, *B. megaterium*, *B. mycoides*, *Bacillus* spp. ^{2}	PEN ^{3}-FOX ^{4}-MER ^{3}-IMI ^{2}-S ^{2}-TOB ^{3}-CLI -GEN-SXT-CIP ^{3}
	7	*Bacillus* spp.	*B. pumilus* ^{3}, *B. cereus*, *B. artrophaeus* ^{2}, *B. safensis*	Susceptible ^{7}
	1	*Bradybacterium*	*Bradybacterium* spp.	Susceptible
	1	*Brevibacillus*	*B. laterosporus*	Susceptible
	1	*Microbacterium*	*M. arborescensis*	Susceptible
	2	*Paenibacillus*	*P. apiarus* ^{2}	PEN-FOX-TOB
	1	*Paenibacillus*	*P. polymyxa*	Susceptible
	1	*Staphylococcus*	*S. hominis*	Susceptible
	1	*Streptomyces*	*S. avidinii*	Susceptible
Gram-negative	1	*Klebsiella*	*K. aerogenes*	AMP-AMC-FOX
	1	*Olivibacter*	*O. soli*	AMP-FOX-CTX-CAZ-C-TOB
	4	*Pseudomonas*	*P. chlororaphis*, *P. caricapapayae*, *P. kilonensis* ^{2}	TIC ^{4}-ATM ^{2}
	3	*Pseudomonas*	*Pseudomonas* spp. ^{2}, *P. kilonensis*	Susceptible

[a] Superscript 2, 3 indicates the number of isolates of each species. [b] Superscript 2, 3, 4, 7 indicates the number of isolates with a specific characteristic, when more than one.

4. Discussion

Soil contains a highly diverse collection of bacteria, making it a very attractive starting point for efforts to discover molecules with antimicrobial activity [17]. In this sense, the present work carried out a massive soil sampling due to the citizen collaboration of professors, teachers, university students and secondary education students, under the MicroMundo project.

Therefore, from a collection of 2600 bacteria, 132 putative antimicrobial producers were obtained in the first screening, which represent 5% of the total isolates recovered. When processing these samples in the laboratory during the second screening, 100 producers were lost, probably due to the stricter criteria of antibacterial effect verification at the university, considering only clear zones of inhibition as putative markers of bacteriocins. However, many other antimicrobial substances have been described apart from antimicrobial peptides with different phenotypes of inhibition halos not considered for this study. On the other hand, bacteriocins are known to be produced in response to signals received from a potential competitor, which then elicits an antagonistic response [18]. Therefore, in this study, the 32 isolates, which showed constant antimicrobial activity throughout the second screenings carried out, were selected for their subsequent characterization.

This work provides information on the soil biodiversity of bacteria with potential inhibitory capacity. Renyi profiles of La Rioja zones reveal a higher diversity in La Rioja Central, although a higher number of antimicrobial-producing isolates among the 132 firstly identified were detected among Logroño samples. In this regard, *Bacillus* (30%) and *Pseudomonas* (20%) were the most predominant genera, in accordance with what was observed by Huang et al., 2021 [19]. However, other genera were found in this study, such as *Paenibacillus* or *Serratia*. These results highlight the potential of soil as a reservoir of bacteria that produce antimicrobial agents; thus, further characterization of isolates could be of interest.

In recent years, bacteria such as *Pseudomonas* spp. and *Bacillus* spp. have been studied and used as biological control agents for plant diseases [20,21], including the antibiosis mechanism for competition for nutrients and niches [22]. *Bacillus* is a genus well known as a producer of antibacterial substances such as lipopeptides, phenols, proteases, and bacteriocins [23]. Species of the genus *Pseudomonas* produce several secondary metabolites that affect other bacteria, fungi, or predators of nematodes and protozoa, such as bacteriocins, ranging from small microcin to large tailocin [24].

Thus, as expected, 14 *Bacillus* spp. and 7 *Pseudomonas* spp. out of the 32 bacteria finally selected as clear producers of antimicrobial substances were identified in this work. According to the *spot-on-lawn* results, higher activity was found against Gram-positive indicator bacteria than against Gram-negative indicator bacteria, being the *Staphylococcus* genera, (including MR-*Staphylococci*), the most susceptible indicator bacteria. It is widely known that most microbial metabolites have specific antimicrobial potential, and they act at the target sites [2]. Seven Gram-positive isolates showed antimicrobial activity against the Gram-negative indicators used. In addition, *Brevibacillus laterosporus* X7262 and *Staphylococcus hominis* X7276 stand out as high producers, which show antimicrobial activity against MS-staphylococci, *L. monocytogenes* and *M. luteus*.

Brevibacillus laterosporus is an aerobic, spore-forming, entomopathogenic microorganism commonly isolated from soil. Some strains have potential activity as biological control agents [25]. In addition, several applications of this bacterium as a biological control agent have been described, highlighting the high toxicity against mosquito larvae among other insects and the activity that promotes growth and improves productivity in bee colonies [26–28].

As for *S. hominis*, it is a normal skin commensal coagulase negative staphylococci (CoNS) described as a bacteriocin producer such as hominicin [29] and nukacin KQU-131 [30], among others. Moreover, recent studies have detected bacteriocin-like-producing staphylococci of environmental origin, including *S. hominis* [31]. Due to their high tolerance to an acidic environment, the resistance to bile, and the capacity to adhering to an epithelial cell line, *S. hominis* has been proposed as a good candidate for probiotic treatments against *S. aureus* [32]. In this sense, Nakatsuji et al., 2017 [33], reported that human *Staphylococcus* commensal species produce antimicrobial peptides that protect us against pathogens that control skin microbiota imbalances, and they demonstrated that a personalized probiotic CoNS cream could alleviate the symptoms of skin dysbiosis such as atopic dermatitis.

In short, advanced and combinatorial therapies that include antibiotics or new molecules with antimicrobial activity could be used as an alternative solution to combat AMR from a biotechnological and biomedical perspective and to solve problems in the agriculture and food industries, among others [34]. The one-health perspective makes clear the need for an ecosystem union to achieve improved objectives in the problem of AMR. Citizens must be integrated into this system, knowing the problem of the urgent need to find antimicrobial molecules, becoming aware of it, and contributing to research through this type of citizen science and service-learning initiatives such as MicroMundo.

Supplementary Materials: The following supporting information can be downloaded at: https://www.mdpi.com/article/10.3390/antibiotics12010057/s1, Figure S1: Coordinates of all soil samples analyzed in this study. Each cluster is differentiated by colors; Figure S2: Renyi diversity profile of the antimicrobial-producing microbial community included in each of the following zones: La Rioja Central, La Rioja East, La Rioja West, Logroño and Outside; Table S1: Characteristics of the 15 indicator bacteria used in this study for the screening of antimicrobial activity production; Table S2: Antibiotics used for disk-diffusion test in Gram-positive and -negative antimicrobial-producer isolates (EUCAST); Table S3: Number of isolates included in each of the established regions based on their sampling location.

Author Contributions: C.T., B.R. and R.F.-F. designed the study; R.F.-F., B.R. and E.N. performed the experiments; R.F.-F. and B.R. made the first analysis of the data and prepared the draft of the paper. All authors have read and agreed to the published version of the manuscript.

Funding: This work was supported by the project MCIN/AEI/10.13039/501100011033 of Spain. Rosa Fernández-Fernández has a predoctoral fellowship FPU from the Ministerio de Ciencia, Innovación y Universidades of Spain (FPU18/05438).

Acknowledgments: Thanks to Agustí Martínez Sancho for his help in statistics and diversity analysis. We appreciate the voluntary participation of the master's students, secondary school teachers and students that allowed us to carry out the sampling of the present study.

Conflicts of Interest: The authors declare no conflict of interest.

References

1. Yount, N.Y.; Weaver, D.C.; De Anda, J.; Lee, E.Y.; Lee, M.W.; Wong, G.C.L.; Yeaman, M.R. Discovery of Novel Type II Bacteriocins Using a New High-Dimensional Bioinformatic Algorithm. *Front. Immunol.* **2020**, *11*, 01873. [CrossRef] [PubMed]
2. Rani, A.; Saini, K.C.; Bast, F.; Varjani, S.; Mehariya, S.; Bhatia, S.K.; Sharma, N.; Funk, C. A Review on Microbial Products and Their Perspective Application as Antimicrobial Agents. *Biomolecules* **2021**, *11*, 1860. [CrossRef] [PubMed]
3. Santajit, S.; Indrawattana, N. Mechanisms of Antimicrobial Resistance in ESKAPE Pathogens. *BioMed Res. Int.* **2016**, *2016*, 1–8. [CrossRef] [PubMed]
4. Soltani, S.; Hammami, R.; Cotter, P.D.; Rebuffat, S.; Said, L.B.; Gaudreau, H.; Bédard, F.; Biron, E.; Drider, D.; Fliss, I. Bacteriocins as a new generation of antimicrobials: Toxicity aspects and regulations. *FEMS Microbiol. Rev.* **2021**, *45*, fuaa039. [CrossRef] [PubMed]
5. Twomey, E.; Hill, C.; Field, D.; Begley, M. Recipe for Success: Suggestions and Recommendations for the Isolation and Characterisation of Bacteriocins. *Int. J. Microbiol.* **2021**, *2021*, 1–19. [CrossRef]
6. Riley, M.A.; Wertz, J.E. Bacteriocins: Evolution, Ecology, and Application. *Annu. Rev. Microbiol.* **2002**, *56*, 117–137. [CrossRef] [PubMed]
7. Telhig, S.; Ben Said, L.; Torres, C.; Rebuffat, S.; Zirah, S.; Fliss, I. Evaluating the Potential and Synergetic Effects of Microcins against Multidrug-Resistant *Enterobacteriaceae*. *Microbiol. Spectr.* **2022**, *10*, e02752-21. [CrossRef]
8. Curtis, T.P.; Sloan, W.T.; Scannell, J.W. Estimating prokaryotic diversity and its limits. *Proc. Natl. Acad. Sci. USA* **2002**, *99*, 10494–10499. [CrossRef]
9. Schloss, P.; Handelsman, J. Toward a Census of Bacteria in Soil. *PLOS Comput. Biol.* **2006**, *2*, e92. [CrossRef]
10. Hover, B.M.; Kim, S.-H.; Katz, M.; Charlop-Powers, Z.; Owen, J.G.; Ternei, M.A.; Maniko, J.; Estrela, A.B.; Molina, H.; Park, S.; et al. Culture-independent discovery of the malacidins as calcium-dependent antibiotics with activity against multidrug-resistant Gram-positive pathogens. *Nat. Microbiol.* **2018**, *3*, 415–422. [CrossRef]
11. Handelsman, J. *Small World Initiative: Research Protocols and Research Guide to Microbial and Chemical Diversity*; XanEdu Publishing Inc.: Ann Arbor, MI, USA, 2015.
12. Valderrama, M.J.; González, E.R.; De Pablo, P.C.; Díez-Orejas, R.; Fernández-Acero, T.; Gil-Serna, J.; De Juan, L.; Martín, H.; Molina, M.; Navarro-García, F.; et al. Educating in antimicrobial resistance awareness: Adaptation of the Small World Initiative program to service-learning. *FEMS Microbiol. Lett.* **2018**, *365*, fny161. [CrossRef] [PubMed]
13. Davis, E.; Sloan, T.; Aurelius, K.; Barbour, A.; Bodey, E.; Clark, B.; Dennis, C.; Drown, R.; Fleming, M.; Humbert, A.; et al. Antibiotic discovery throughout the Small World Initiative: A molecular strategy to identify biosynthetic gene clusters involved in antagonistic activity. *MicrobiologyOpen* **2017**, *6*, e00435. [CrossRef] [PubMed]
14. Robredo, B.; Fernández-Fernández, R.; Torres, C. Antimicrobial resistance as a nexus between teaching and research. *J. Biol. Educ.* **2021**, 1–17. [CrossRef]
15. Ghosh, A.; Dey, N.; Bera, A.; Tiwari, A.; Sathyaniranjan, K.; Chakrabarti, K.; Chattopadhyay, D. Culture independent molecular analysis of bacterial communities in the mangrove sediment of Sundarban, India. *Saline Syst.* **2010**, *6*, 1. [CrossRef]
16. EUCAST. The European Committee on Antimicrobial Susceptibility Testing. *Breakpoint Tables for Interpretation of MICs and Zone Diameters. Version 12.0.* 2022. Available online: http://www.eucast.org (accessed on 26 November 2022).
17. Reddy, B.V.B.; Kallifidas, D.; Kim, J.H.; Charlop-Powers, Z.; Feng, Z.; Brady, S.F. Natural Product Biosynthetic Gene Diversity in Geographically Distinct Soil Microbiomes. *Appl. Environ. Microbiol.* **2012**, *78*, 3744–3752. [CrossRef] [PubMed]
18. Hibbing, M.E.; Fuqua, C.; Parsek, M.R.; Peterson, S.B. Bacterial competition: Surviving and thriving in the microbial jungle. *Nat. Rev. Microbiol.* **2010**, *8*, 15–25. [CrossRef] [PubMed]
19. Huang, Z.; Liu, B.; Yin, Y.; Liang, F.; Xie, D.; Han, T.; Liu, Y.; Yan, B.; Li, Q.; Huang, Y.; et al. Impact of biocontrol microbes on soil microbial diversity in ginger (*Zingiber officinale* Roscoe). *Pest Manag. Sci.* **2021**, *77*, 5537–5546. [CrossRef]
20. Chowdhury, S.P.; Hartmann, A.; Gao, X.; Borriss, R. Biocontrol mechanism by root-associated Bacillus amyloliquefaciens FZB42—A review. *Front. Microbiol.* **2015**, *6*, 780. [CrossRef]
21. Balthazar, C.; Novinscak, A.; Cantin, G.; Joly, D.L.; Filion, M. Biocontrol Activity of *Bacillus* spp. and *Pseudomonas* spp. Against *Botrytis cinerea* and Other Cannabis Fungal Pathogens. *Phytopathology* **2022**, *112*, 549–560. [CrossRef]

22. Dimkić, I.; Janakiev, T.; Petrović, M.; Degrassi, G.; Fira, D. Plant-associated Bacillus and Pseudomonas antimicrobial activities in plant disease suppression via biological control mechanisms—A review. *Physiol. Mol. Plant Pathol.* **2021**, *117*, 101754. [CrossRef]
23. Zhao, J.; Zhou, Z.; Bai, X.; Zhang, D.; Zhang, L.; Wang, J.; Wu, B.; Zhu, J.; Yang, Z. A novel of new class II bacteriocin from Bacillus velezensis HN-Q-8 and its antibacterial activity on Streptomyces scabies. *Front. Microbiol.* **2022**, *13*, 943232. [CrossRef] [PubMed]
24. Ghequire, M.G.; De Mot, R. Ribosomally encoded antibacterial proteins and peptides from *Pseudomonas*. *FEMS Microbiol. Rev.* **2014**, *38*, 523–568. [CrossRef] [PubMed]
25. de Oliveira, E.J.; Rabinovitch, L.; Monnerat, R.G.; Passos, L.K.J.; Zahner, V. Molecular Characterization of *Brevibacillus laterosporus* and Its Potential Use in Biological Control. *Appl. Environ. Microbiol.* **2004**, *70*, 6657–6664. [CrossRef] [PubMed]
26. Barbieri, G.; Ferrari, C.; Mamberti, S.; Gabrieli, P.; Castelli, M.; Sassera, D.; Ursino, E.; Scoffone, V.C.; Radaelli, G.; Clementi, E.; et al. Identification of a Novel *Brevibacillus laterosporus* Strain with Insecticidal Activity Against *Aedes albopictus* Larvae. *Front. Microbiol.* **2021**, *12*, 624014. [CrossRef] [PubMed]
27. Javed, K.; Qiu, D. Protein Elicitor PeBL1 of *Brevibacillus laterosporus* Enhances Resistance Against *Myzus persicae* in Tomato. *Pathogens* **2020**, *9*, 57. [CrossRef]
28. Khaled, J.M.; Al-Mekhlafi, F.A.; Mothana, R.A.; Alharbi, N.S.; Alzaharni, K.E.; Sharafaddin, A.H.; Kadaikunnan, S.; Alobaidi, A.S.; Bayaqoob, N.I.; Govindarajan, M.; et al. *Brevibacillus laterosporus* isolated from the digestive tract of honeybees has high antimicrobial activity and promotes growth and productivity of honeybee's colonies. *Environ. Sci. Pollut. Res.* **2017**, *25*, 10447–10455. [CrossRef]
29. Kim, J.Y.; Kwon, J.H.; Ahn, S.H.; Lee, S.I.; Han, Y.S.; Choi, Y.O.; Lee, S.Y.; Ahn, K.M.; Ji, G.E. Effect of probiotic mix (*Bifidobacterium bifidum, Bifidobacterium lactis, Lactobacillus acidophilus*) in the primary prevention of eczema: A double-blind, randomized, placebo-controlled trial. *Pediatr. Allergy Immunol.* **2010**, *21*, e386–e393. [CrossRef]
30. Wilaipun, P.; Zendo, T.; Okuda, K.-I.; Nakayama, J.; Sonomoto, K. Identification of the Nukacin KQU-131, a New Type-A(II) Lantibiotic Produced by *Staphylococcus hominis* KQU-131 Isolated from Thai Fermented Fish Product (Pla-ra). *Biosci. Biotechnol. Biochem.* **2008**, *72*, 2232–2235. [CrossRef]
31. Fernández-Fernández, R.; Lozano, C.; Eguizábal, P.; Ruiz-Ripa, L.; Martínez-Álvarez, S.; Abdullahi, I.N.; Zarazaga, M.; Torres, C. Bacteriocin-Like Inhibitory Substances in Staphylococci of Different Origins and Species With Activity Against Relevant Pathogens. *Front. Microbiol.* **2022**, *13*, 870510. [CrossRef]
32. Sung, C.; Kim, B.; Kim, S.; Joo, H.; Kim, P. Probiotic potential of *Staphylococcus hominis* MBBL 2–9 as anti- *Staphylococcus aureus* agent isolated from the vaginal microbiota of a healthy woman. *J. Appl. Microbiol.* **2010**, *108*, 908–916. [CrossRef]
33. Nakatsuji, T.; Chen, T.H.; Narala, S.; Chun, K.A.; Two, A.M.; Yun, T.; Shafiq, F.; Kotol, P.F.; Bouslimani, A.; Melnik, A.V.; et al. Antimicrobials from human skin commensal bacteria protect against *Staphylococcus aureus* and are deficient in atopic dermatitis. *Sci. Transl. Med.* **2017**, *9*, eaah4680. [CrossRef] [PubMed]
34. Sumi, C.D.; Yang, B.W.; Yeo, I.-C.; Hahm, Y.T. Antimicrobial peptides of the genus *Bacillus*: A new era for antibiotics. *Can. J. Microbiol.* **2015**, *61*, 93–103. [CrossRef] [PubMed]

antibiotics

MDPI

Article

Integrating Network Pharmacology Approaches to Decipher the Multi-Target Pharmacological Mechanism of Microbial Biosurfactants as Novel Green Antimicrobials against Listeriosis

Mohd Adnan [1], Arif Jamal Siddiqui [1], Emira Noumi [1], Sami Hannachi [1], Syed Amir Ashraf [2], Amir Mahgoub Awadelkareem [2], Mejdi Snoussi [1], Riadh Badraoui [1,3], Fevzi Bardakci [1], Manojkumar Sachidanandan [4], Mirav Patel [5] and Mitesh Patel [5,*]

1 Department of Biology, College of Science, University of Hail, Hail P.O. Box 2440, Saudi Arabia
2 Department of Clinical Nutrition, College of Applied Medial Sciences, University of Hail, Hail P.O. Box 2440, Saudi Arabia
3 Section of Histology-Cytology, Medicine Faculty of Tunis, University of Tunis El Manar, La Rabta 1007, Tunis, Tunisia
4 Department of Oral Radiology, College of Dentistry, University of Hail, Hail P.O. Box 2440, Saudi Arabia
5 Department of Biotechnology, Parul Institute of Applied Sciences and Centre of Research for Development, Parul University, Vadodara 391760, India
* Correspondence: patelmeet15@gmail.com

Abstract: *Listeria monocytogenes* (*L. monocytogenes*) is a serious food-borne pathogen that can cause listeriosis, an illness caused by eating food contaminated with this pathogen. Currently, the treatment or prevention of listeriosis is a global challenge due to the resistance of bacteria against multiple commonly used antibiotics, thus necessitating the development of novel green antimicrobials. Scientists are increasingly interested in microbial surfactants, commonly known as "biosurfactants", due to their antimicrobial properties and eco-friendly nature, which make them an ideal candidate to combat a variety of bacterial infections. Therefore, the present study was designed to use a network pharmacology approach to uncover the active biosurfactants and their potential targets, as well as the signaling pathway(s) involved in listeriosis treatment. In the framework of this study, 15 biosurfactants were screened out for subsequent studies. Among 546 putative targets of biosurfactants and 244 targets of disease, 37 targets were identified as potential targets for treatment of *L. monocytogenes* infection, and these 37 targets were significantly enriched in a Gene Ontology (GO) analysis, which aims to identify those biological processes, cellular locations, and molecular functions that are impacted in the condition studied. The obtained results revealed several important biological processes, such as positive regulation of MAP kinase activity, protein kinase B signaling, ERK1 and ERK2 cascade, ERBB signaling pathway, positive regulation of protein serine/threonine kinase activity, and regulation of caveolin-mediated endocytosis. Several important KEGG pathways, such as the ERBBB signaling pathway, TH17 cell differentiation, HIF-1 signaling pathway, *Yersinia* infection, Shigellosis, and C-type lectin receptor signaling pathways, were identified. The protein–protein interaction analysis yielded 10 core targets (IL2, MAPK1, EGFR, PTPRC, TNF, ITGB1, IL1B, ERBB2, SRC, and mTOR). Molecular docking was used in the latter part of the study to verify the effectiveness of the active biosurfactants against the potential targets. Lastly, we found that a few highly active biosurfactants, namely lichenysin, iturin, surfactin, rhamnolipid, subtilisin, and polymyxin, had high binding affinities towards IL2, MAPK1, EGFR, PTPRC, TNF, ITGB1, IL1B, ERBB2, SRC, and mTOR, which may act as potential therapeutic targets for listeriosis. Overall, based on the integrated network pharmacology and docking analysis, we found that biosurfactants possess promising anti-listeriosis properties and explored the pharmacological mechanisms behind their effect, laying the groundwork for further research and development.

Keywords: *Listeria monocytogenes*; listeriosis; network pharmacology; biosurfactants; antimicrobial

Citation: Adnan, M.; Siddiqui, A.J.; Noumi, E.; Hannachi, S.; Ashraf, S.A.; Awadelkareem, A.M.; Snoussi, M.; Badraoui, R.; Bardakci, F.; Sachidanandan, M.; et al. Integrating Network Pharmacology Approaches to Decipher the Multi-Target Pharmacological Mechanism of Microbial Biosurfactants as Novel Green Antimicrobials against Listeriosis. *Antibiotics* **2023**, *12*, 5. https://doi.org/10.3390/antibiotics12010005

Academic Editor: Helena P. Felgueiras

Received: 7 November 2022
Revised: 15 December 2022
Accepted: 16 December 2022
Published: 20 December 2022

1. Introduction

More than 200 diseases can be caused in humans by food-borne contaminations, which are caused by a variety of factors that are involved with the cause and development of disease related to food consumption [1]. In this regard, we can point to the increasing population of the world, which has led to the subsequent rise in the demand for food, as well as microbial genomic diversification and selection pressures, resulting in the emergence of new pathogens as a result of the growing popularity of eating outside the home [2]. An infection caused by the bacterium *Listeria monocytogenes* (*L. monocytogenes*) is called "listeriosis" and is usually a result of eating food that has been contaminated with this food pathogen. In a wide range of food products, such as dairy products, raw vegetables, and raw meat, as well as ready-to-eat products, this bacterium has been found to be present [3]. The *L. monocytogenes* are a Gram-positive, rod-shaped, non-spore forming, non-capsule forming bacteria, which are motile at 10 to 25 °C [4]. They can infect a wide range of human and animals cell types. Few populations of humans are reported to carry the bacterium without showing symptoms in the intestinal tract [5]. Following the ingestion of bacterium by the host, *L. monocytogenes* first encounters epithelial cells of the gut lining and then enters the host's monocytes, macrophages, or polymorphonuclear leukocytes. The bacterium becomes blood-borne and multiplies both intracellularly and extracellularly. In pregnant women, it can migrate through the placenta to reach the fetus intracellularly [6]. When *L. monocytogenes* is infected in mice, the bacteria first appear in macrophages before spreading to liver hepatocytes [7]. Several outbreaks have been associated with the consumption of ready-to-eat food, because *L. monocytogenes* is capable of growing at refrigerated temperatures [8].

There are several high-risk populations that are susceptible to listeriosis, including the elderly, pregnant women, newborns, and immunocompromised patients due to kidney transplant, cancer, HIV/AIDS, and steroid therapy [8]. Around the world, there are approximately 1600 cases of listeriosis each year, and approximately 260 people die from it [9]. Despite the fact that there are a small number of cases of listeriosis in the world, the high rate of death associated with this infection makes it an important public health concern. Due to this, there is a need to implement effective medical management for listeriosis. Therefore, alternative measures are needed to control *L. monocytogenes* in the food industry.

Over the past few years, natural products and their derivatives have been gaining more and more attention as insights into research and possible drug sources for targeted therapy, owing to their variety of structural features, multi-target action, and low toxicity [10]. There have been a great number of dramatic advances in high-throughput screening techniques over the past few decades that have greatly contributed to the discovery of novel drugs based on natural products [11]. Hence, a new discovery of a potential bioactive compound that can affect the pathophysiology of diseases and disorders will be considered a thunderbolt of this new era of pharmaceuticals.

Biological surfactants (biosurfactants) are surface active compounds which are synthesized by the microbes (bacteria and fungi) on their cell surface or excreted that can reduce surface and interfacial tension [12]. There is no doubt that biosurfactants are becoming more and more popular among scientists because of their eco-friendliness properties, scalability, durability under harsh environmental conditions, specificity, and versatility, which make them appealing for their application in various fields [13]. There are numerous applications for these compounds as antimicrobials, anti-adhesives, and anticancer agents, in addition to being extensively used for the purposes of recovery of oil, bioremediation, and emulsification in industry [14].

In previous studies, biosurfactants have been demonstrated to have antimicrobial, antibiofilm, and anti-listeriosis properties [13–15], suggesting that they could potentially be useful for preventing and treating listeriosis. In spite of this, very few studies have been published that have examined the use of biosurfactants in the prevention and treatment of listeriosis, and no research has examined the mechanisms behind their action [15].

Insights into the mechanisms of action of biosurfactants against listeriosis will be possible if studies focusing on molecular targets and their related signal pathways are conducted. To accomplish this purpose, we utilized network pharmacology [16,17] and a molecular docking methodology [18] approach in the present study to construct a multidimensional network of "component–target–pathway–disease" that is able to explain the biological mechanisms underlying biosurfactants for the prevention and treatment of listeriosis. It is intended that the results of the present study will provide a scientific foundation for clinical trial research and the development of biosurfactant products in the future. Figure 1 illustrates the flowchart of this study.

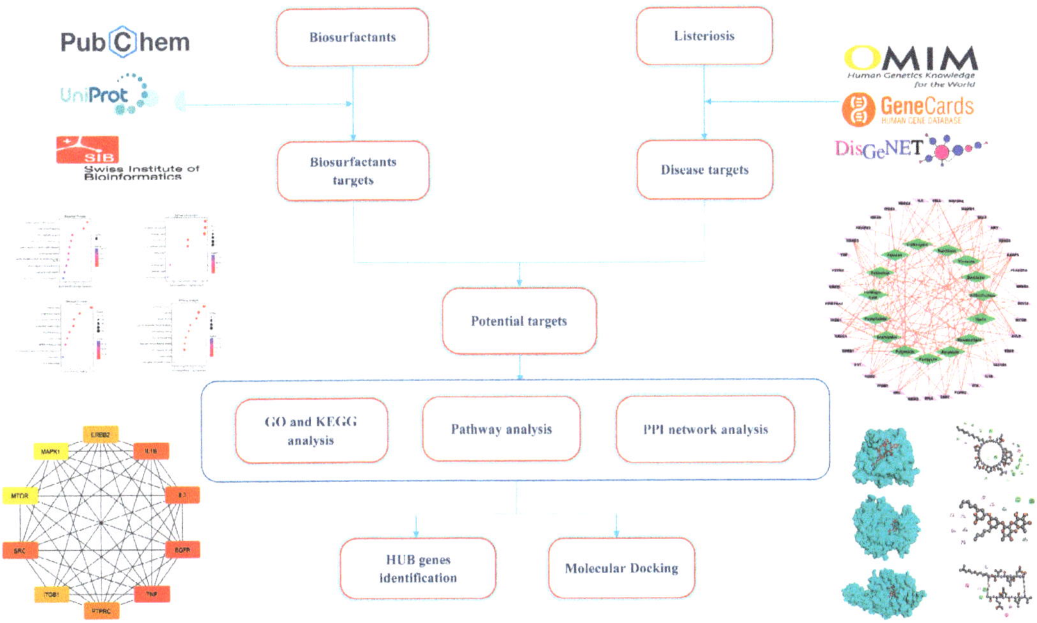

Figure 1. Framework based on an integration strategy of network pharmacology.

2. Results

2.1. Identification of Active Components of Biosurfactants

In total, 15 biosurfactants were selected, and their detailed information was retrieved from the PubChem database in order to be analyzed, using the SwissTargetPrediction database (Table 1). We predicted the potential protein targets of each biosurfactant by using SwissTargetPrediction (Figure 2A–F, Figure 3A–F, Figure 4A–F, and Figure 5A–C). Following the removal of duplicate targets from the target prediction, screening of 546 potential targets was conducted for further evaluation. A visual compound–target network was subsequently constructed by using Cytoscape 3.9.1 in order to construct a visual network with 546 nodes and 545 edges (Figure 6A). The nodes represent ingredients and their corresponding targets. The higher the degree corresponding to the node, the greater the pharmacological effects of this ingredient or target. The calculated average shortest path length, betweenness centrality, closeness centrality, and degree of nodes in the network are shown in Table 2.

Table 1. List of biosurfactants.

Sr. No.	Biosurfactant	Microbial Origin	References	Molecular Formula	PubChem	Canonical SMILE
1	Surfactin	*Bacillus subtilis* *Bacillus siamensis*	[19,20]	$C_{53}H_{93}N_7O_{13}$	5066078	CC(C)CCCCCCCCCC1CC(=O)NC(C(=O)NC(C(=O)NC(C(=O)NC(C(=O)NC(C(=O)NC(C(=O)NC(C(=O)O1)CC(C)C)CC(C)C)CC(=O)O)C(C)C)CC(C)C)CC(C)C)CCC(=O)O
2	Rhamnolipid	*Pseudomonas aeruginosa*	[21]	$C_{32}H_{58}O_{13}$	5458394	CCCCCCCC(CC(=O)O)OC(=O)CC(CCCCCCC)OC1C(C(C(C(O1)C)O)O)OC2C(C(C(C(O2)C)O)O)O
3	Viscosin	*Pseudomonas fluorescens*	[22]	$C_{54}H_{95}N_9O_{16}$	72937	CCCCCCCC(CC(=O)NC(CC(C)C)C(=O)NC(CCC(=O)O)C(=O)NC1C(OC(=O)C(NC(=O)C(NC(=O)C(NC(=O)C(NC(=O)C(NC(=O)C(NC1=O)C(C)C)CC(C)C)CO)CC(C)C)CO)C(C)CC)C)O
4	Liposan	*Candida lipolytica*	[23]	$C_8H_{14}O_2S_2$	864	C1CSSC1CCCCC(=O)O
5	Lichenysin	*Bacillus licheniformis*	[24]	$C_{51}H_{90}N_8O_{12}$	11804102	CC(C)CCCCCCCCCC1CC(=O)NC(C(=O)NC(C(=O)NC(C(=O)NC(C(=O)NC(C(=O)NC(C(=O)NC(C(=O)O1)C(C)C)CC(C)C)CC(=O)O)C(C)C)CC(C)C)CC(C)C)CCC(=O)N
6	Iturin	*Bacillus subtilis* *Bacillus amyloliquefaciens*	[25]	$C_{48}H_{74}N_{12}O_{14}$	158570	CCCCCCCCCCCC1CC(=O)NC(C(=O)NC(C(=O)NC(C(=O)NC(C(=O)N2CCCC2C(=O)NC(C(=O)NC(C(=O)N1)CO)CC(=O)N)CCC(=O)N)CC(=O)N)CC3=CC=C(C=C3)O)CC(=O)N
7	Arthrofactin	*Arthrobacter* sp. strain MIS38	[26]	$C_{64}H_{111}N_{11}O_{20}$	23724538	CCCCCCCC1CC(=O)NC(C(=O)NC(C(=O)NC(C(=O)NC(C(=O)NC(C(=O)NC(C(=O)NC(C(=O)NC(C(=O)NC(C(=O)NC(C(=O)O1)CC(=O)O)C(C)CC)C(C)CC)CO)CC(C)C)CO)CC(C)C)CC(C)C)C(C)O)CC(=O)O)CC(C)C
8	Amphisin	*Pseudomonasfluorescens*	[27]	C66H114N12O20	101134740	CCCCCCCC(CC(=O)NC(CC(C)C)C(=O)NC(CC(=O)O)C(=O)NC1C(OC(=O)C(NC(=O)C(NC(=O)C(NC(=O)C(NC(=O)C(NC(=O)C(NC(=O)C(NC1=O)CC(C)C)CC(C)C)CO)CC(C)C)CCC(=O)N)CC(C)C)C(C)CC)CC(=O)O)C)O
9	Putisolvin	*Pseudomonas putida*	[28]	C65H113N13O19	139588800	CCCCCC(=O)NC(CC(C)C)C(=O)NC(CCC(=O)O)C(=O)NC(CC(C)C)C(=O)NC(C(C)CC)C(=O)NC(CCC(=O)N)C(=O)NC(CO)C(=O)NC(C(C)C)C(=O)NC(C(C)CC)C(=O)NC1COC(=O)C(NC(=O)C(NC(=O)C(NC1=O)CC(C)C)C(C)C)CO

Table 1. *Cont.*

Sr. No.	Biosurfactant	Microbial Origin	References	Molecular Formula	PubChem	Canonical SMILE
10	Ustilagic Acid	*Ustilago maydis*	[29]	C36H64O18	52922086	CCCC(CC(=O)OC1C(C(C(OC1OC2C(OC(C(C2O)O)OCC(CCCCCCCCCCCCC(C(=O)O)O)O)COC(=O)C)CO)O)O)O
11	Pumilacidin	*Bacillus pumilus*	[30]	C55H99N7O12	101174694	CCC(C)C1C(=O)OC(CC(=O)NC(C(=O)NC(C(=O)NC(CNC(C(=O)NC(C(=O)NC(C(=O)N1)CC(C)C)CC(=O)O)CC(C)C)CC(C)C)CC(C)C)CCC(=O)O)CCCCCCCCCCC(C)C
12	Fengycin	*Bacillus subtilis*	[25]	C72H110N12O20	443591	CCCCCCCCCCCCCC(CC(=O)NC(CCC(=O)O)C(=O)NC(CCCN)C(=O)NC1CC2=CC=C(C=C2)OC(=O)C(NC(=O)C(NC(=O)C(NC(=O)C3CCCN3C(=O)C(NC(=O)C(NC(=O)C(NC1=O)C(C)O)CCC(=O)O)C)CCC(=O)N)CC4=CC=C(C=C4)O)C(C)CC)O
13	Subtilisin	*Bacillus subtilis*	[31]	C18H25N3O6	92174084	CC(C(=O)NOC(=O)C1=CC=CC=C1)NC(=O)C(C)NC(=O)OC(C)(C)C
14	Gramicidin S	*Brevibacillus brevis*	[32]	C60H92N12O10	73357	CC(C)CC1C(=O)NC(C(=O)N2CCCC2C(=O)NC(C(=O)NC(C(=O)NC(C(=O)NC(C(=O)N3CCCC3C(=O)NC(C(=O)NC(C(=O)N1)CCCN)C(C)C)CC4=CC=CC=C4)CC(C)C)CCCN)C(C)C)CC5=CC=CC=C5
15	Polymyxin	*Paenibacillus polymyxa*	[33]	C48H82N16O14	3083714	CC(C)CC1C(=O)NC(C(=O)NC(C(=O)NC(C(=O)NCCC(C(=O)NC(C(=O)NC(C(=O)N1)CC2=CC=CC=C2)CCN)NC(=O)C(CCN)NC(=O)C(C(C)O)NC(=O)C(CCN)NC(=O)O)C(C)O)CCN)CCN

Table 2. Topological parameters of the compound.

Sr. No.	Biosurfactant	Degree	Betweenness	Closeness
1	Putisolvin	12	218.86357	0.44347826
2	Surfactin	11	260.51117	0.42857143
3	Lichenysin	11	180.90701	0.43589744
4	Arthrofactin	10	295.14645	0.4214876
5	Amphisin	10	98.94803	0.4214876
6	Iturin	10	312.37665	0.39534885
7	Pumalicidin	10	114.981255	0.42857143
8	Subtilisin	10	543.5165	0.39534885
9	Polymyxin	9	186.33307	0.4015748
10	Viscosin	9	126.53945	0.4214876
11	Fengycin	8	143.53413	0.38345864
12	Gramicidin	8	158.42682	0.38345864
13	Ustilagic Acid	6	220.03003	0.3167702
14	Rhamnolipid	6	151.10707	0.37226278
15	Liposan	3	112.77881	0.3090909

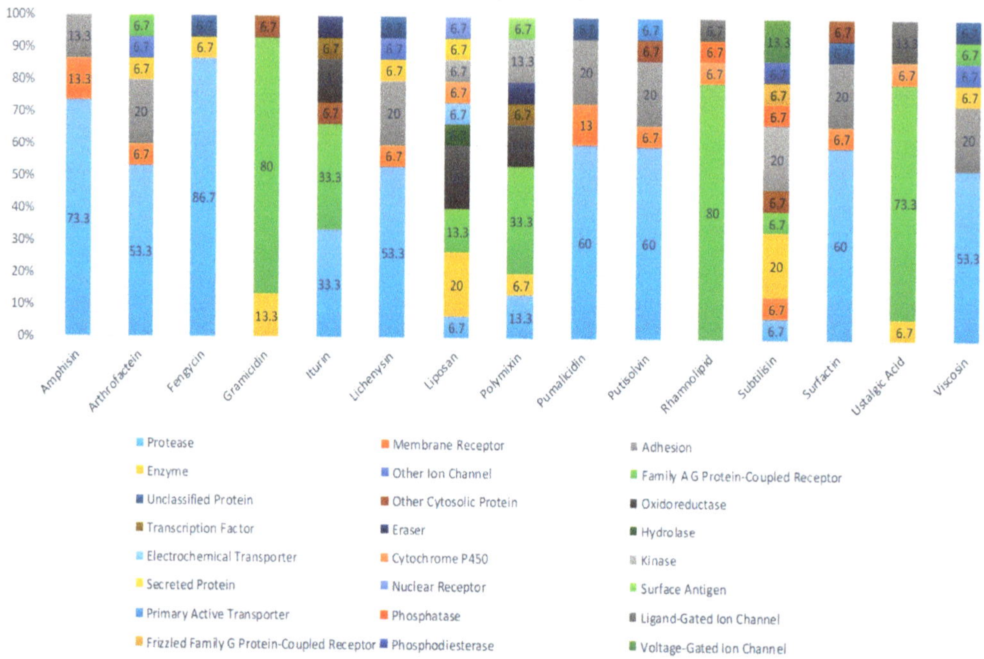

Figure 2. Biosurfactants and their potential proteins interaction, as retrieved from the SwissTargetPrediction server.

2.2. Listeriosis and Intersection Target

The human genome database was used to collect the targets that are related to listeriosis. A total of 197, 276, and 211 targets were identified in the OMIM, DisGeNET, and GeneCard databases, respectively. As a result of removing duplicate entries from these three kinds of databases, a total of 244 listeriosis targets were obtained (Figure 6B). By intersecting these targets with component targets, a total of 37 intersection targets were obtained, as shown in Figure 7A. Figure 7B,C show a diagram of component intersection targets that has 52 nodes and 133 edges that were created with the help of Cytoscape.

2.3. Construction of Protein–Protein Interaction Network (PPI) and Key Targets

Utilizing the GeneMANIA tool, we imported 37 target genes in order to obtain a PPI network that demonstrates the relationships between these 37 target genes and other genes in the network. In the results, the percentage represents the weight that is given to interaction relationships in the network. According to our results, 29.46% of the interactions between the targets in the network resulted in co-expressions, and 35.77% of them resulted in physical interactions. Furthermore, there was a relationship between co-localization and shared protein domains (Figure 8A). In Table 3, we provide the calculated average length of shortest paths to the three central nodes, betweenness centrality, closeness centrality, and degree of each node in the network. There were ten targets in the network which are organized in the order of high to low, according to the topology properties of the network, corresponding to EGFR, SRC, IL1B, IL2, PTPRC, ERBB2, ITGB1, MAPK1, MTOR, and TNF (Figure 8B). Biosurfactants may be able to prevent and treat listeriosis by targeting these ten targets, as they may be the key targets for biosurfactants.

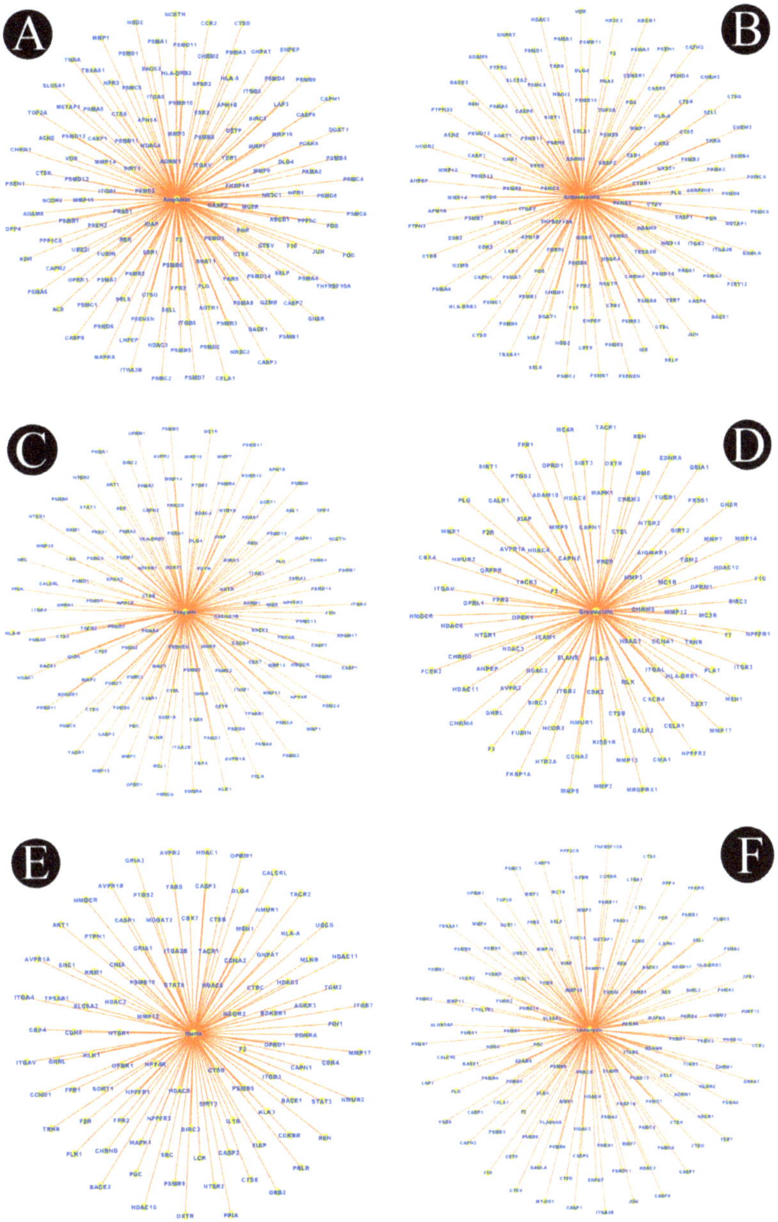

Figure 3. Biosurfactants and their potential target networks: (**A**) amphisin, (**B**) arthrofactin, (**C**) fengycin, (**D**) gramicidin, (**E**) iturin, and (**F**) lichenysin. Edges (orange color) represent respective protein targets.

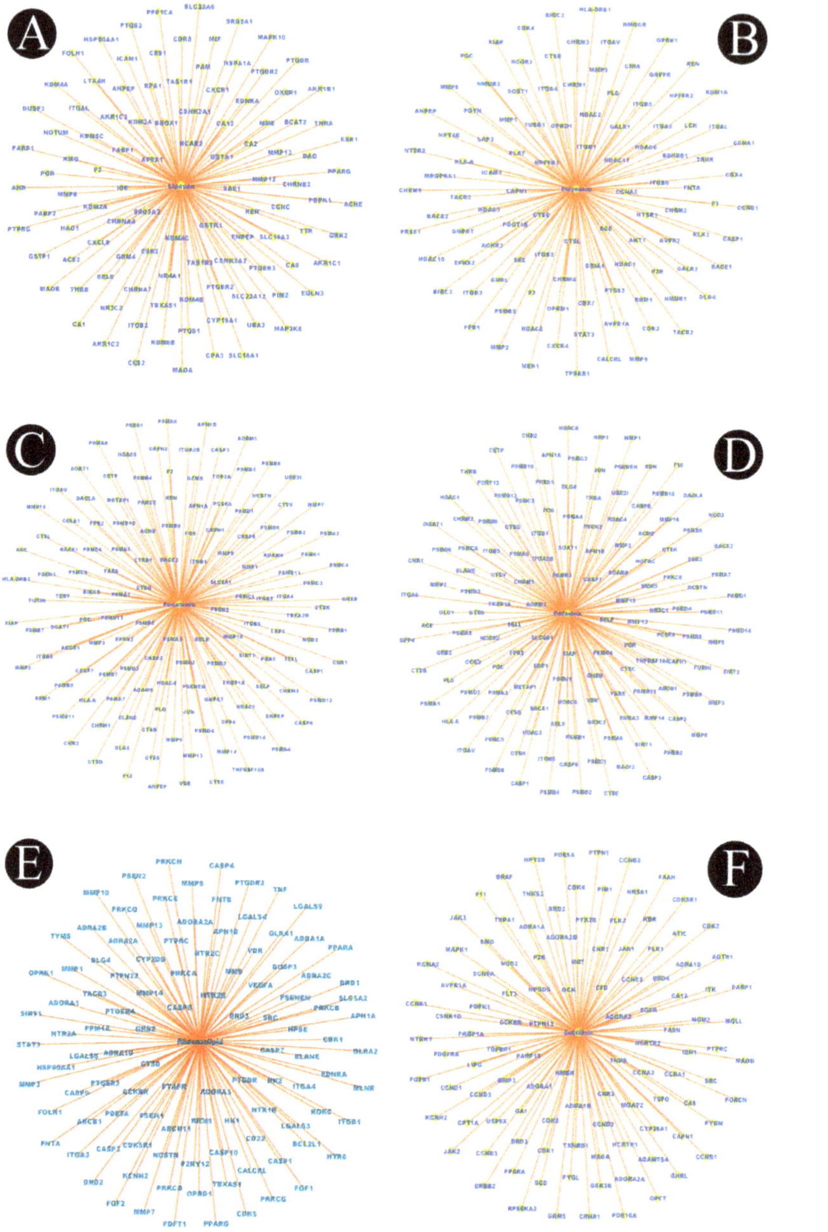

Figure 4. Biosurfactants and their potential target networks. (**A**) liposan, (**B**) polymyxin, (**C**) pumalicidin, (**D**) putisolvin, (**E**) rhamnolipid, and (**F**) subtilisin. Edges (orange color) represent respective protein targets.

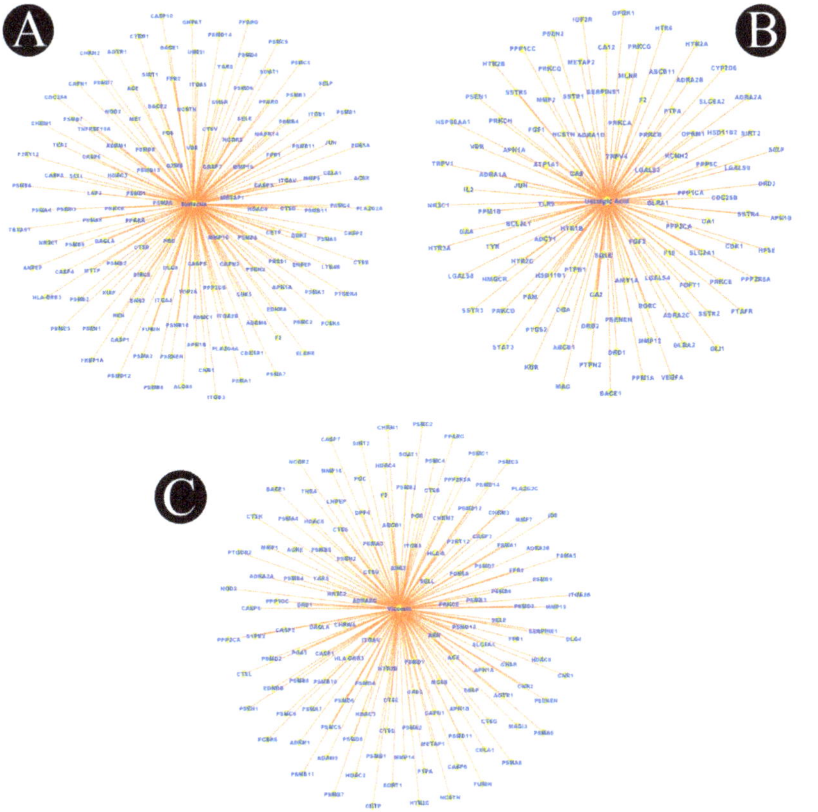

Figure 5. Biosurfactants and their potential target networks. (**A**) surfactin, (**B**) ustilagic acid, and (**C**) viscosin. Edges (orange color) represent respective protein targets.

2.4. Functional GO and KEGG Pathways

By using the Shiny GO 0.76.2 database analytical tool, the 37 intersected genes were enriched by GO and KEGG analysis. As a result of incorporating biological process (BP), molecular function (MF) and cellular component (CC) (Figure 9A–C), along with a p-value < 0.05, as screening conditions, a total of 1255 items were obtained pertaining to biological process, 149 items were obtained pertaining to molecular function, and 94 items were obtained pertaining to cellular composition. The hypothesis was put forth that biosurfactants could be involved in inhibiting listeriosis through the positive regulation of MAP kinase activity, protein kinase B signaling, ERK1 and ERK2 cascade, and the ERBB signaling pathway; positive regulation of protein serine/threonine kinase activity and leukocyte cell–cell adhesion; positive regulation of establishment of protein localization; and regulation of caveolin-mediated endocytosis via molecular functions such as integrin binding, phosphoprotein binding, protein tyrosine kinase activity, growth factor receptor binding, phosphatase binding, cytokine activity, NEDD8 transferase activity, and cadherin binding in cellular compartments such as membrane raft, membrane microdomain, focal adhesion, cell–substrate junction, myelin sheath, basal plasma membrane, and basal part of the cell. A total of 190 enrichment results were obtained from the KEGG pathway enrichment analysis. Among them Shigellosis, *Yersinia* infection, the ERBB signaling pathway, Th17 cell differentiation, the HIF-1 signaling pathway, the C-type lectin receptor signaling pathway, and bladder cancer pathways are closely associated with listeriosis and are in accordance with the enrichment results of GO. There was a significant abundance of KEGG pathways and gene pathways with p-values ≤ 0.05. Based on the Shiny GO platform, the first ten components were analyzed (Figure 9D). Based on the statistical analysis, ten proteins exhibited a high frequency of participation in the first 10 pathways, indicating that they played a major role in the enrichment pathway. The ten core proteins are EGFR, SRC, IL1B, IL2, PTPRC, ERBB2, ITGB1, MAPK1, MTOR, and TNF.

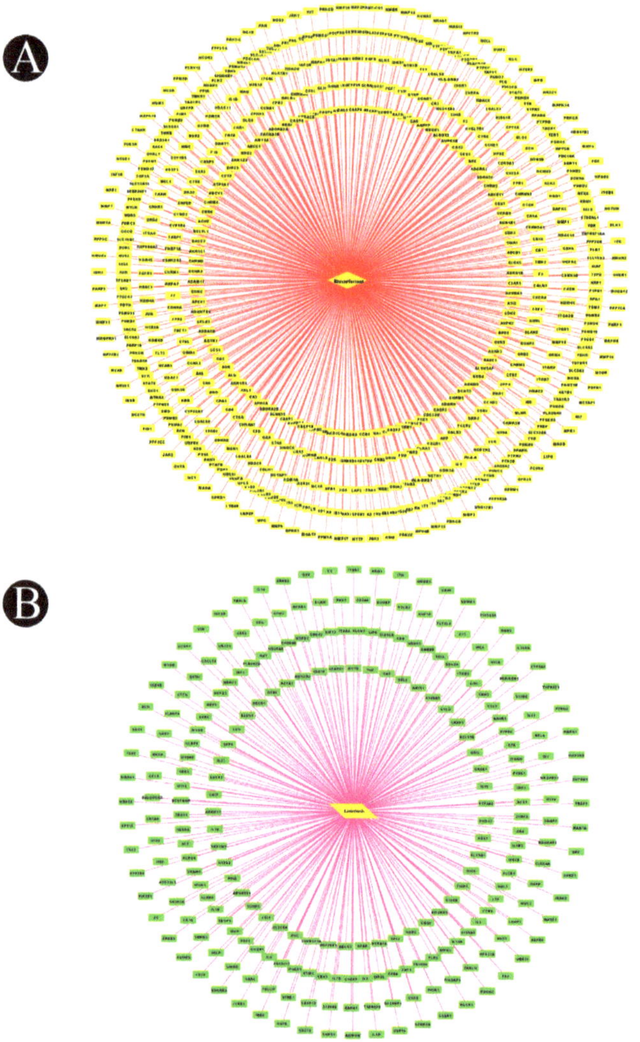

Figure 6. (**A**) Biosurfactant–gene network after removing duplication of genes (diamond indicates biosurfactants and rectangle indicates target proteins). (**B**). Disease–gene network after removing duplication of genes (parallelogram indicates disease, and rectangle indicates target proteins).

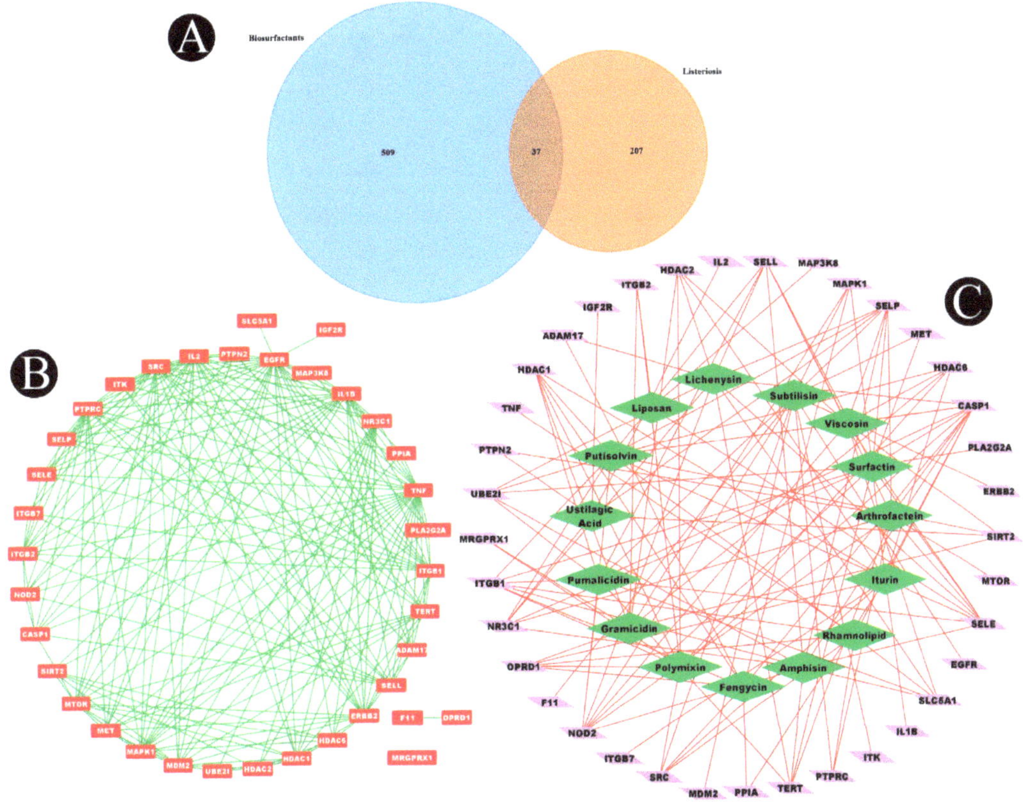

Figure 7. (**A**) Venn diagram showing common genes between listeriosis and biosurfactants. (**B**) Interconnected common genes network constructed by using Cytoscape. (**C**) Biosurfactants–common-genes target network (diamond indicates biosurfactants and parallelogram indicates common target proteins).

2.5. Molecular Docking

Virtual screening using molecular docking is a computational method for identifying potential leads against predefined targets. By employing this method, compounds with appreciable binding affinities and specific interactions with target proteins were identified. A docking analysis of all the biosurfactants revealed the presence of several compounds with a significant affinity for the respective target proteins (Figure 10). The highest binding affinity was found between IL2–lichenysin (−6.0 kJ/mol), MAPK1–polymyxin (−7.2 kJ/mol), EGFR–rhamnolipid (−6.7 kJ/mol), PTPRC–surfactin (−6.2 kJ/mol), TNF–subtilisin (−6.0 kJ/mol), ITGB1–lichenysin (−7.8 kJ/mol), IL1B–iturin (−7.4 kJ/mol), ERBB2–iturin (−6.2 kJ/mol), SRC–surfactin (−7.0 kJ/mol), and mTOR–subtilisin (−6.5 kJ/mol). These results suggest that the few selected biosurfactants have a significant level of binding efficiency with respective proteins, which may contribute to the development of a potential binding partner for selective proteins that could be used for drug development. The best biosurfactants observed occupying the active site in different ways can be seen in Figures 11–15 and Table 4.

Table 4. *Cont.*

Sr. No.	Protein	Receptor–Ligand	Interaction Type	Distance
		A:TYR268:HH - N:UNK1:O	Conventional Hydrogen Bond	2.03597
		N:UNK1:H - A:ASP286:OD1	Conventional Hydrogen Bond	2.62308
		N:UNK1:H - N:UNK1:O	Conventional Hydrogen Bond	2.48544
		N:UNK1:H - N:UNK1:O	Conventional Hydrogen Bond	2.1777
		N:UNK1:H - N:UNK1:O	Conventional Hydrogen Bond	2.83049
		N:UNK1:H - N:UNK1:O	Conventional Hydrogen Bond	1.55546
		N:UNK1:H - N:UNK1:O	Conventional Hydrogen Bond	2.51507
2	3WLW	N:UNK1:H - A:THR285:O	Conventional Hydrogen Bond	2.22765
		N:UNK1:H - A:SER284:O	Conventional Hydrogen Bond	3.005
		N:UNK1:H - N:UNK1:O	Conventional Hydrogen Bond	2.72822
		A:LYS309:CE - N:UNK1:O	Carbon Hydrogen Bond	3.69856
		A:LEU250:CB - N:UNK1	Pi-Sigma	3.92346
		A:ALA249 - N:UNK1	Alkyl	4.39901
		A:ALA249 - N:UNK1:C	Alkyl	4.10378
		A:VAL251 - N:UNK1	Alkyl	5.23527
		A:VAL251 - N:UNK1	Alkyl	5.01063
		N:UNK1 - A:LEU250	Alkyl	4.96966
		N:UNK1:C - A:VAL287	Alkyl	5.09767
		A:MET341:HN - N:UNK1:O	Conventional Hydrogen Bond	1.97267
		A:SER345:HN - N:UNK1:O	Conventional Hydrogen Bond	2.85586
		A:ASN391:HD21 - N:UNK1:O	Conventional Hydrogen Bond	2.986
		A:ASN391:HD22 - N:UNK1:O	Conventional Hydrogen Bond	2.8606
		N:UNK1:H - A:LEU273:O	Conventional Hydrogen Bond	2.23507
		N:UNK1:H - A:LEU273:O	Conventional Hydrogen Bond	2.69853
		N:UNK1:H - A:GLN275:O	Conventional Hydrogen Bond	2.83421
3	4MXO	A:GLY274:CA - N:UNK1:O	Carbon Hydrogen Bond	3.14193
		A:GLY344:CA - N:UNK1:O	Carbon Hydrogen Bond	3.14994
		N:UNK1:C - N:UNK1:O	Carbon Hydrogen Bond	3.54737
		A:VAL281 - N:UNK1	Alkyl	4.86428
		A:VAL281 - N:UNK1	Alkyl	5.14814
		A:ALA293 - N:UNK1	Alkyl	4.50016
		A:LYS295 - N:UNK1	Alkyl	5.25134
		A:ALA403 - N:UNK1:C	Alkyl	3.80138
		N:UNK1:C - A:MET314	Alkyl	4.94124
		N:UNK1:C - A:VAL323	Alkyl	3.63942

Table 4. *Cont.*

Sr. No.	Protein	Receptor–Ligand	Interaction Type	Distance
4	5FMV	N:UNK1 - A:LEU273	Alkyl	4.79106
		N:UNK1:C - A:LEU273	Alkyl	4.84683
		A:PHE278 - N:UNK1	Pi-Alkyl	5.36838
		N:UNK1:H - A:ASP508:OD2	Salt Bridge	2.60084
		N:UNK1:H - A:ASP508:OD2	Conventional Hydrogen Bond	2.42554
		N:UNK1:H - A:ASP508:OD1	Conventional Hydrogen Bond	2.71193
		A:LYS448 - N:UNK1	Alkyl	4.81167
		A:PRO449 - N:UNK1	Alkyl	4.98618
		A:HIS404 - N:UNK1	Pi-Alkyl	4.73078
		A:TRP487 - N:UNK1	Pi-Alkyl	4.80377
		A:TRP487 - N:UNK1	Pi-Alkyl	4.83494
		A:TRP487 - N:UNK1	Pi-Alkyl	4.37643
5	9ILB	N:UNK1:H - A:THR79:OG1	Conventional Hydrogen Bond	2.04809
		N:UNK1:H - N:UNK1:O	Conventional Hydrogen Bond	2.48545
		N:UNK1:H - A:GLU25:OE2	Conventional Hydrogen Bond	3.01317
		N:UNK1:H - N:UNK1:O	Conventional Hydrogen Bond	2.08891
		N:UNK1:H - N:UNK1:O	Conventional Hydrogen Bond	1.5553
		N:UNK1:H - A:LEU134:O	Conventional Hydrogen Bond	2.12807
		N:UNK1:H - A:VAL132:O	Conventional Hydrogen Bond	2.61684
		N:UNK1:H - A:LEU80:O	Conventional Hydrogen Bond	2.59836
		N:UNK1:C - N:UNK1:O	Carbon Hydrogen Bond	3.5902
		A:PHE133 - N:UNK1	Pi-Pi Stacked	3.75995
		A:TYR24 - N:UNK1	Pi-Alkyl	4.38175
		A:TYR24 - N:UNK1:C	Pi-Alkyl	4.05141
		N:UNK1 - A:PRO131	Pi-Alkyl	5.36192
6	4G6O	A:ASN152:HD22 - :UNL1:N	Conventional Hydrogen Bond	2.68599
		UNL1:H - :UNL1:O	Conventional Hydrogen Bond	1.97392
		UNL1:H - :UNL1:O	Conventional Hydrogen Bond	1.72687
		UNL1:H - :UNL1:O	Conventional Hydrogen Bond	2.6197
		UNL1:H - A:ASP109:OD2	Conventional Hydrogen Bond	2.68182
		UNL1:H - :UNL1:O	Conventional Hydrogen Bond	2.16901
		UNL1:H - A:CYS164:SG	Conventional Hydrogen Bond	3.02337
		UNL1:H - :UNL1:O	Conventional Hydrogen Bond	2.8299
		UNL1:H - A:ASP104:O	Conventional Hydrogen Bond	2.77312

Table 4. *Cont.*

Sr. No.	Protein	Receptor–Ligand	Interaction Type	Distance
		UNL1:H - :UNL1:O	Conventional Hydrogen Bond	2.25806
		UNL1:H - :UNL1:O	Conventional Hydrogen Bond	2.61859
		UNL1:H - A:GLU31:OE1	Conventional Hydrogen Bond	2.17908
		UNL1:H - A:ASN152:OD1	Conventional Hydrogen Bond	2.75093
		UNL1:H - :UNL1:O	Conventional Hydrogen Bond	2.45343
		UNL1:H - A:ASP165:OD1	Conventional Hydrogen Bond	2.30444
		UNL1:H - :UNL1:O	Conventional Hydrogen Bond	2.29142
		UNL1 - A:ARG65	Pi-Alkyl	5.08193
7	4WKQ	UNL1:H - A:ASN842:OD1	Conventional Hydrogen Bond	2.34513
		UNL1:H - A:ASP837:OD2	Conventional Hydrogen Bond	2.64181
		UNL1:H - :UNL1:O	Conventional Hydrogen Bond	2.7793
		UNL1:H - :UNL1:O	Conventional Hydrogen Bond	2.61272
		A:ARG841:CD - :UNL1:O	Carbon Hydrogen Bond	3.48781
		UNL1:C - A:ASP855:OD2	Carbon Hydrogen Bond	3.36649
		UNL1:C - A:ASP855:OD2	Carbon Hydrogen Bond	2.93591
		A:LEU718 - :UNL1	Alkyl	4.75503
		A:LEU718 - :UNL1	Alkyl	4.63701
		A:VAL726 - :UNL1	Alkyl	4.56822
		A:VAL726 - :UNL1	Alkyl	5.49427
		A:ALA743 - :UNL1	Alkyl	4.47905
		A:LEU844 - :UNL1	Alkyl	5.36797
		A:LEU844 - :UNL1	Alkyl	5.0695
		UNL1:C - A:LYS745	Alkyl	4.04982
		UNL1:C - A:MET766	Alkyl	5.00922
		UNL1:C - A:LEU788	Alkyl	4.51178
8	5WBU	UNL1:H - A:MET2345:SD	Conventional Hydrogen Bond	2.81877
		A:ILE2356:CG2 - :UNL1	Pi-Sigma	3.70617
		A:TYR2225 - :UNL1	Pi-Pi T-shaped	4.91263
		UNL1:C - A:PRO2169	Alkyl	4.70722

Table 4. *Cont.*

Sr. No.	Protein	Receptor–Ligand	Interaction Type	Distance
9	7CEB	A:TYR295:HH - :UNL1:O	Conventional Hydrogen Bond	2.30825
		A:TYR411:HH - :UNL1:O	Conventional Hydrogen Bond	2.74757
		UNL1:H - :UNL1:O	Conventional Hydrogen Bond	1.65568
		UNL1:H - :UNL1:O	Conventional Hydrogen Bond	1.6278
		UNL1:C - A:TYR234	Pi-Sigma	3.72179
		UNL1:C - A:ILE356	Alkyl	4.29736
		UNL1:C - A:PRO185	Alkyl	4.4092
		A:TRP91 - :UNL1	Pi-Alkyl	5.11969
		A:TRP91 - :UNL1	Pi-Alkyl	5.3081
		A:HIS110 - :UNL1	Pi-Alkyl	5.20706
		A:PHE237 - :UNL1	Pi-Alkyl	5.3189
		A:PHE237 - :UNL1	Pi-Alkyl	4.60088
		A:TYR295 - :UNL1	Pi-Alkyl	4.91972
		A:TYR295 - :UNL1:C	Pi-Alkyl	5.27292
		A:TYR411 - :UNL1:C	Pi-Alkyl	4.32441
10	2AZ5	A:GLN61:HE12 - :UNL1:O	Conventional Hydrogen Bond	2.61257
		A:TYR119:HH - :UNL1:O	Conventional Hydrogen Bond	2.86757
		A:TYR151:HH - :UNL1:O	Conventional Hydrogen Bond	2.59785
		A:LEU63:CD1 - :UNL1	Pi-Sigma	3.80396
		A:LEU63:CD2 - :UNL1	Pi-Sigma	3.73556
		UNL1:C - A:TYR119	Pi-Sigma	3.95427
		UNL1 - A:PRO117	Pi-Alkyl	4.8335

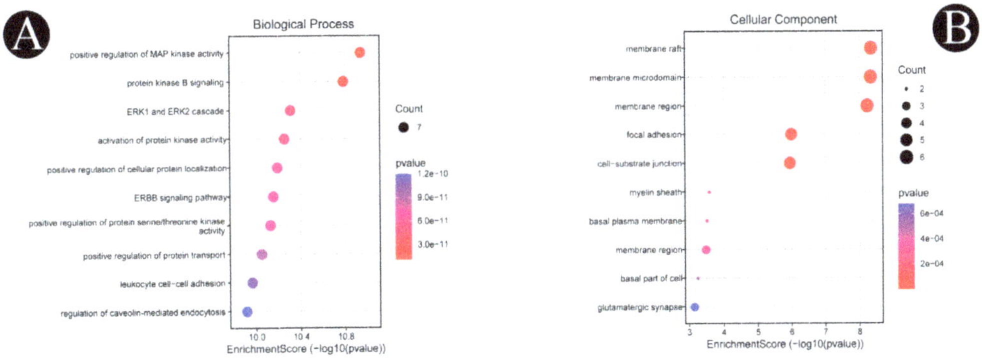

Figure 9. *Cont.*

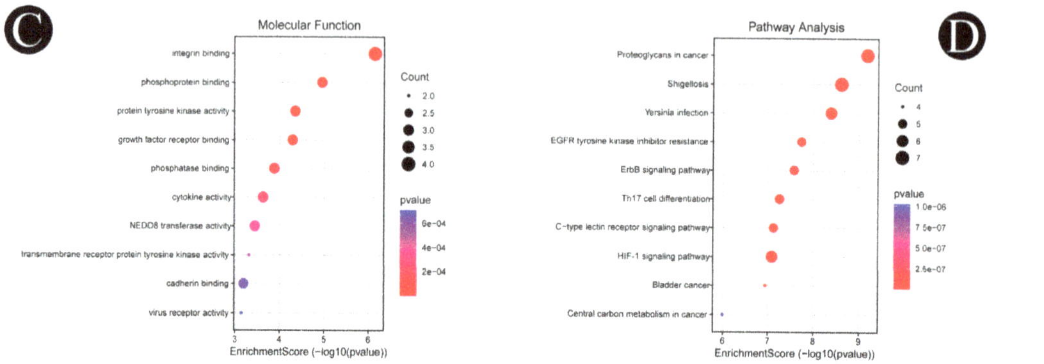

Figure 9. GO enrichment and KEGG pathway analyses of 37 target proteins (p-value ≤ 0.05). (**A**) The top 10 biological processes. (**B**) The top 10 cellular components. (**C**) The top 10 molecular functions. (**D**) The top 10 KEGG pathways. The color scales indicate the different thresholds for the p-values, and the sizes of the dots represent the number of genes corresponding to each term.

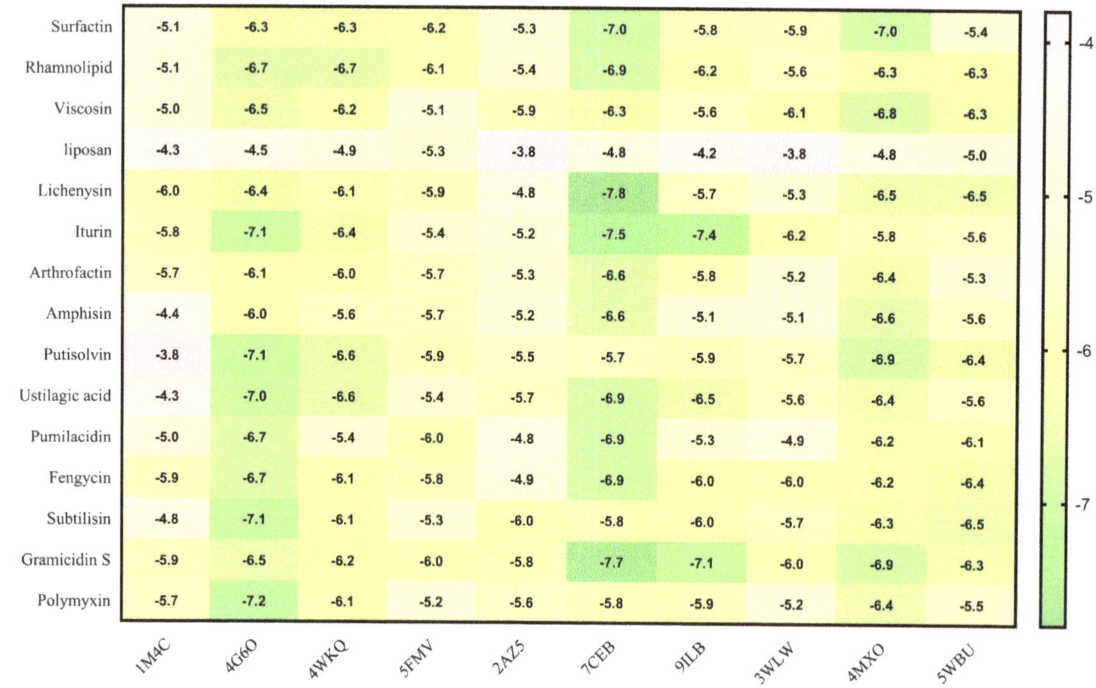

Figure 10. Binding affinities of top-rated pose of ligand–receptor complex.

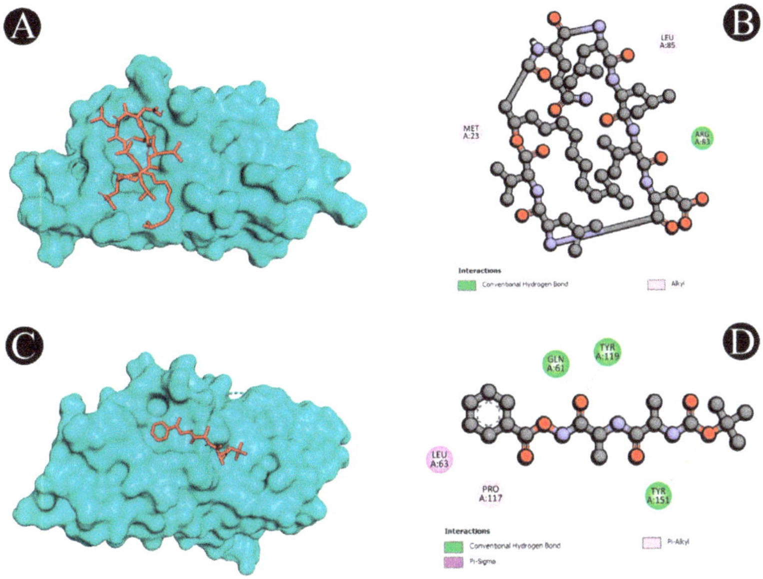

Figure 11. (**A**,**B**) Visualization of docking analysis of IL2 and lichenysin. (**C**,**D**) Visualization of docking analysis of TNF and subtilisin.

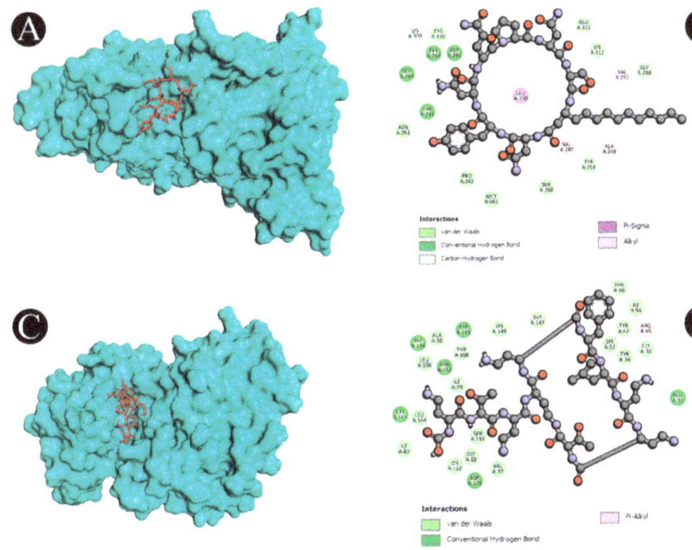

Figure 12. (**A**,**B**). Visualization of docking analysis of ERBB2 and iturin. (**C**,**D**) Visualization of docking analysis of MAPK1 and polymyxin.

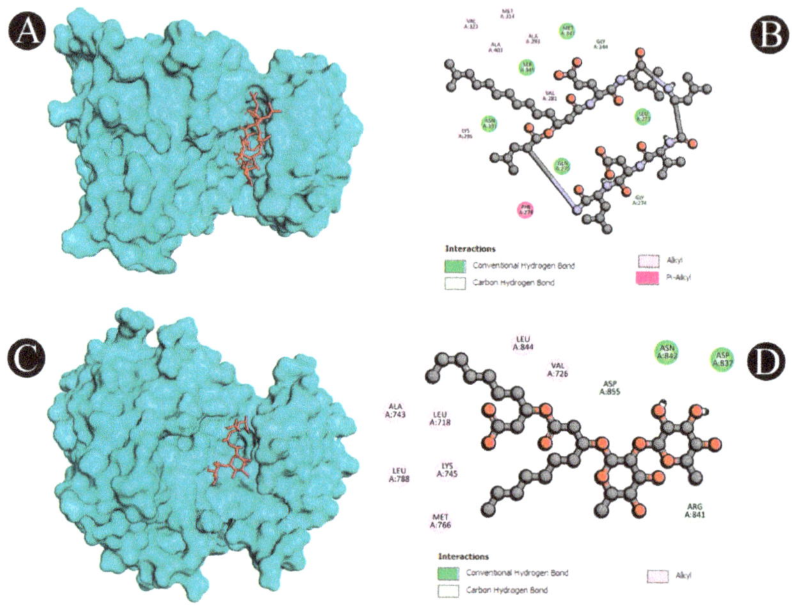

Figure 13. (**A**,**B**) Visualization of docking analysis of SRC and surfactin. (**C**,**D**) Visualization of docking analysis of EGFR and rhamnolipid.

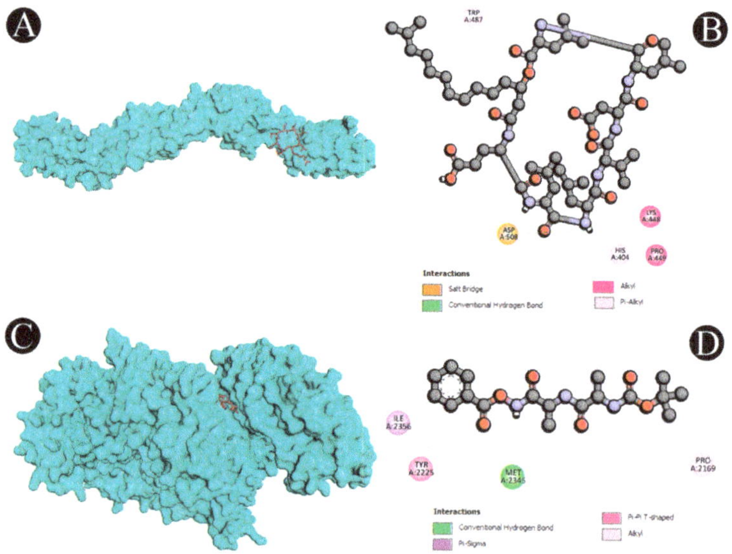

Figure 14. (**A**,**B**) Visualization of docking analysis of PTPRC and surfactin. (**C**,**D**) Visualization of docking analysis of MTOR and subtilisin.

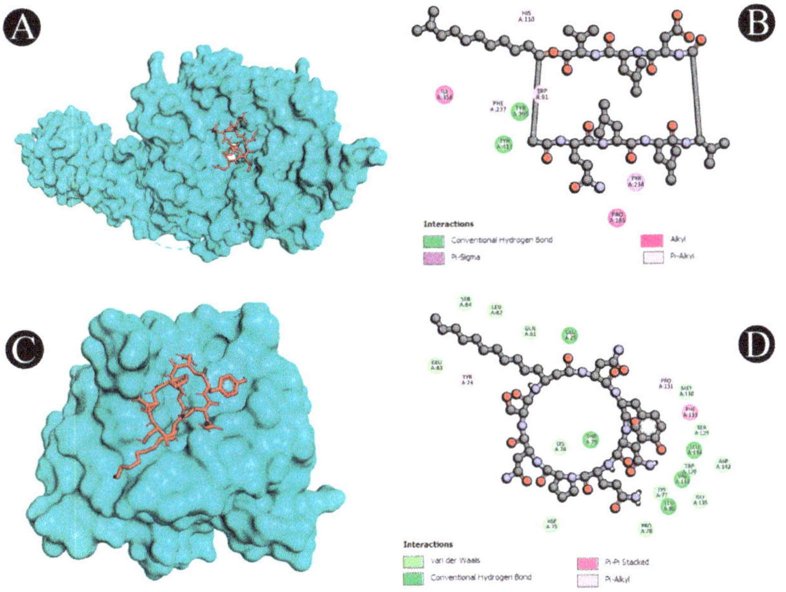

Figure 15. (**A**,**B**) Visualization of docking analysis of ITGB1 and lichenysin. (**C**,**D**) Visualization of docking analysis of IL1B and iturin.

3. Discussion

Over the past 80 years, *L. monocytogenes* has been identified as a human pathogen that has the potential to cause disease. There has been a demographic shift in the last few decades, and there has been an explosion of immunosuppressive medications used for treating malignancies and managing organ transplants. This has led to an increasing number of immunocompromised individuals who are at an increased risk for listeriosis [34]. There is also the factor of changing consumer lifestyles which has resulted in less time available for food preparation, as well as an increase in the use of ready-to-eat food and take-away food. Food production and technology have drastically changed in recent years, resulting in foods with longer shelf-lives that are considered to be "Listeria-risk foods"; the bacteria multiply for a longer period of time, so the food does not undergo a listericidal process before consumption [35]. A high case-fatality rate of between 20 and 30% has been reported for listeriosis, compared to other common food-borne pathogens. During the past three decades, an epidemiological investigation has suggested that epidemic and sporadic listeriosis are primarily linked with the ingestion of foods or food products that are contaminated. While listeriosis is a rare food-borne illness in comparison to other food-borne illnesses, it is a very serious one. There is a high mortality rate associated with the disease even with adequate antibiotic treatment. Approximately, 90% of patients who have listeriosis are hospitalized, and many of them are in intensive care units. As a result, listeriosis is a serious problem around the world. Presently, listeriosis remains a significant challenge, and current treatment options are not adequate to combat the disease [36].

A number of biosurfactants have been tested for their antimicrobial activity, which has shown to be effective against different types of bacterial pathogens, such as *Clostridium perfringens*, *Bacillus subtilis*, *Staphylococcus aureus*, etc. (Gram-positive bacteria); *Escherichia coli*, *Enterobacter aerogenes*, *Salmonella Typhimurium*, etc. (Gram-negative bacteria); and *Mucor* sp., *Phytophthora capsici*, *Fusarium graminearum*, *Botrytis cinerea*, and *Phytophthora infestans* (pathogenic fungi) [37,38]. There is no complete understanding of how these compounds exert their antimicrobial activity; however, one of their proposed sites of action is the cell membrane since they are amphipathic and thus can interact with phospholipids [39].

Moreover, to date only one study is published on the anti-listeriosis activity of biosurfactants. According to the report published by de Araujo et al. [40], there is evidence that *P. aeruginosa* PA1 produces rhamnolipids that have antibacterial activity against *L. monocytogenes* ATCC 19112 and ATCC 7644. In addition to screening microbial surfactants for the treatment of listeriosis, the present study identifies a new therapeutic concept for further investigation into the mechanism of biosurfactants. For complex diseases such as listeriosis, in terms of predictive analysis, network pharmacology offers unique advantages [41]. Analyzing the PPI network, 10 core targets, namely EGFR, SRC, IL1B, IL2, PTPRC, ERBB2, ITGB1, MAPK1, MTOR, and TNF, for biosurfactants against listeriosis were screened out in the present study. EGF (epidermal growth factor) receptors are tyrosine kinases that bind ligands from the EGF family and activate signaling cascades in order to convert extracellular signals into appropriate cellular responses. In order to induce endocytosis, *L. monocytogenes* interacts with this tyrosine kinase receptor and E-cadherin, which might be a common pathogen invasion mechanism for the entry of *L. monocytogenes* [42]. SRC (proto-oncogene tyrosine-protein kinase Src) is reported as closely related with the pathogenicity of *L. monocytogenes*. As a result of *L. monocytogenes* infection, the heavy chain of non-muscle myosin IIA (NMHC-IIA) is phosphorylated at a specific tyrosine residue [43]. The pro-inflammatory cytokine interleukin-1beta (IL-1β) has been known to have a protective function against a variety of bacterial, fungal, and viral infections [44]. Bacterial pathogens are capable of exploiting host cell signaling pathways in order to adhere to or internalize host cells. A frequent molecular alteration involves the phosphorylation of tyrosine kinases on host non-receptors and receptors [45].

Bacteria can induce phosphorylation through direct contact with host cells or via soluble factors [45]. In order to infect host cells, *L. monocytogenes* reported activating the ERBB2/ERBB3 heterodimer pathway [46]. Pathogens such as *Yersinia pseudotuberculosis*, *Staphylococcus aureus*, *Neisseria* species, and enteroaggregative *Escherichia coli* exploit the Integrin subunit beta 1 (ITGβ1) receptor for adhesion to or invasion of mammalian cells [47–51]. Additionally, another protein, mTOR (mammalian target of rapamycin), plays a crucial role in *Listeria* entry. Cell growth, autophagy, and actin cytoskeleton development are controlled by mTOR, a serine/threonine kinase that responds to growth factor stimulation and nutrient, energy, or oxygen availability. Further to this, MAPK family proteins are also reported to play a crucial role in the infection of *L. monocytogenes*, and therefore, the treatment of infection with MAPK inhibitors is reported to affect the inhibition of bacterial internalization towards the host cells during infection [52]. Additionally, tumor necrosis factor (TNF) is a cytokine that has also been reported to play an active role in the susceptibility to *L. monocytogenes* infection. The dysregulation of TNF production and function has been reported to be associated with *L. monocytogenes* pathogenesis since it plays a crucial role in inflammation [53]. The production of TNF is shown to contribute to the protection against *L. monocytogenes* infection in an experimental model, and it can also stimulate the production of IFN-γ [54]. In experiments on severe combined immunodeficiency mice infected with *L. monocytogenes*, it is also found to be involved in a T-cell-independent pathway that leads to macrophage activation [55]. Moreover, the contribution of TNF to the pathogenesis of *L. monocytogenes* has been shown when TNF- or TNF Receptor 1 (R1)-deficient mice succumbed to the *L. monocytogenes* infection relatively quickly instead of recovering after a few days like the control mice [56]. All of these protein targets which come out as a result of the present study can therefore be considered to be most important targets for treating *L. monocytogenes* infections in the future.

Based on the GO analysis, possible targets for biosurfactants against listeriosis are involved in multiple important GO processes. Human epidermal growth factor receptor tyrosine kinases have a size of around 180 kDa and are a family of candidate tyrosine kinase receptors [57]. This family of receptors (EGFR, also termed ERBB1/HER1, ERBB2 or neu/HER2, ERBB3 or HER3, and ERBB4 or HER4) is characterized by dimerization with other receptors that are either of the same nature (homodimerization) or of a different nature (heterodimerization) [58–60]. Interestingly, ERBB receptors play important roles in cancer development [61] and are also found in signaling between bacteria and their hosts. It has been shown that the binding of *Neisseria meningitidis* to endothelial cells leads to the clustering of ERBB2 receptors, followed by phosphorylation of receptor tyrosine and activation of downstream signaling molecules, leading to actin polymerization and bacterial internalization [62]. The envelope glycoprotein B of the human cytomegalovirus (HCMV) binds to EGFR and promotes its tyrosine phosphorylation upon heterodimerization with ErbB3, resulting in virus entry and viral protein synthesis [63]. Likewise, *L. monocytogenes* and other bacteria also trigger the activation of the ErbB2/ErbB3 heterodimer signaling pathway in order to invade host cells [46]. One more key protein that is utilized in classical endocytic mechanisms to allow various particles to be internalized is clathrin or caveolin [64–66]. In order to move between epithelial cells, *L. monocytogenes* hijacks the caveolin–endocytic machinery. The activation of these processes is mediated by a subset of caveolar proteins (caveolin-1, cavin-2, and EHD2). Moreover, it is well-known that pathogens manipulate the post-translational modifications (PTMs) of host proteins to interfere with the normal functioning of host cells in various ways. A key target among these modifications is ubiquitin (UBI), ubiquitin-like proteins (UBLs), and neural precursor cell expressed developmentally downregulated protein 8 (NEDD8), which regulate pathways necessary for the host cell. The PTM modifiers, for instance, regulate the pathways that are crucial to the spread of infection, such as the entry, replication, propagation, or detection of the pathogen by the host, which have all been linked to these PTM modifiers. Different enzymes are involved in this biological process, as well as molecular functions, such as protein kinase binding activity, and the reactions are occurring in a variety of locations, such as the membrane and cytoplasm [67]. There are several biological processes involved in this process that are mediated by different enzymes, along with molecular functions such as protein kinase activity, and all of these reactions occur in multiple locations, such as the membrane and the cytoplasm of the cell. Based on these findings, we suggested that biosurfactants might have an impact on these processes as a result obtained from the GO analysis in this study.

According to the KEGG pathways analysis, potential targets of biosurfactants against listeriosis are significantly enriched in several important pathways, such as the ERBB signaling pathway, C-type lectin receptor signaling, Th17 cell differentiation and HIF-1 signaling pathway, etc. As described above, the ERBB signaling pathway plays an important role in listeriosis to invade bacteria in host cells. A key role that dendritic cells play in tailoring immune responses to pathogens is the expression of C-type lectin receptors (CLRs). Different signaling pathways are triggered by CLRs following the binding of pathogens, which are responsible for triggering the expression of specific cytokines that determine the fate of T cells during polarization. The activation of certain CLRs can be accompanied by the activation of nuclear factor-kappa B, while other CLRs can influence the activation of Toll-like receptors via signaling pathways. Depending on what signaling motifs are present in the cytoplasmic domains of CLRs, they can induce many different types of responses, including pro-inflammatory, antimicrobial, endocytic, phagocytic, and anti-inflammatory responses [68]. Th17 cells have been found to belong to a subgroup of cells that secrete IL-17, or IL-17A, a component of the inflammatory response. Together with Thl, Th2, and Tregs, Th17 cells make up four subsets of CD4+T cells. Under the stimulation of IL-6 and TGF-β, Th17 cells are differentiated by Th0 cells. A key role that they play is in the regulation of the immune system and in the defense of the host [69]. One of the most important transcription factors in maintaining oxygen homeostasis is hypoxia-inducible factor 1 (HIF-1), which is one of the many transcription factors involved in the process. There are two subunits of this protein: an inducibly expressed HIF-1alpha subunit and a constitutively expressed HIF-1beta subunit. In the presence of normoxia, HIF-1 alpha undergoes a process of hydroxylation at specific prolyl residues in order to undergo an immediate ubiquitination and subsequently be degraded by the proteasome. Contrary to this, under hypoxia, the alpha subunit of HIF-1 becomes stable and begins to interact with coactivators such as p300/CBP in order to modulate its transcriptional activity. As a master regulator of hypoxia-inducible genes, HIF-1 regulates a number of hypoxia-inducible genes under hypoxic conditions. The HIF-1 gene family encodes proteins that play a key role in improving oxygen delivery and enhancing cells' adaptive responses to oxygen deprivation. It is important to note that nitric oxide and several growth factors are also stimulatory factors that can induce HIF-1, so it is not only in response to decreased oxygen availability that it is induced but also in response to other stimulants [70].

Furthermore, we performed docking experiments for the biosurfactants and the ten Hub genes in accordance with the "compounds-targets networks". Additionally, the results of docking analyses confirmed our results and showed that lichenysin, iturin, surfactin, rhamnolipid, subtilisin, and polymyxin bind stably to the active pockets of target proteins. Therefore, these compounds could be considered for use as a potential treatment for listeriosis by inhibiting proteins such as, IL2, MAPK1, SRC, EGFR, PTPRC, TNF, IL1B, and ERBB2. Taking into account the role of network pharmacology, the present study examines the active biosurfactants, their potential targets, and their associated pathways, as they pertain to the treatment of listeriosis, which provides a theoretical foundation for further experimental studies. In consideration of the limitations of network pharmacology, it is only through data mining that the basic pharmacological mechanisms for the treatment of listeriosis can be identified. Currently, network pharmacology relies on a variety of databases to support the analysis of bioactive properties. Due to the fact that there are many different information sources and experimental data in databases, it is inevitable that they will show discrepancies. In spite of the fact that we have presented some interesting results, further research and clinical trials are required to evaluate the potential of biosurfactants to validate their usage as a prevention measure against *Listeria* and other food-borne diseases.

4. Materials and Methods

4.1. Biosurfactants Target Prediction

In the present study, we selected biosurfactants that have been reported to possess antimicrobial activity in the literature. The information about their structure, molecular weights, and canonical smiles and the corresponding sdf files were obtained from the PubChem database (http://pubchem.ncbi.nlm.nih.gov/), accessed on 13 September 2022. Using a public database, SwissTargetPrediction, and a STITCH database, we predicted the target of bioactive microbial biosurfactants based on the species of *Homo sapiens* as the only species targeted for the study. To complete the process of standardizing the target names, UniProtKB database (https://www.uniprot.gov/) was used [71,72].

4.2. Network Construction for Compound–Targets

The biosurfactants that were collected and the effective targets that were identified were analyzed by using Cytoscape 3.9.1 software (http://www.cytoscape.org/) for the creation of a compound–target network. To measure the topology scores of the nodes in the network, we used the CytoNCA plugin (v2.1.6), which measures betweenness, closeness, and the centrality of subgraphs of the nodes in each graph. Accordingly, the option "without weight" was selected [73].

4.3. Protein Targets Associated with Listeriosis

A search for targets related to listeriosis was conducted by using keywords such as "listeriosis" and "listeriosis infection" in the GeneCards database (https://www.genecards.org/) [74], Online Mendelian Inheritance in Man database (OMIM, https://omim.org/), and gene–disease associations database (DisGeNET, http://www.disgenet.org/) accessed on 20 September 2022 [75,76], and the Universal Protein database (UniProt, https://www.UniProt.org/) was used to convert the target protein name to a gene name [77]. All listeriosis targets were acquired after repetitive targets were removed.

4.4. Target Screening and Network Construction for Biosurfactants and Listeriosis

In order to detect the core target of the biosurfactants for the treatment of listeriosis, the prediction results of the biosurfactants were matched with the search results of the listeriosis related targets, and the target with the most overlap was selected as the core target. Using the FunRich Tool version 3.1.3, we mapped the targets that biosurfactants and listeriosis share. A Venn diagram was drawn in order to visualize the process. In order to construct a common target network, the Cytoscape software version 3.9.1 was used.

4.5. Protein–Protein Interaction Network (PPI) Construction and Target Identification

The GeneMANIA tool, in addition to constructing a PPI network, is able to find a series of genes related to the input gene based on a large volume of function-related data and analyze the interaction between these genes, based on their co-localization and co-expression [78]. In the present study, GeneMANIA was used to build a protein–protein interaction network related to the cross-gene interactions between biosurfactants and listeriosis based on the analysis of the cross-gene analysis. As a result of the GeneMANIA analysis, we were able to obtain not only information about the relationships between the input cross genes, but also information about the relationships between other closely related targets as well. Accordingly, we label this new set of genes predicted to be biosurfactant targets for listeriosis in the following analysis. The topology parameters of the PPI network were calculated by using Network Analyzer in order to identify the main nodes of the network and the key proteins across the network, while the degree of centrality of the network (betweenness, closeness, and subgraph) was calculated by using CytoNCA.

4.6. Analysis of Gene Ontology (GO) Function and Kyoto Encyclopedia of Genes and Genomes (KEGG) Pathway Enrichment

Based on the results obtained from the above screening, the target of biosurfactants that shared a common target with listeriosis was imported into the Enrichr database (https://maayanlab.cloud/Enrichr/). An analysis was conducted to explore the enrichment of GO functions and pathways within the human genome based on the species *H. sapiens*. As part of the functional analysis of GO, we considered biological processes (BPs), cellular components (CCs), and molecular functions (MFs). The data were visualized as histograms and bubble charts, using the SRPLOT application (http://bioinformatics.com.cn/srplot), as well as the ShinyGO 0.76.2 database (https://bioinformatics.sdstate.edu/go/).

4.7. Construction of Target-Path/Functional Networks

In order to perform a deeper analysis of the signal pathways, biological processes, and molecular functions, ten representative pathways were screened. As part of the construction of the target pathway/functional network, the ShinyGO 0.76.2 database was used (https://bioinformatics.sdstate.edu/go/). Through the use of enrichment analysis, potential targets of biosurfactants for treating listeriosis, biological processes, and signaling pathways were defined by nodes in the network, and the interactions between these nodes were defined by edges.

4.8. Findings of Hub Genes

We tested the PPI network obtained from STRING, using the CytoHubba plugin of Cytoscape. This plugin was used to analyze the core regulatory genes of the PPI network, as well as the identification of key targets within the network. As part of the screening process, the core compounds were tested under the assumption that the "Degree" parameter of the node in the "active ingredient target-disease" network was above the mean. A virtual screening approach based on molecular docking was carried out between the biosurfactants and identified hub genes.

4.9. Molecular Docking Analysis

4.9.1. Protein and Ligand Structures

To study the interaction of selected biosurfactants with the identified protein targets in the present study, a molecular docking analysis was performed. Crystal structures of identified hub target proteins such as TNF (PDB ID: 2AZ5), EGFR (PDB ID: 4WKQ), IL1B (PDB ID: 9ILB), IL2 (PDB ID: 1M4C), SRC (PDB ID: 4MXO), PTPRC (PDB ID: 5FMV), ITGB1 (PDB ID: 7CEB), ERBB2 (PDB ID: 3WLW), MTOR (PDB ID: 5WBU), and MAPK1 (PDB ID: 4G6O) were retrieved from RCSB PDB. Two-dimensional structures of selected biosurfactants were retrieved from the well-known organic compound database PubChem in SDF format. These compounds were then converted into three-dimensional structures, using Avogadro, and saved in PDB format [79].

4.9.2. Ligand Preparation

For the preparation of input files for docking, Autodock software 1.5.7 [80] was used. The structures were minimized with MMFF94 force field. Steepest Descent algorithm was used for optimization with a total of 5000 steps. During minimization, the structure was updated every 1 step, and minimization was terminated when the energy difference is less than 0.1. Energy-minimized structures were saved in PDB format.

4.9.3. Prediction of Binding Site

The binding sites of all protein structures were predicted by using Discovery Studio v. 21.1.0.20298 [81]. The pocket with the highest score was considered to be the most probable binding site of the proteins.

4.9.4. Molecular Docking

Three-dimensional structures of proteins derived from RCSB PDB were prepared for molecular docking, using AutoDock Tools [80]. Before the docking experiment, we used AutoDock Tools software to preprocess the crystal structure of the target proteins, including removing excess protein chains, ligands, and water molecules, and structures were optimized by adding missing hydrogen atoms. Structure files (PDB format) of all biosurfactants were docked separately against the protein structures, using molecular docking software AutoDock 4.2.6. [80]. All the parameters used for the docking of biosurfactants with the proteins were kept the same, except for the grid center differed for each protein inside the grid box. Auto Grid was used for the preparation of the grid map, using a grid box. The grid size was set to 90 × 90 × 90xyz points for all proteins. Grid spacing was kept to 0.500 Å for all the proteins. The grid center for 2AZ5 was designated at dimensions (x, y, and z), −14.888, 68.771, and 32.730; for 4WKQ at (x, y, and z), −2.761, 201.806, and 26.195; for 9ILB at (x, y and z), −13.592, 13.466, and 0.200; for 1M4C at (x, y, and z), 14.170, −10.108, and 20.056; for 4MXO at (x, y, and z), 9.193, −33.735, and −7.984; for 5MFV at (x, y, and z), 30.052, −18.946, and 32.074; for 7CEB at (x, y, and z), 43.112, 46.089, and −1.124; for 3WLW at (x, y, and z), 36.088, 26.277, and −20.954; for 5WBU at (x, y, and z), 11.014, −18.240, and −30.383; and for 4G6O at (x, y, and z), 14.790, 5.794, and 17.485. The grid box is cantered in such a way that it encloses the entire binding site of each protein and provides enough space for the translation and rotation of ligands. The generated docked conformation was ranked by predicted binding energy, and the topmost binding energy docked conformation was analyzed through the use of the PyMOL and Discovery Studio Visualizer [81]. By using the Discovery Studio Visualizer, it was possible to explore the types of interactions, the participating residuals, and the atomic coordinates involved.

5. Conclusions

The purpose of this study was to investigate the molecular mechanisms of biosurfactants to treat listeriosis, using a network pharmacology approach and molecular docking. As a result of the current study, biosurfactants have been found to be capable of targeting multiple proteins and regulating multiple signaling pathways induced by *L. monocytogenes* infection, indicating that biosurfactants may have a regulatory effect on listeriosis caused by *L. monocytogenes*. Furthermore, our findings indicate that IL2, MAPK1, EGFR, PTPRC, TNF, ITGB1, IL1B, ERBB2, and mTOR genes may be viable therapeutic targets for the reduction of listeriosis. In addition to providing an alternative or complementary therapy for the treatment of listeriosis, these findings lay the foundation for future studies. There are, however, some limitations to this study, as pharmacological and clinical research still needs to be conducted to verify our findings. A groundwork has been laid for further study of biosurfactants' protective mechanisms and drug discovery applications based on network pharmacology.

Author Contributions: Conceptualization, M.A. and M.P. (Mitesh Patel); methodology, M.S. (Mejdi Snoussi), E.N., A.J.S., S.H. and S.A.A.; validation, M.P. (Mirav Patel), M.S. (Manojkumar Sachidanandan), M.S. (Mejdi Snoussi), F.B. and R.B.; formal analysis, M.P. (Mitesh Patel), M.P. (Mirav Patel), M.A., M.S. (Mejdi Snoussi), F.B. and R.B.; investigation, S.A.A., A.J.S., M.S. (Manojkumar Sachidanandan) and A.M.A.; data curation, M.P. (Mitesh Patel), F.B., A.M.A., S.A.A., S.H. and E.N.; writing—original draft preparation, M.P. and M.A.; writing—review and editing, F.B., M.S. (Mejdi Snoussi), M.S. (Manojkumar Sachidanandan) and E.N.; software, M.P. (Mirav Patel), M.A. and M.P. (Mitesh Patel); visualization, M.P. (Mirav Patel), M.P. (Mitesh Patel) and A.J.S.; supervision, M.A. and M.P. (Mitesh Patel); project administration, M.A. All authors have read and agreed to the published version of the manuscript.

Funding: This research has been funded by Scientific Research Deanship at University of Ha'il-Saudi Arabia through project number RG-21093.

Institutional Review Board Statement: Not applicable.

Informed Consent Statement: Not applicable.

Data Availability Statement: All data generated or analyzed during this study are included in this article.

Acknowledgments: Authors are thankful to Scientific Research Deanship at University of Ha'il-Saudi Arabia for supporting this study through project number RG-21093.

Conflicts of Interest: The authors declare no conflict of interest. The funders had no role in the design of the study; in the collection, analyses, or interpretation of data; in the writing of the manuscript, or in the decision to publish the results.

References

1. Ge, H.; Wang, Y.; Zhao, X. Research on the Drug Resistance Mechanism of Foodborne Pathogens. *Microb. Pathog.* **2022**, *162*, 105306. [CrossRef] [PubMed]
2. Nyachuba, D.G. Foodborne Illness: Is It on the Rise? *Nutr. Rev.* **2010**, *68*, 257–269. [CrossRef] [PubMed]
3. White, D.G.; Zhao, S.; Simjee, S.; Wagner, D.D.; McDermott, P.F. Antimicrobial Resistance of Foodborne Pathogens. *Microbes Infect.* **2002**, *4*, 405–412. [CrossRef]
4. Jones, G.S.; Bussell, K.M.; Myers-Morales, T.; Fieldhouse, A.M.; Bou Ghanem, E.N.; D'Orazio, S.E. Intracellular *Listeria monocytogenes* comprises a minimal but vital fraction of the intestinal burden following foodborne infection. *Infect. Immun.* **2015**, *83*, 3146–3156. [PubMed]
5. Rouquette, C.; Berche, P. The pathogenesis of infection by Listeria *monocytogenes*. *Microbiologia* **1996**, *12*, 245–258.
6. Kazmierczak, M.J.; Wiedmann, M.; Boor, K.J. Alternative sigma factors and their roles in bacterial virulence. *Microbiol. Mol. Biol. Rev.* **2005**, *69*, 527–543. [CrossRef]
7. Mansfield, B.E.; Freitag, N.E. Listeria monocytogenes pathogenesis: Exploration of alternative hosts. In *Abstracts of the General Meeting of the American Society for Microbiology*; American Society for Microbiology: Washington, DC, USA, 2003; Volume 103, p. B-186.
8. Gandhi, M.; Chikindas, M.L. Listeria: A Foodborne Pathogen That Knows How to Survive. *Int. J. Food Microbiol.* **2007**, *113*, 1–15.
9. Voronina, O.L.; Tartakovsky, I.S.; Yuyshuk, N.D.; Ryzhova, N.N.; Aksenova, E.I.; Kunda, M.S.; Kutuzova, A.V.; Melkumyan, A.R.; Karpova, T.I.; Gruzdeva, O.A.; et al. Analysis of Sporadic Cases of Invasive Listeriosis in a Metropolis. *J. Microbiol. Epidemiol. Immun.* **2020**, *97*, 546–555. [CrossRef]
10. Noor, F.; Noor, A.; Ishaq, A.R.; Farzeen, I.; Saleem, M.H.; Ghaffar, K.; Aslam, M.F.; Aslam, S.; Chen, J.-T. Recent Advances in Diagnostic and Therapeutic Approaches for Breast Cancer: A Comprehensive Review. *Curr. Pharm. Des.* **2021**, *27*, 2344–2365. [CrossRef]

11. Noor, F.; Saleem, M.H.; Chen, J.-T.; Javed, M.R.; Al-Megrin, W.A.; Aslam, S. Integrative Bioinformatics Approaches to Map Key Biological Markers and Therapeutic Drugs in Extramammary Paget's Disease of the Scrotum. *PLoS ONE* **2021**, *16*, e0259408.
12. Kashif, A.; Rehman, R.; Fuwad, A.; Shahid, M.K.; Dayarathne, H.N.P.; Jamal, A.; Aftab, M.N.; Mainali, B.; Choi, Y. Current advances in the classification, production, properties and applications of microbial biosurfactants—A critical review. *Adv. Colloid Inter. Sci.* **2022**, *306*, 102718. [CrossRef] [PubMed]
13. Patel, M.; Siddiqui, A.J.; Hamadou, W.S.; Surti, M.; Awadelkareem, A.M.; Ashraf, S.A.; Alreshidi, M.; Snoussi, M.; Rizvi, S.M.D.; Bardakci, F. Inhibition of Bacterial Adhesion and Antibiofilm Activities of a Glycolipid Biosurfactant from *Lactobacillus Rhamnosus* with Its Physicochemical and Functional Properties. *Antibiotics* **2021**, *10*, 1546. [CrossRef] [PubMed]
14. Adnan, M.; Siddiqui, A.J.; Hamadou, W.S.; Ashraf, S.A.; Hassan, M.I.; Snoussi, M.; Badraoui, R.; Jamal, A.; Bardakci, F.; Awadelkareem, A.M. Functional and Structural Characterization of *Pediococcus pentosaceus*-Derived Biosurfactant and Its Biomedical Potential against Bacterial Adhesion, Quorum Sensing, and Biofilm Formation. *Antibiotics* **2021**, *10*, 1371. [CrossRef]
15. Magalhães, L.; Nitschke, M. Antimicrobial Activity of Rhamnolipids against *Listeria monocytogenes* and Their Synergistic Interaction with Nisin. *Food Control* **2013**, *29*, 138–142. [CrossRef]
16. Boezio, B.; Audouze, K.; Ducrot, P.; Taboureau, O. Network-based Approaches in Pharmacology. *Mol. Inform.* **2017**, *36*, 1700048. [CrossRef] [PubMed]
17. Zhou, Z.; Chen, B.; Chen, S.; Lin, M.; Chen, Y.; Jin, S.; Chen, W.; Zhang, Y. Applications of Network Pharmacology in Traditional Chinese Medicine Research. *Evid. Based. Complement. Altern. Med.* **2020**, *2020*, 1646905. [CrossRef] [PubMed]
18. Park, M.; Park, S.-Y.; Lee, H.-J.; Kim, C.E. A Systems-Level Analysis of Mechanisms of *Platycodon grandiflorum* Based on a Network Pharmacological Approach. *Molecules* **2018**, *23*, 2841. [CrossRef]
19. Ongena, M.; Jacques, P. *Bacillus* lipopeptides: Versatile weaponsfor plant disease biocontrol. *Trends. Microbiol.* **2007**, *16*, 115–125. [CrossRef]
20. Pathak, K.V.; Keharia, H. Application of extracellular lipopeptide bio-surfactant produced by endophytic *Bacillus subtilis* K1 isolated from aerial rootsof banyan (*Ficus benghalensis*) in microbially enhanced oil recovery (MEOR). *3 Biotech* **2014**, *4*, 41–48. [CrossRef]
21. Rahman, K.S.M.; Thahira-Rahman, J.; McClean, S.; Marchant, R.; Banat, I.M. Rhamnolipid biosurfactants production by strains of *Pseudomonas aeruginosa* using low cost materials. *Biotechnol. Prog.* **2002**, *18*, 1277–1281. [CrossRef]
22. Alsohim, A.S.; Taylor, T.B.; Barrett, G.A.; Gallie, J.; Zhang, X.X.; Altamirano-Junqueira, A.E.; Johnson, L.J.; Rainey, P.B.; Jackson, R.W. The biosurfactant viscosin produced by P seudomonas fluorescens SBW 25 aids spreading motility and plant growth promotion. *Environ. Microbiol.* **2014**, *16*, 2267–2281. [CrossRef] [PubMed]
23. Rufino, R.D.; Luna, J.M.; Sarubbo, L.A.; Rodrigues, L.R.M.; Teixeira, J.A.C.; Campos-Takaki, G.M. Antimicrobial and anti-adhesive potential of a biosurfactant Rufisan produced by Candida lipolytica UCP 0988. *Colloids Surf. B Biointerfaces* **2011**, *84*, 1–5. [CrossRef] [PubMed]
24. Yeak, K.Y.C.; Perko, M.; Staring, G.; Fernandez-Ciruelos, B.M.; Wells, J.M.; Abee, T.; Wells-Bennik, M.H. Lichenysin Production by *Bacillus licheniformis* Food Isolates and Toxicity to Human Cells. *Front. Microbiol.* **2022**, *13*, 831033. [CrossRef] [PubMed]
25. Arrebola, E.; Jacobs, R.; Korsten, L. Iturin A is the principal inhibitor inthe biocontrol activity of Bacillus amyloliquefaciens PPCB004 against posthar-vest fungal pathogens. *J. Appl. Microbiol.* **2010**, *108*, 386–395. [CrossRef]
26. Morikawa, M.; Daido, H.; Takao, T.; Murata, S.; Shimonishi, Y.; Imanaka, T. A new lipopeptide biosurfactant produced by *Arthrobacter* sp. strain MIS38. *J. Bacteriol.* **1993**, *175*, 6459–6466. [CrossRef]
27. Groboillot, A.; Portet-Koltalo, F.; Le Derf, F.; Feuilloley, M.J.; Orange, N.; Poc, C.D. Novel application of cyclolipopeptide amphisin: Feasibility study as additive to remediate polycyclic aromatic hydrocarbon (PAH) contaminated sediments. *Int. J. Mol. Sci.* **2011**, *12*, 1787–1806. [CrossRef]
28. Kuiper, I.; Lagendijk, E.L.; Pickford, R.; Derrick, J.P.; Lamers, G.E.; Thomas-Oates, J.E.; Lugtenberg, B.J.; Bloemberg, G.V. Characterization of two Pseudomonas putida lipopeptide biosurfactants, putisolvin I and II, which inhibit biofilm formation and break down existing biofilms. *Mol. Microbiol.* **2004**, *51*, 97–113. [CrossRef]
29. Hewald, S.; Josephs, K.; Bölker, M. Genetic analysis of biosurfactant production in *Ustilago maydis*. *Appl. Environ. Microbiol.* **2005**, *71*, 3033–3040. [CrossRef]
30. Saggese, A.; Culurciello, R.; Casillo, A.; Corsaro, M.M.; Ricca, E.; Baccigalupi, L. A marine isolate of *Bacillus pumilus* secretes a pumilacidin active against *Staphylococcus aureus*. *Mar. Drugs* **2018**, *16*, 180. [CrossRef]
31. Gudiña, E.J.; Fernandes, E.C.; Rodrigues, A.I.; Teixeira, J.A.; Rodrigues, L.R. Biosurfactant production by *Bacillus subtilis* using corn steep liquor as culture medium. *Front. Microbiol.* **2015**, *6*, 59.
32. Abdelhamid, H.N.; Khan, M.S.; Wu, H.F. Graphene oxide as a nanocarrier for gramicidin (GOGD) for high antibacterial performance. *RSC Adv.* **2014**, *4*, 50035–50046. [CrossRef]
33. Eras-Muñoz, E.; Farré, A.; Sánchez, A.; Font, X.; Gea, T. Microbial biosurfactants: A review of recent environmental applications. *Bioengineered* **2022**, *13*, 12365–12391. [CrossRef] [PubMed]
34. Bodro, M.; Paterson, D.L. Listeriosis in Patients Receiving Biologic Therapies. *Eur. J. Clin. Microbiol. Infect. Dis.* **2013**, *32*, 1225–1230. [CrossRef] [PubMed]
35. Allerberger, F.; Wagner, M. Listeriosis: A Resurgent Foodborne Infection. *Clin. Microbiol. Infect.* **2010**, *16*, 16–23. [CrossRef] [PubMed]

36. Melo, J.; Andrew, P.W.; Faleiro, M.L. *Listeria monocytogenes* in Cheese and the Dairy Environment Remains a Food Safety Challenge: The Role of Stress Responses. *Food. Res. Int.* **2015**, *67*, 75–90. [CrossRef]
37. Benincasa, M.; Abalos, A.; Oliveira, I.; Manresa, A. Chemical Structure, Surface Properties and Biological Activities of the Biosurfactant Produced by *Pseudomonas aeruginosa* LBI from Soapstock. *Antonie Van Leeuwenhoek* **2004**, *85*, 1–8. [CrossRef]
38. Haba, E.; Pinazo, A.; Jauregui, O.; Espuny, M.J.; Infante, M.R.; Manresa, A. Physicochemical Characterization and Antimicrobial Properties of Rhamnolipids Produced by *Pseudomonas aeruginosa* 47T2 NCBIM 40044. *Biotechnol. Bioeng.* **2003**, *81*, 316–322. [CrossRef]
39. Ortiz, S.; López-Alonso, V.; Rodríguez, P.; Martínez-Suárez, J.V. The Connection between Persistent, Disinfectant-Resistant Listeria Monocytogenes Strains from Two Geographically Separate Iberian Pork Processing Plants: Evidence from Comparative Genome Analysis. *Appl. Environ. Microbiol.* **2016**, *82*, 308–317. [CrossRef]
40. De Araujo, L.V.; Guimarães, C.R.; da Silva Marquita, R.L.; Santiago, V.M.J.; de Souza, M.P.; Nitschke, M.; Freire, D.M.G. Rhamnolipid and Surfactin: Anti-Adhesion/Antibiofilm and Antimicrobial Effects. *Food Control* **2016**, *63*, 171–178. [CrossRef]
41. Chen, W.; Ding, C.; Yu, J.; Wang, C.; Wan, L.; Hu, H.; Ma, J. Wuzi Yanzong Pill—Based on Network Pharmacology and In Vivo Evidence—Protects Against Spermatogenesis Disorder via the Regulation of the Apoptosis Pathway. *Front. Pharmacol.* **2020**, *11*, 592827. [CrossRef]
42. Yap, P.-C.; MatRahim, N.-A.; AbuBakar, S.; Lee, H.Y. Antilisterial Potential of Lactic Acid Bacteria in Eliminating *Listeria monocytogenes* in Host and Ready-to-Eat Food Application. *Microbiol. Res.* **2021**, *12*, 17. [CrossRef]
43. Almeida, M.T.; Mesquita, F.S.; Cruz, R.; Osório, H.; Custódio, R.; Brito, C.; Vingadassalom, D.; Martins, M.; Leong, J.M.; Holden, D.W. Src-Dependent Tyrosine Phosphorylation of Non-Muscle Myosin Heavy Chain-IIA Restricts Listeria Monocytogenes Cellular Infection. *J. Biol. Chem.* **2015**, *290*, 8383–8395. [CrossRef] [PubMed]
44. Sahoo, M.; Ceballos-Olvera, I.; del Barrio, L.; Re, F. Role of the Inflammasome, IL-1. *Sci. World J.* **2011**, *11*, 2037–2050. [CrossRef] [PubMed]
45. Cossart, P.; Pizarro-Cerdá, J.; Lecuit, M. Invasion of Mammalian Cells by *Listeria monocytogenes*: Functional Mimicry to Subvert Cellular Functions. *Trends. Cell. Biol.* **2003**, *13*, 23–31. [CrossRef] [PubMed]
46. Oliveira, M.J.; Lauwaet, T.; de Bruyne, G.; Mareel, M.; Leroy, A. *Listeria monocytogenes* Produces a Pro-Invasive Factor That Signals via ErbB2/ErbB3 Heterodimers. *J. Cancer Res. Clin. Oncol.* **2005**, *131*, 49–59. [CrossRef]
47. Isberg, R.R.; Barnes, P. Subversion of Integrins by Enteropathogenic *Yersinia*. *J. Cell. Sci.* **2001**, *114*, 21–28. [CrossRef]
48. Fowler, T.; Wann, E.R.; Joh, D.; Johansson, S.; Foster, T.J.; Höök, M. Cellular Invasion by Staphylococcus Aureus Involves a Fibronectin Bridge between the Bacterial Fibronectin-Binding MSCRAMMs and Host Cell B1 Integrins. *Eur. J. Cell. Biol.* **2000**, *79*, 672–679. [CrossRef]
49. Hauck, C.R.; Ohlsen, K. Sticky Connections: Extracellular Matrix Protein Recognition and Integrin-Mediated Cellular Invasion by Staphylococcus Aureus. *Curr. Opin. Microbiol.* **2006**, *9*, 5–11. [CrossRef]
50. Van Putten, J.P.M.; Duensing, T.D.; Cole, R.L. Entry of OpaA+ Gonococci into HEp-2 Cells Requires Concerted Action of Glycosaminoglycans, Fibronectin and Integrin Receptors. *Mol. Microbiol.* **1998**, *29*, 369–379. [CrossRef]
51. Izquierdo, M.; Alvestegui, A.; Nataro, J.P.; Ruiz-Perez, F.; Farfan, M.J. Participation of Integrin A5β1 in the Fibronectin-Mediated Adherence of Enteroaggregative *Escherichia coli* to Intestinal Cells. *Biomed. Res. Int.* **2014**, *2014*, 781246. [CrossRef]
52. Hashino, M.; Tachibana, M.; Nishida, T.; Hara, H.; Tsuchiya, K.; Mitsuyama, M.; Watanabe, K.; Shimizu, T.; Watarai, M. Inactivation of the MAPK Signaling Pathway by Listeria Monocytogenes Infection Promotes Trophoblast Giant Cell Death. *Front. Microbiol.* **2015**, *6*, 1145. [CrossRef] [PubMed]
53. Sedgwick, J.D.; Riminton, D.S.; Cyster, J.G.; Körner, H. Tumor necrosis factor: A master-regulator of leukocyte movement. *Immunol. Today* **2000**, *21*, 110–113. [CrossRef] [PubMed]
54. Tripp, C.S.; Wolf, S.F.; Unanue, E.R. Interleukin 12 and tumor necrosis factor alpha are costimulators of interferon gamma production by natural killer cells in severe combined immunodeficiency mice with listeriosis, and interleukin 10 is a physiologic antagonist. *Proc. Natl. Acad. Sci. USA* **1993**, *90*, 3725–3729. [CrossRef] [PubMed]
55. Bancroft, G.J.; Sheehan, K.C.; Schreiber, R.D.; Unanue, E.R. Tumor necrosis factor is involved in the T cell-independent pathway of macrophage activation in scid mice. *J. Immunol.* **1989**, *143*, 127–130. [PubMed]
56. Li, X.; Körner, H.; Liu, X. Susceptibility to Intracellular Infections: Contributions of TNF to Immune Defense. *Front. Microbiol.* **2020**, *11*, 1643. [CrossRef] [PubMed]
57. Yarden, Y.; Sliwkowski, M.X. Untangling the ErbB Signalling Network. *Nat. Rev. Mol. Cell Biol.* **2001**, *2*, 127–137. [CrossRef]
58. Garrett, T.P.J.; McKern, N.M.; Lou, M.; Elleman, T.C.; Adams, T.E.; Lovrecz, G.O.; Zhu, H.-J.; Walker, F.; Frenkel, M.J.; Hoyne, P.A. Crystal Structure of a Truncated Epidermal Growth Factor Receptor Extracellular Domain Bound to Transforming Growth Factor α. *Cell* **2002**, *110*, 763–773. [CrossRef]
59. Ogiso, H.; Ishitani, R.; Nureki, O.; Fukai, S.; Yamanaka, M.; Kim, J.-H.; Saito, K.; Sakamoto, A.; Inoue, M.; Shirouzu, M. Crystal Structure of the Complex of Human Epidermal Growth Factor and Receptor Extracellular Domains. *Cell* **2002**, *110*, 775–787. [CrossRef]
60. Cho, H.-S.; Mason, K.; Ramyar, K.X.; Stanley, A.M.; Gabelli, S.B.; Denney, D.W.; Leahy, D.J. Structure of the Extracellular Region of HER2 Alone and in Complex with the Herceptin Fab. *Nature* **2003**, *421*, 756–760. [CrossRef]
61. Olayioye, M.A.; Neve, R.M.; Lane, H.A.; Hynes, N.E. The ErbB Signaling Network: Receptor Heterodimerization in Development and Cancer. *EMBO J.* **2000**, *19*, 3159–3167. [CrossRef]

62. Hoffmann, I.; Eugène, E.; Nassif, X.; Couraud, P.-O.; Bourdoulous, S. Activation of ErbB2 Receptor Tyrosine Kinase Supports Invasion of Endothelial Cells by *Neisseria meningitidis*. *J. Cell Biol.* **2001**, *155*, 133–144. [CrossRef] [PubMed]
63. Wang, X.; Huong, S.-M.; Chiu, M.L.; Raab-Traub, N.; Huang, E.S. Epidermal Growth Factor Receptor Is a Cellular Receptor for Human Cytomegalovirus. *Nature* **2003**, *424*, 456–461. [CrossRef] [PubMed]
64. Conner, S.D.; Schmid, S.L. Regulated Portals of Entry into the Cell. *Nature* **2003**, *422*, 37–44. [CrossRef]
65. Nabi, I.R.; Le, P.U. Caveolae/Raft-Dependent Endocytosis. *J. Cell Biol.* **2003**, *161*, 673–677. [CrossRef] [PubMed]
66. McMahon, H.T.; Boucrot, E. Molecular Mechanism and Physiological Functions of Clathrin-Mediated Endocytosis. *Nat. Rev. Mol. Cell Biol.* **2011**, *12*, 517–533. [CrossRef]
67. Jubelin, G.; Taieb, F.; Duda, D.M.; Hsu, Y.; Samba-Louaka, A.; Nobe, R.; Penary, M.; Watrin, C.; Nougayrède, J.-P.; Schulman, B.A. Pathogenic Bacteria Target NEDD8-Conjugated Cullins to Hijack Host-Cell Signaling Pathways. *PLoS Pathog.* **2010**, *6*, e1001128. [CrossRef] [PubMed]
68. Hoving, J.C.; Wilson, G.J.; Brown, G.D. Signalling C-type Lectin Receptors, Microbial Recognition and Immunity. *Cell. Microbiol.* **2014**, *16*, 185–194. [CrossRef]
69. Schmidt-Weber, C.B.; Akdis, M.; Akdis, C.A. TH17 Cells in the Big Picture of Immunology. *J. Allergy Clin. Immunol.* **2007**, *120*, 247–254. [CrossRef]
70. Galanis, A.; Pappa, A.; Giannakakis, A.; Lanitis, E.; Dangaj, D.; Sandaltzopoulos, R. Reactive Oxygen Species and HIF-1 Signalling in Cancer. *Cancer Lett.* **2008**, *266*, 12–20. [CrossRef]
71. Daina, A.; Michielin, O.; Zoete, V. SwissTarget Prediction: Updated Data and New Features for Efficient Prediction of Protein Targets of Small Molecules. *Nucleic Acids Res.* **2019**, *47*, W357–W364. [CrossRef]
72. Gfeller, D.; Michielin, O.; Zoete, V. Shaping the Interaction Landscape of Bioactive Molecules. *Bioinformatics* **2013**, *29*, 3073–3079. [CrossRef]
73. Zhang, J.; Li, H.; Zhang, Y.; Zhao, C.; Zhu, Y.; Han, M. Uncovering the Pharmacological Mechanism of Stemazole in the Treatment of Neurodegenerative Diseases Based on a Network Pharmacology Approach. *Int. J. Mol. Res.* **2020**, *21*, 427.
74. Safran, M.; Dalah, I.; Alexander, J.; Rosen, N.; Iny Stein, T.; Shmoish, M.; Nativ, N.; Bahir, I.; Doniger, T.; Krug, H. GeneCards Version 3: The Human Gene Integrator. *Database* **2010**, *2010*, baq020. [PubMed]
75. Piñero, J.; Bravo, À.; Queralt-Rosinach, N.; Gutiérrez-Sacristán, A.; Deu-Pons, J.; Centeno, E.; García-García, J.; Sanz, F.; Furlong, L.I. DisGeNET: A Comprehensive Platform Integrating Information on Human Disease-Associated Genes and Variants. *Nucleic Acids Res.* **2016**, *45*, D833–D839. [CrossRef] [PubMed]
76. Piñero, J.; Ramírez-Anguita, J.M.; Saüch-Pitarch, J.; Ronzano, F.; Centeno, E.; Sanz, F.; Furlong, L.I. The DisGeNET Knowledge Platform for Disease Genomics: 2019 Update. *Nucleic Acids Res.* **2020**, *48*, D845–D855.
77. UniProt Consortium. Uniport: A Hub for Protein Information. *Nucleic Acids Res.* **2015**, *43*, D204–D212. [CrossRef]
78. Franz, M.; Rodriguez, H.; Lopes, C.; Zuberi, K.; Montojo, J.; Bader, G.D.; Morris, Q. GeneMANIA Update 2018. *Nucleic Acids Res.* **2018**, *46*, W60–W64. [CrossRef]
79. Hanwell, M.D.; Curtis, D.E.; Lonie, D.C.; Vandermeersch, T.; Zurek, E.; Hutchison, G.R. Avogadro: An advanced semantic chemical editor, visualization, and analysis platform. *J. Cheminformatics* **2012**, *4*, 17. [CrossRef]
80. Morris, G.M.; Huey, R.; Olson, A.J. Using Autodock for Ligand-receptor Docking. *Curr. Protoc. Bioinformatics* **2008**, *24*, 8.14.1–8.14.40. [CrossRef]
81. Accelrys. *Discovery Studio, 2.1*; Accelrys: San Diego, CA, USA, 2008.

antibiotics

MDPI

Article

Biosynthesised Silver Nanoparticles Loading onto Biphasic Calcium Phosphate for Antibacterial and Bone Tissue Engineering Applications

Varun Prasath Padmanabhan [1,†], Pugalmani Sivashanmugam [2,†], Ravichandran Kulandaivelu [1,*], Suresh Sagadevan [3,*], Balu Sridevi [4], Rajakumar Govindasamy [5] and Muthu Thiruvengadam [5,*]

1 Department of Analytical Chemistry, University of Madras, Guindy Campus, Tamil Nadu, Chennai 600025, India
2 Department of Orthodontics, Saveetha Dental College and Hospitals, Saveetha Institute of Medical and Technical Sciences, Saveetha University, Tamil Nadu, Chennai 600077, India
3 Nanotechnology & Catalysis Research Centre, University of Malaya, Kuala Lumpur 50603, Malaysia
4 Department of Electronics and Communication Engineering, Velammal Institute of Technology, Ponneri, Tamil Nadu, Thiruvallur 601204, India
5 Department of Applied Bioscience, College of Life and Environmental Sciences, Konkuk University, Seoul 05029, Republic of Korea
* Correspondence: raavees@gmail.com (R.K.); drsureshnano@gmail.com (S.S.); muthu@konkuk.ac.kr (M.T.)
† These authors contributed equally to this work.

Abstract: Biphasic calcium phosphate (BCP) serves as one of the substitutes for bone as it consists of an intimate mixture of beta-tricalcium phosphate (β-TCP) and hydroxyapatite (HAP) in different ratios. BCP, because of its inbuilt properties such as osteoconductivity, biocompatibility, and biostability in several clinical models serves as a bone substituent for orthopedic applications. Therefore, the present study aimed to assess the effectiveness of silver (Ag) nanoparticles (NPs) combined with BCP composites for the orthopedic sector of bone tissue regeneration and growth. In this regard, we first synthesized Ag-BCP microclusters by the double-emulsion method and then characterized the composite for various physicochemical properties, including the crystallinity and crystal structure, bonding and functionality, porosity, morphology, surface charges, topography, and thermal stability. In addition, the antibacterial activity of Ag-BCP was tested against gram-positive and gram-negative microorganisms such as *Staphylococcus aureus*, *Candida albicans*, and *Escherichia coli*. Finally, the cytocompatibility of Ag-BCP was confirmed against the fibroblast cells in vitro.

Keywords: biphasic calcium phosphate; Ag nanoparticles; antibacterial activity; hydroxyapatite; β-tricalcium phosphate; MTT assay

Citation: Padmanabhan, V.P.; Sivashanmugam, P.; Kulandaivelu, R.; Sagadevan, S.; Sridevi, B.; Govindasamy, R.; Thiruvengadam, M. Biosynthesised Silver Nanoparticles Loading onto Biphasic Calcium Phosphate for Antibacterial and Bone Tissue Engineering Applications. *Antibiotics* **2022**, *11*, 1780. https://doi.org/10.3390/antibiotics11121780

Academic Editor: Helena P. Felgueiras

Received: 7 November 2022
Accepted: 2 December 2022
Published: 8 December 2022

1. Introduction

In recent years, research relating to stem cells and tissue engineering has produced efficacious therapeutic strategies for the treatment of damaged bones and their cells through the regeneration/remodeling pathways [1]. For such applications, the composite scaffolds containing the ingredients such as polymers [2,3], ceramics [4], metal nanoparticles (NPs), and their composites [5] are highly suitable because of their inbuilt properties such as porosity, conductivity, resistance, and biocompatibility. These synthetic composites with their porous and biocompatible nature provide a suitable environment for regeneration with complete functionality and effective proliferation of cells that eventually replace the diseased bone cells. Furthermore, the synthetic bone scaffolds made up of said ingredients have the capacity to encapsulate the therapeutic drug molecules that are useful for the treatment of commonly attacking orthopedic diseases like bone and bone marrow tumors, osteoporosis, and to avert an infection [6]. Of the various ingredients, the bioceramic material, biphasic calcium phosphate (BCP) has several interesting properties and the most important is its mineral portion which is relatively easy to process and has an excellent cell–cell attachment capacity. All these properties support its incorporation as a bone substituent, biocement, surface coating, drug-delivery platform, and tissue engineering scaffold [7–9]. Additional advantages of BCP material are the low cost, unlimited availability, biocompatibility, predictability, biosafety, and lower morbidity to the patient over autografts and allografts. Hence, this material serves as an attractive option for bone tissue engineering, dental replacements, craniofacial surgeries, spinal surgery, and neurosurgeries [10].

The properties of BCP materials (similar to many different composites) can be strongly influenced by altering the production parameters such as the solution pH, sintering temperature, and purification processes. The calcium phosphates (Ca_3PO_4) thus formed consist of unique physicochemical characteristics like an altered surface area, porosity, surface energy, charges, and roughness [11–13]. In this direction, to control the pore sizes of bioceramic compounds, one approach involves the incorporation of porogens and pore-formers. An alternative method to this approach is the application of heat treatment to generate macropores (diameter >100 μm) and micropores (with <10 μm diameter) [14]. Therefore, taking advantage of this heat-induced method for the formation of high-surface bioceramics having a macroporous and microporous nature to suit osteoconductivity, many researchers have demonstrated the role of BCP as a bone substituent [15,16]. Moreover, the adjacent concavities and nearby walls of the macropores serve as a salient point and favor the formation of geometric-dependent of bone [17,18]. Furthermore, the surface dissolution leads to the supersaturation of calcium (Ca) and phosphate (PO_4^{3-}) ions, causing reprecipitation and the generation of a biocompatible surface layer that permits an easy bonding of bone with the synthetic bioceramic. This process has an impact on the potential of osteoinduction [19,20]. Of various kinds of Ca_3PO_4, the BCP kind is made up of stable and soluble phases of ions with varying concentrations. Furthermore, the other form, hydroxyapatite (HAP; chemical formula $Ca_{10}(PO_4)_6(OH)_2$), would be advantageous on top of other calcium phosphates because of the guided bioactivity in linking the resorption/solubilization and biomaterial stability towards the promotion of bone growth [21,22].

Silver (Ag) nanoparticles (NPs) are widely used in the treatment of bacterial infections associated with injuries, wounds, tissue engineering and in the water treatment sector [23]. Ag NPs have high surface charges, surface area, surface-to-volume ratio, and surface oxygen defects that promote antibacterial activity in contrast to the other Ag-salts and organometallics [24,25]. Nevertheless, the stability and dispersion of Ag NPs curb their biological efficiency by aggregation that leads to the formation of larger-sized crystals and decreases the cumulative surface area. In some instances, this aggregation is overcome to stabilize the NPs on substrates, leading to stability enhancement and an associated antibacterial effect.

By considering the potential advantages of BCP to serve as a bone substituent and Ag NPs for the impending antibacterial activity, the present work aimed to develop a nanocomposite that has multiple functions to suit bone tissue engineering applications. For that, we fabricated a hybrid composite consisting of a BCP matrix decorated in situ with that of Ag NPs and for the formation of the composite, we employed a facile double-emulsion method. The nanocomposite was analyzed for its physicochemical properties by making use of various instrumental techniques like FTIR, powder XRD, SEM, zeta potential, surface topology and TGA-DTG. Further, we tested the controlled drug release, antibacterial activity, and in vitro cell viability capacity of as-synthesized Ag-BCP nanocomposite.

2. Materials and Methods

2.1. Formation of Ag NPs

About 50 mL of 0.01 mM $AgNO_3$ in an aqueous solution was mixed with 50 mL of 0.1 mM glucose solution and the reaction was maintained at a pH of 11 using ammonia. The reaction was kept for aging for 6 h and until the solution color changed from colorless to yellow, confirming the formation of Ag NPs. The Ag NPs were separated by centrifugation, washed with ethanol 2–3 times, dried, and stored for loading onto BCP.

$$C_6H_{12}O_6 + 2AgNO_3 \xrightarrow{2NH_4OH} \text{Gluconic acid} + 2Ag + 2NH_4NO_3 + H_2O \quad (1)$$

2.2. Synthesis of Ag-Decorated BCP

For the fabrication of the Ag-BCP nanocomposite, we first formed BCP by the combination of hydroxyapatite (HAP) and β-tricalcium phosphate (β-TCP). Both were prepared individually. HAP was first prepared by mixing 50 mL of 1 M Calcium Nitrate as Ca precursor solution (adjusted to a pH of 11 using NH_4OH) with 50 mL of 0.66 M phosphate solution in a dropwise manner. After the complete addition, the milky-white-colored solution was stirred constantly for another 2 h to generate a white precipitate which was kept for 24 h. Then, the precipitate was separated by filtration and washed with a solvent mixture containing a 1:2 ratio of ethanol to water. The precipitate was kept in an oven furnace for sintering at 800 °C overnight to finally generate a white powder of HAP.

For the formation of β-TCP, a simple co-precipitation method was used. To this end, 50 mL of an aqueous solution of calcium nitrate (0.9 M) was added dropwise to 50 mL of ammonium dihydrogen phosphate (0.6 M) at a pH of 8 maintained by using concentrated ammonia. Here, the Ca/P ratio of 1.5 was retained manually and the magnetic stirring (set in the range of 200–250 rpm) was continued for 2 h even after the complete addition. After that, the precipitate was separated, washed with an ethanol-water mixture, dried in an oven, crushed, and sintered at 900 °C in a muffle furnace for 1 h. The final product was stored in an airtight container.

For the synthesis of BCP implants [26], HAP and β-TCP powders in a 60:40 ratio were grounded homogeneously using a mortar and pestle. The finely powdered mixture was initially dried at 55 °C overnight and further subjected to 100 °C for 5 h. The fully dried powder was collected and sealed with a polythene cover until its use. Finally, for the loading of Ag NPs onto BCP, individual aqueous solutions containing equal amounts of Ag NPs (25 mL) and BCP (25 mL slurry) were added together. The mixture was subjected to ultrasonication to undergo homogenous mixing for a period of 15 min. This resulted in the formation of a light-yellow-colored viscous solution that was kept in an oven at 110 °C for drying. The dried powder of the Ag-BCP composite was collected and used for further analysis.

2.3. Instrumentation

Powdered X-ray diffraction (XRD) analysis was carried out to understand the crystal structure and crystallinity. The powder samples were run in the 2θ range of 20–80° (Model: Smart Lab se X-ray, Rigaku, Japan; k = 1.5418 Å). The morphology of the samples was investigated using a field emission scanning electron microscopy (FESEM) connected with an energy-dispersive X-ray diffractometer (Model: JOEL JFM 6390 Scientific, Peabody, MA, USA). The functionality and bonding of samples were studied using Fourier transform infrared spectroscopy (FTIR) in the wavenumber range of 4000–400 cm^{-1} (Spectrum 2, PerkinElmer, Waltham, MA, USA) and confocal Raman spectroscopy (Alpha 300r, Witech, Braunschweig, Germany). Furthermore, atomic force microscopy (AFM) studies were employed to investigate the surface nature of the samples (Park xe7, Park system, Suwon, Republic of Korea). The elemental composition of the samples was analyzed using X-ray photoelectron spectroscopy (XPS, ULVAC-PHI, Inc; Model: PHI5000 Version Probe III). For the thermal stability and phase changes, thermogravimetric analysis (TGA) and differential thermal analysis (DTA) were performed (Netzsch sta 2500 was measured under N_2 environment between 30–800 °C). The zeta potential and dynamic light scattering analysis were used to determine the particle size and surface charges (Horiba Scientific Sz-100 instrument).

2.4. Measurement of Porosity

To investigate the porosity of pelletized BCP and Ag-BCP samples, the liquid displacement technique was used. The pelletized samples of BCP and Ag-BCP are not soluble in ethanol. Thus, the penetrating ability of ethanol into the sample pores restricts the occurrence of any swelling or shrinkage. For the testing, the sample pellet with a known weight (W) was first immersed in a graduated cylinder that already had a known volume (V_1) of ethanol. The emigration (of ethanol) followed by depressurization for the undisturbed samples can be seen. This diffusion of ethanol into the pores can be continued until we see the halting of air bubbles and at this stage, the volume of ethanol (i.e., sample pellet soaked in ethanol) was noted as V_2 and we measured the difference in two volumes (V_2–V_1). Further, the sample pellet was removed from the cylinder containing ethanol and we measured the residual volume (V_3) of ethanol. By making use of W, V_1, V_2, and V_3 in the following formula, the porosity was calculated.

$$\text{Porosity, } \varepsilon = \frac{(V1 - V3)}{(V2 - V3)} \quad (2)$$

2.5. In-Vitro Bioactivity and Biodegradation Studies

To investigate the extent of bioactivity for our materials in stimulated fluid (SBF) in vitro, we first prepared a solution consisting of NaCl (7.9 g), $NaHCO_3$ (0.3 g), KCl (0.2 g), $K_2HPO_4.3H_2O$ (0.2 g), $MgCl_2.6H_2O$ (0.3 g), $CaCl_2$ (0.2 g), Na_2SO_4 (0.07 g) and $(CH_2OH)_3CNH_2$ (6.0 g) in double-distilled water (added one after the other). The formed solution mixture pH was set to 7.4 with the help of HCl and maintained at 37 °C. For the testing, the pelletized samples of BCP and Ag-BCP were soaked in SBF for 21 days and after that, removed, rinsed with de-ionized water, and further subjected to SEM analysis to investigate the extent of biomineralization (formation of any mass) at the surface.

The in vitro biodegradation/biodissolution studies were performed by investigating the amount of Ag and Ca ions that were released into the buffer solution and by recording the morphological changes linked to the release. The pellet (with a weight of 0.5 g as W_0) made from the granules was first immersed in a 20 mL tris-buffer solution maintained at a pH of 7.4 at 37 °C. At the end of 3 weeks of incubation, we rinsed the sample pellets with ethanol, dried, and measured its final mass (W_t). The weight loss can be calculated using the equation:

$$\text{Weight decrease} = (W_t/W_0) \times 100\% \quad (3)$$

2.6. Studies of Drug Loading and Release

The drug loading capacity and release efficiency from the BCP and Ag-BCP matrices were evaluated using the typical drug, Ciprofloxacin (CIP). Typically, 10 mg of CIP was dispersed in 100 mL of double-distilled water containing 90 mg either of BCP or Ag-BCP. The mixture was allowed to stir for 24 h at room temperature. Then, the precipitate was separated by centrifugation, rinsed with distilled water, and dried. The drug-loaded sample in the form of a pellet was collected. Further, for the drug release studies, a known weight of bioceramic pellet loaded with CIP was placed in phosphate-buffered saline (PBS; pH 7.4) and subjected to horizontal agitation on a shaking water bath set at 37 °C. After each specified interval of time, about 5 mL of the sample (containing the released CIP) was collected and replaced with an equal amount of fresh medium. UV–Vis spectrometry was used for the qualitative and quantitative investigation of released CIP at various time intervals.

2.7. Antibacterial and Antifungal Activity

The antibacterial activity of our bioceramic samples was measured by the agar disc diffusion method involving Muller Hinton Agar (MHA) medium. The stock cultures were maintained at 4 °C on the slant of the nutrient agar. The active cultures were prepared by transferring a loop full of bacterial cells from the cultured stocks to nutrient broth-loaded test tubes and subjected to incubation at 37 °C for 24 h. The strains were used for their antibacterial activity against the Gram-positive *S. aureus*, Gram-negative *E. coli*, and antifungal activity against *C. albicans*. The discs were prepared with 20 μL of each of the samples (Ag-BCP, HAP, β-TCP, and BCP), a negative control of dimethylsulfoxide (DMSO), and a Standard 1 mg/mL of Streptomycin as positive control). The plates were incubated for another 24 h at 37 °C and, finally, the growth of microbes was investigated by measuring the diameter of the zone of inhibition (ZoI).

For the antifungal activity, the assay procedure is almost the same. We used the agar disc diffusion method, where the stock cultures were maintained at 4 °C on the slant of potato dextrose agar (PDA). Briefly, 4.4 g of PDA was weighed and dissolved in 100 mL of distilled water followed by the addition of 1 g of agar. Then, we subjected the media to sterilization, solidified the media for 1 h, and spread the inoculums on solid plates with a sterile swab moistened with the fungal suspension. The discs contained 20 μL of each of the testing samples (Ag-BCP, HAP, β-TCP, and BCP), a negative control (DMSO), and a positive control (1 mg/mL of Ketoconazole). The extent of antifungal activity was measured by incubating the sample-treated plates at 37 °C for 24 h and finally recording the diameter of ZoI in mm.

2.8. In Vitro Cell Viability Assay

To investigate the cytocompatibility performance of as-synthesized bioceramic composites, in vitro cell viability studies were carried out on the L929 mouse fibroblast cell line over a 24 h period. Briefly, 1×10^5 cells per well were added to a 96-well plate containing Dulbeccos Modified Eagle Medium (DMEM) and 10% fetal bovine serum (FBS). The cells were allowed to grow until reaching their confluency level. Then, they were washed with a fresh serum-free medium 2–3 times, followed by starvation for 1 h at 37° C. Subsequently, the cells were treated with different concentrations (31.2–1000 mg/mL) of bioceramics, BCP, and Ag-BCP over a 24-h period. Then, the old medium was replaced with a fresh serum-free medium comprising 0.5 mg/mL of MTT (3-[4,5-dimethylthiazol-2-yl] 2,5-diphenyl tetrazolium bromide) and incubated for another 4 h at 37 °C in a CO_2 incubator. The medium containing MTT was removed and the cells were washed with PBS to eliminate any unreacted reagent. Then, we added DMSO while thoroughly mixing by pipetting up and down to dissolve the formed formazan crystals. Finally, the purple-blue-colored formazan crystals were analyzed spectrophotometrically by recording the absorbance at 570 nm (Biorad 680). Using these readings, the cytotoxicity was determined using the Graph pad prism 5 software. Furthermore, the viable L929 cells were observed using inverted phase-contrast microscopy.

$$\text{Percentage (\%) of cell viability} = (\text{Sample's OD}/\text{Control's OD}) \times 100 \qquad (4)$$

2.9. Statistical Analysis

All the statistical analyses were performed using a one-way analysis of variance (ANOVA) and the data presented are the mean ± standard deviation of at least three individual experiments with the value of $p < 0.05$ as statistically significant.

3. Results and Discussion

3.1. Physicochemical Analysis

Figure 1 shows the FTIR spectral analysis of (a) HAP, (b) β-TCP, (c) BCP, and (d) Ag-BCP samples. All spectra indicate the presence of characteristic bands at 1032, 1098, and 1133 cm^{-1} due to the triply degenerated (υ_3) asymmetric stretching vibrations of P-O bonds. Furthermore, the observation of bands at 602 and 560 cm^{-1} signifies the υ_4 vibration of -PO_4 group and the band at 926 cm^{-1} indicates the υ_1 vibration of phosphate bond. The bending vibrational band observed at 630 cm^{-1} infers the liberational -OH group (due to surface adsorbed water vapor/moisture) and the bands at 498 and 452 cm^{-1} are assigned to the υ_2 vibration of the $PO_4{}^{3-}$ group. The asymmetric bending vibrations of the phosphate group present in HAP were evident through the observation of a band at 608 cm^{-1}. Similarly, for the β-TCP sample, some prominent sharp bands were observed at 560 and 602 cm^{-1} and can be linked to the bending vibrations of PO_4^{3-} group. The band at 960 cm^{-1} (visible as a minor hump) stems from the υ_1 frequency of vibration, and the band at 1037 cm^{-1} from the P-O stretching vibration of PO_4^{3-} ions in β-TCP [27]. Further, for each sample, we observed some prominent bands at 3572 and 630 cm^{-1} owing to the presence of hydroxyl groups, i.e., the sharp band at 730 cm^{-1} is due to H_2O (surface adsorbed) is available in all the tested samples, indicating the presence of moisture. For the BCP sample, the υ_1 and υ_4 absorption bands were observed at 926 and 567 cm^{-1}, respectively. The intensity of the band observed at 926 cm^{-1} was least pronounced for BCP and Ag-BCP samples and was due to the composition of the 60:40 ratio of HAP and β-TCP. We observed no evidence for the occurrence of any chemical bonding between BCP and Ag, meaning that the Ag NPs were physically embedded in the BCP. From the analysis, the band patterns that appeared in all the samples (a-d) are best correlated with the earlier reports. Table 1 shows the composition of FT-IR spectra of all the samples.

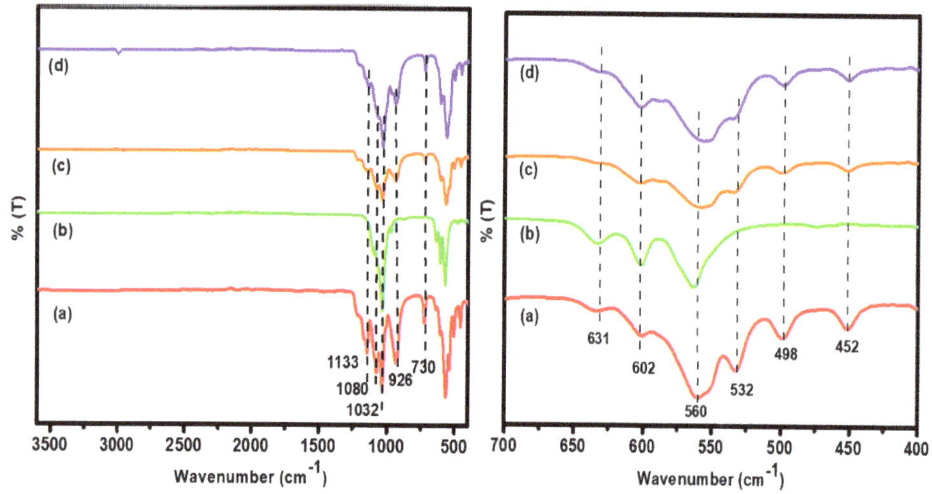

Figure 1. Comparison of the FTIR spectra of (**a**) HAP, (**b**) β-TCP, (**c**) BCP, and (**d**) Ag-BCP samples.

Table 1. FTIR spectral data for (a) HAP, (b) β-TCP, (c) BCP, and (d) Ag-BCP samples.

Vibrational Frequency (cm^{-1})	Band Assignment			
	HAP	β-TCP	BCP	Ag-BCP
3572, 630	-OH group	-OH group	-OH group	-OH group
2920	-	-	-	Glucose-assisted Ag NPs (C-H) stretching
1032, 1098, 1133	Asymmetric stretching vibrations of the P–O bonds	1133 is absent	Asymmetric stretching of the P–O bonds	Asymmetric stretching of the P–O bonds
1037	-	PO_4^{3-} ions found in β-TCP	PO_4^{3-} ions found in β-TCP	PO_4^{3-} ions found in β-TCP
926	Symmetric stretching (υ_1) of P–O bond from PO_4^{3-} group	-	Symmetric stretching (υ_1) of P–O bond of PO_4^{3-} group	Symmetric stretching (υ_1) of P–O bond of PO_4^{3-} group
960	-	Symmetric stretching (υ_1) of P–O bond of PO_4^{3-} group	Symmetric stretching (υ_1) of P–O bond of PO_4^{3-} group	Symmetric stretching (υ_1) of P–O bond of PO_4^{3-} group
730	Owing to H_2O	-	-	Owing to H_2O
631	Liberational OH group	Liberational OH group	Liberational OH group	Liberational OH group
602, 560	Phosphate bands (υ_4)	Vibrational bands of PO_4^{3-}	Phosphate bands (υ_4)	Phosphate bands (υ_4)
498, 452	Phosphate bands (υ_2)	Phosphate bands (υ_2)	Phosphate bands (υ_2)	Phosphate bands (υ_2)

Figure 2 compares the Raman spectroscopic analysis of (a) HAP, (b) β-TCP, (c) BCP, and (d) Ag-BCP samples. Table 2 shows the Raman spectral composition of all the samples. The band with the highest intensity around 966 cm^{-1} relates to the symmetrical stretching vibrations (υ_1) of the phosphate (PO_4^{-3}) confirming the formation of HAP. Furthermore, for the same HAP sample, the symmetrical (υ_2) and antisymmetrical bending (υ_4) vibrations of the PO_4^{-3} groups appeared at 432, 445, 572, and 598 cm^{-1}, respectively. Additionally, the asymmetric stretching vibrations (υ_3) of PO_4^{-3} ions in the HAP were observed as weak intensity bands around 1056 and 1090 cm^{-1} [28]. The high-intensity band at 964 cm^{-1} besides a weak shoulder at 948 cm^{-1} indicates the internal vibrations of β-TCP in the BCP sample. For the BCP sample shown in Figure 2c, the spectrum has the same major band at 966 cm^{-1} and a shoulder band at 948 cm^{-1} corresponding to the β-TCP phase, as the BCP is composed of HAP and β-TCP in a 60:40 ratio. Finally, for the Ag-BCP sample (Figure 2d), the Ag NPs were loaded onto the BCP composite. The Ag particles were synthesized using glucose and thus in this reaction gluconic acid is formed as a by-product which is observed as a minor band at 1364 cm^{-1} [29]. Furthermore, the minor band at 226 cm^{-1} corresponds to the presence of Ag NPs.

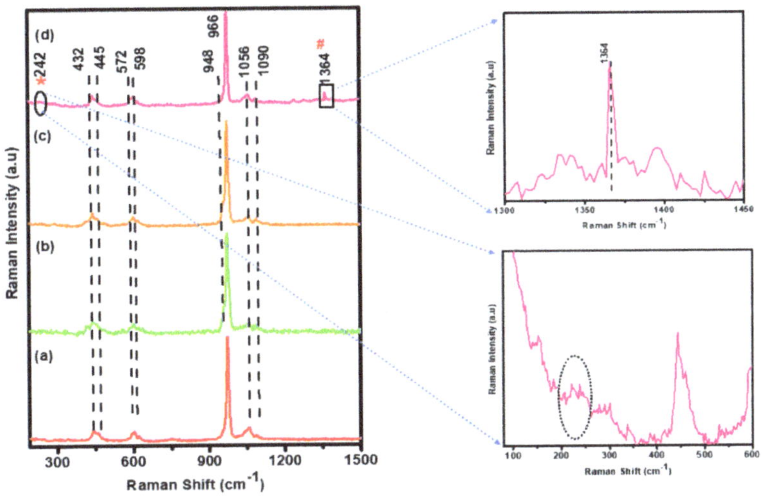

Figure 2. Comparison of the Raman spectra of (a) HAP, (b) β-TCP, (c) BCP, and (d) Ag-BCP samples.

Table 2. Raman spectral data of (a) HAP, (b) β-TCP, (c) BCP, and (d) Ag-BCP.

Frequency of Vibration (cm^{-1})	Band Assignments			
	HAP	β-TCP	BCP	Ag-BCP
1364	-	-	-	Gluconic acid
1056 1090	Asymmetric stretching vibrations (υ_3) of PO_4^{-3}	Asymmetric stretching vibrations (υ_3) of PO_4^{-3}	Asymmetric stretching vibrations (υ_3) of PO_4^{-3}	Asymmetric stretching vibrations (υ_3) of PO_4^{-3}
966	Symmetric stretching vibrations (υ_1) of PO_4^{-3} group	-	Symmetric stretching vibrations (υ_1) of PO_4^{-3} group	Symmetric stretching vibrations (υ_1) of PO_4^{-3} group
964, 948	-	Internal vibrations of β-TCP in BCP	Internal vibrations of β-TCP in BCP	Internal vibrations of β-TCP in BCP
432, 445	Symmetrical bending (υ_2)	Symmetrical bending (υ_2)	Symmetrical bending (υ_2)	Symmetrical bending (υ_2)
572, 598	Asymmetric bending (υ_4) vibrations of PO_4^{-3}	Asymmetric bending (υ_4) vibrations of PO_4^{-3}	Asymmetric bending (υ_4) vibrations of PO_4^{-3}	Asymmetric bending (υ_4) vibrations of PO_4^{-3}

The powder XRD reflection patterns of the four bioceramics (HAP, β-TCP, BCP, and Ag-BCP) provided in Figure 3 confirm the formation of highly crystalline phases in all samples, as shown by the narrow and sharp pattern. The XRD patterns provided in Figure 3a,b show the calcium phosphate precursors derived from the HAP sintered at 800 °C and β-TCP at 900 °C, respectively. From the comparative analysis, the HAP patterns have perfectly matched with the parent HAP, as provided by the JCPDS card No. 09–0432. The crystalline nature of the HAP sample was confirmed by the patterns observed at 2θ of 25.9° (002), 28.6° (210), 31.7° (211), 32.2° (112), 34.0° (202), and 39.8° (310). These diffraction patterns indicate the presence of HAP in BCP (JCPDS No. 9-432). The calcium to phosphorous (Ca/P) ratio was found to be 1.6. Similarly, the reflection patterns of the β-TCP sample (Figure 3b) appeared at 2θ of 21.8° (024), 25.8° (1010), 27.8° (214), 31.0° (0210), 32.4° (128), and 34.3° (220) (JCPDS No. 9-169), with the Ca/P ratio being 1:5. Hence, all the diffraction patterns of HA and β-TCP in BCP were found to be more distinct and thereby indicate the crystalline nature. Additionally, the XRD patterns of BCP (Figure 3c) and Ag-BCP (Figure 3d) were observed at 2θ of 31.76° (211), 32.15° (112), 32.89° (300), and 34.02° (202), with the Ca/P ratio of 1:6. These data also confirm the formation of BCP in its highly crystalline phase. It is evident from the FTIR spectrum (Figure 1a–c) and the XRD phase composition (Figure 3c,d) that there is no formation of a Calcite (calcium carbonate) pattern at 29.4°. Moreover, it can be observed from the XRD pattern that the characteristic patterns of HAP in the BCP sample are much greater than in β-TCP [30]. Finally, the XRD pattern of the Ag BCP sample (Figure 3d) is in good agreement with that of BCP (Figure 3c). Furthermore, the observation of no additional patterns for Ag NPs indicates its lower concentration.

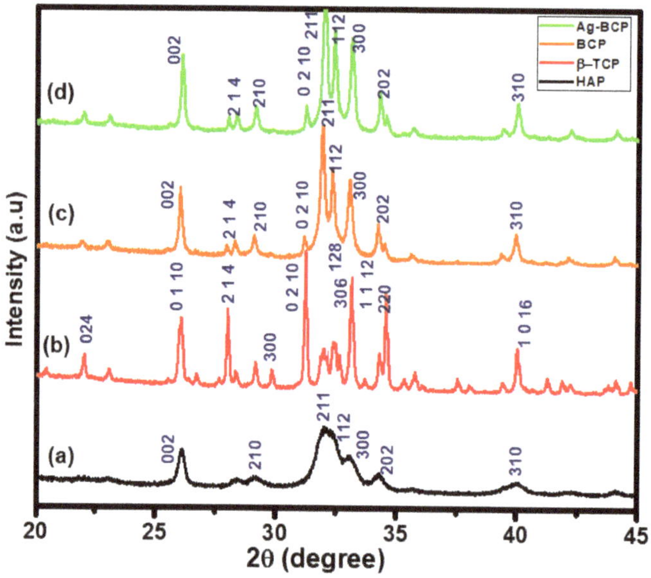

Figure 3. Powder XRD analysis for (a) HAP, (b) β-TCP, (c) BCP, and (d) Ag-BCP.

Figure 4 represents the FESEM analysis of (a) HAP, (b) β-TCP, (c) BCP, and (d) Ag-BCP at three different magnifications. The surfaces of all samples appear to be rough and granular and the maximum effect can be seen in β-TCP. The FESEM of HAP (Figure 4a1) showed that there is a decreased surface roughness. Furthermore, the FESEM of β-TCP (Figure 4b1) shows the micro- and macropores with a well-organized pore network, which supports its enhanced solubility effects. Furthermore, this kind of pore arrangement is predicted to permit the uptake of fluid, cell accommodation, and a greater surface area. The FESEM of BCP (Figure 4c1) showed less roughness and the surface is uneven with its patterns similar to the earlier report [27]. Finally, the morphology of Ag-BCP provided in Figure 4d1 shows some small decorating particles and are referred to as Ag NPs.

Figure 4. FESEM analysis of (**a**) HAP, (**b**) β-TCP, (**c**) BCP, and (**d**) Ag-BCP.

Figure 5 shows the zeta potential analysis of HAP, β-TCP, BCP, and Ag-BCP samples in an aqueous solution. These studies were used to estimate the colloidal stability and dispersions of bio-ceramics in solution. From the analysis, the zeta potential values of HAP, β-TCP, BCP, and Ag-BCP were obtained to be −35.1 mV, −36.9 mV, −40.3 mV, and −44.1 mV, respectively (Figure 5a–d). The zeta potential value obtained for pure HAP was greater than −30 mV compared to the literature report [31] and in the same way, the β-TCP sample with a value of −36.9 mV changed to −40.3 mV for BCP, indicating that the system arrived at a state of moderate stability. Further, for the Ag-BCP sample, the potential value of −44.1 mV was observed indicating the influential stability of the Ag NPs onto the surface of BCP.

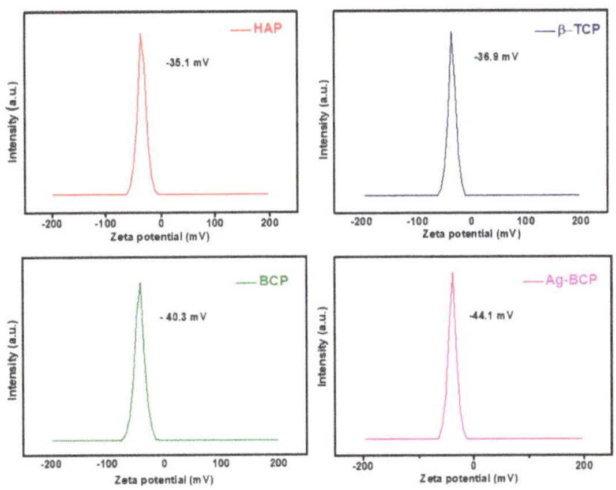

Figure 5. Zeta potential analysis of HAP, β-TCP, BCP, and Ag-BCP.

The AFM provided topographic analyses of BCP and Ag-BCP samples and are shown in Figure 6. For the topographical examination, both samples in their powdered form were coated with Aluminum foil. As shown in Figure 6a, the topography of the BAP sample confirms the formation of homogeneously arranged agglomerated globules. Furthermore, the 2D image of Ag-BCP (Figure 6b) shows that the triangular shape is embedded on the globules representing Ag and globules for the BCP. Similarly, the 3D image of Ag-BCP shows the formation of highly crystalline homogenous peaks corresponding to Ag and are in good agreement with the 2D image peaks [32] (Table S1).

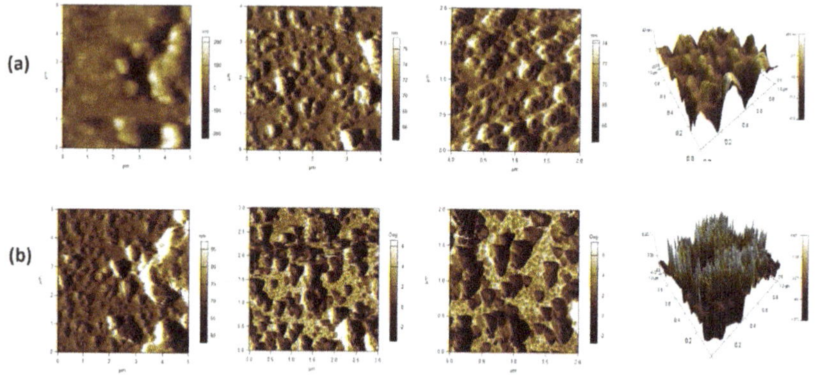

Figure 6. Surface topographical analysis for (**a**) BCP and (**b**) Ag-BCP.

The XPS analysis was used to determine the elemental composition of Ag-BCP samples and the results are provided in Figure 7. The image provides the XPS spectrum along with the elemental peaks of C, Ca, O, Ag, and P. The spectrum of the elements O1s, P2p, Ca2p, Ag3d, and C1s are due to the adsorption of hydrocarbon impurities. The C-C component at 285 eV is used to calibrate the energy level. The spectrum of C1s shows two major peaks inclined at 284.95 eV for the non-oxygenated sp2 carbon ring (C-C) and at 286.5 eV for the sp3 (C-O) oxygenated functional group of carbon. Furthermore, the XPS of Ca2p shows two distinct peaks 347.35 and 351.23 eV corresponding to the $Ca2p_{3/2}$ and $Ca2p_{1/2}$, respectively. The XPS of O1s shows binding energies of 530.8 eV and 533.2 eV corresponding to -OH and P-O-P, respectively. Similarly, the prominent spectrum for Ag arises at 369.87 eV and 372.18 eV corresponding to $3d_{5/2}$ and $3d_{3/2}$. The observation of this Ag peak confirms the successful decoration of Ag NPs onto the surface of the BAP sample [33]. The phosphate spectra of P2p peaks at 132.8 eV and 135 eV correspond to $P2p_{3/2}$ and $P2p_{1/2}$ and confirm the success of HAP formation.

Figure 8 provides a comparison of the thermal stability of BCP and Ag-BCP samples as analyzed by TGA and DTA. The initial weight loss of up to ≈210 °C for both samples was due to the loss of moisture/adsorbed water (up to 200 °C) and lattice water (up to 650 °C). Above this, the occurrence of weight loss was seen in several stages between 200 °C and 900 °C, confirming the transformation of the HAP phase into a β-TCP phase in the BCP sample [34]. For the BCP and Ag-BCP samples, the total weight loss measured around 1000 °C was only 1% and 0.5%, respectively, thereby indicating that the Ag decoration reduced the thermal stability of BCP.

Figure 9 shows the comparison of porosity measurements of BCP and Ag-BCP samples. The porosity of bare BCP was observed around 15.20%. BCP samples decorated with Ag NPs showed an increase in porosity value at 32%. Such an observation of increased porosity values is due to the occurrence of chemical interactions of Ag NPs with the BCP matrices. This is the most useful factor in bone tissue engineering applications, as it can facilitate cell growth and migration, protein delivery to the cells, and preserve tissue volume with temporary mechanical function [35].

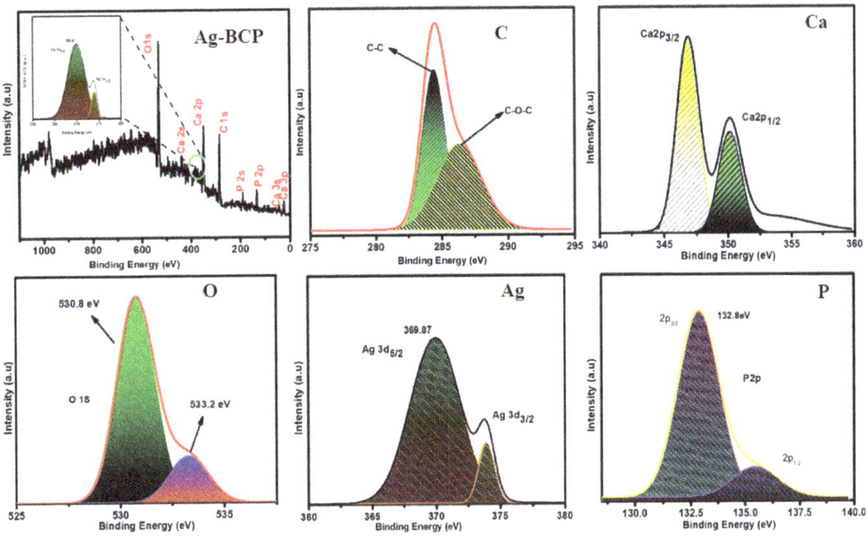

Figure 7. XPS spectroscopic analysis of Ag-BCP samples.

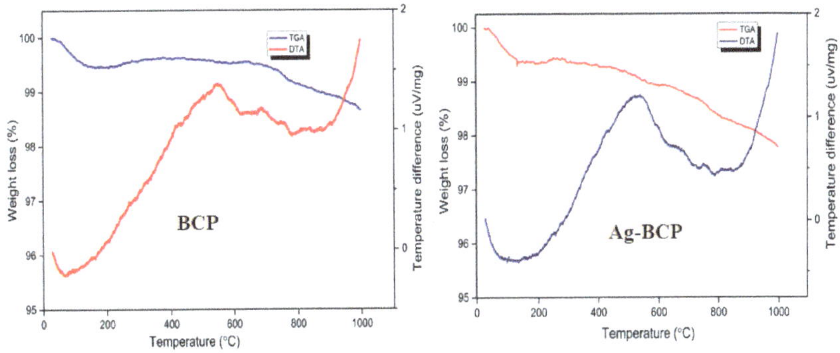

Figure 8. TGA and DTA analysis of BCP and Ag-BCP samples.

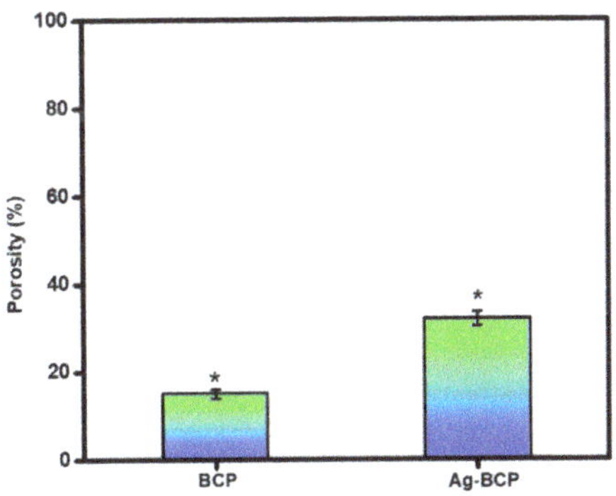

Figure 9. Porosity evaluation of BCP and Ag-BCP samples by the liquid displacement method. * denoted as statistical analyses were performed using a one-way analysis of variance (ANOVA) and the data presented are the mean ± standard deviation.

3.2. Studies of Bioactivity, Biodegradation, and Drug Release

Figure 10 shows the in vitro bioactivity studies of BCP and Ag-BCP samples as investigated by the amount of apatite formed on the sample's surface. The sample's morphological changes associated with the formation of surface masses were recorded by SEM. The materials were maintained in SBF for 14 days at 37 °C. The optical and SEM images of BCP (Figure 10a,c) show fibers with an anisotropic aspect of ~5–10 μm in length and 0.5–1.0 μm in width. However, the Ag-BCP sample (Figure 10b,d) witnessed the formation of an apatite layer at the surface of pellets and the newly formed layer comprised of tiny spherical particles of calcium phosphate crystals. This indicates that the Ag-BCP composite serves as a bioactive material with the ability to generate an apatite layer that can bond bones with implant materials [36].

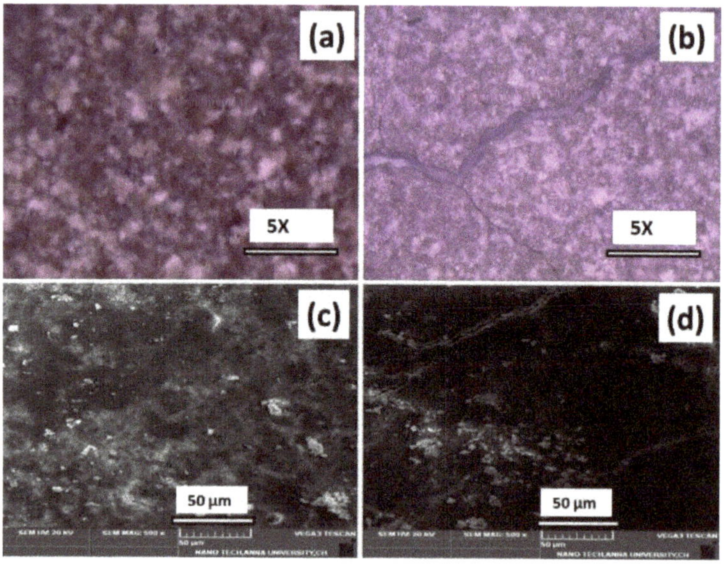

Figure 10. Bioactivity investigation recorded by the optical microscopic images for BCP (**a**) and Ag-BCP s (**b**) samples and the corresponding SEM images of BCP (**c**) and Ag-BCP (**d**).

Optical microscopic and SEM images of BCP and Ag-BCP samples are employed to understand the biodegradation efficiencies linked to surface morphological changes followed by the incubation in SBF (14 days, 37 °C, see Figure 11(i)) For both samples, the surface roughness increased due to the degradation of material into the SBF medium. Furthermore, after 21 days in SBF, 79.45% of BCP and 65.25% of the Ag-BCP composite degraded (Figure 11(ii)). Further, the biodegradation behavior of BCP and Ag-BCP samples in tris-buffer is compared in Figure 11(iii). We observed an overall increase in the pH value after the sample's immersion. This indicates that the BCP sample had an increased pH compared to Ag-BCP. Over 3 weeks, Ag-BCP had the slowest bio-dissolution, while for the BCP, the fastest mass decrease and high dissolution rates with pH increase were noted. This result indicates that both samples are highly biodegradable. Pure BCP (as against Ag-BCP) degrades fastest and the differences can be linked to the availability of solid Ag NPs in the tri-component system. This difference in biodegradation is well suited for bone tissue engineering applications as it helps to withstand mechanical stress and creates an encouraging environment for cell attachment and growth.

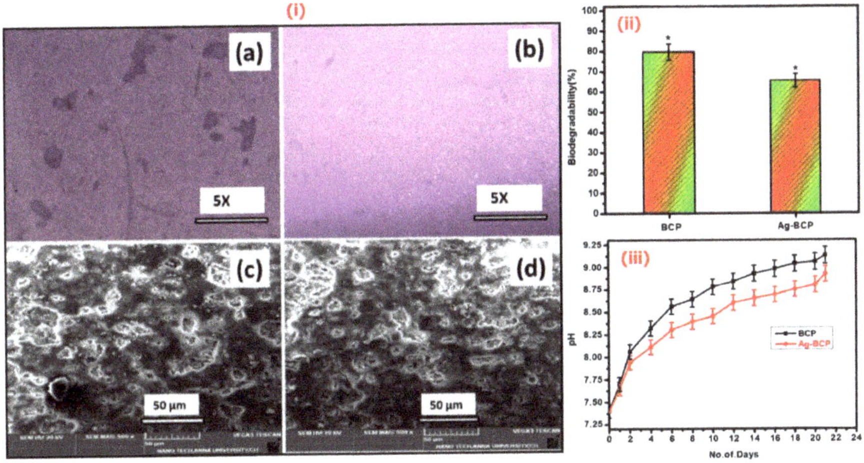

Figure 11. (**i**) Biodegradability investigation using optical microscopy (**a**,**b**) and SEM analysis (**c**,**d**) for the BCP and Ag-BCP samples. The degradability efficiency (%) evaluation is determined by the weight-loss method (* denoted as statistical analyses were performed using a one-way analysis of variance (ANOVA) and the data presented are the mean ± standard deviation) (**ii**) and a pH meter investigation (**iii**).

Figure 12 shows the pattern of CIP release from BCP and Ag-BCP samples under physiological conditions (PBS; pH 7.4). The analysis was carried out by measuring the optical absorption as a function of time. BCP exhibited an initial burst release of ~43% within the first 5 mins. In contrast, the Ag-BCP sample showed a relatively lower release of only 30%. Over 45 mins, BCP released ±92.65%, whereas Ag-BCP released only ±72.37% in a slow and controlled way. This difference is linked to the presence of Ag which supports the occurrence of heterogeneous oxidation reactions requiring the combined effects of dissolved oxygen and protons [36]. From the CIP release studies, we observed that the Ag-BCP has controlled release behavior due to its capacity to maintain the heterogenous particles of varying sizes, shapes, and phases.

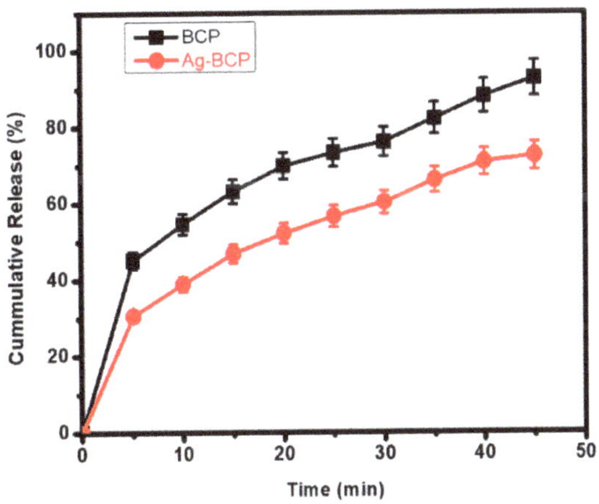

Figure 12. Comparison of CIP release behavior of BCP and Ag-BCP.

3.3. *In Vitro Antimicrobial and Cytocompatibility Studies*

HAP, β-TCP, BCP, and Ag-BCP samples were tested for their antibacterial activity against the Gram-positive *S. aureus*, Gram-negative *E. coli*, and the antifungal activity against *C. albicans* (shown in Figure 13a–c). The reason for selecting *S. aureus* for the studies is that it is responsible for the biofilm formation on bone implants. *E. coli* strains have a reducing capability towards BCP. From the comparison of results provided in Figure 13, the Ag-BCP sample performed almost equal to the standard (Std) in both antibacterial and antifungal activities. Ag-BCP had a moderate inhibitory effect against all bacteria or fungi during a 12 h culturing period. Among the two different bacterial cultures, Ag-BCP had the highest activity towards gram-positive bacteria (*S. aureus*) and fungi (*C. albicans*) where the ZoI was about 10 mm. Furthermore, gram-negative bacteria (*E. coli*) were inhibited in their growth with ZoI of 8 mm. The ZoI of all the samples (HAP, β-TCP, BCP, and Ag-BCP) against the tested microbial cultures are provided in Table 3. Ag-BCP shows better antimicrobial (antibacterial and antifungal) activity than that of the other three samples and thereby confirms the role played by the Ag NPs impregnated onto the BCP composite.

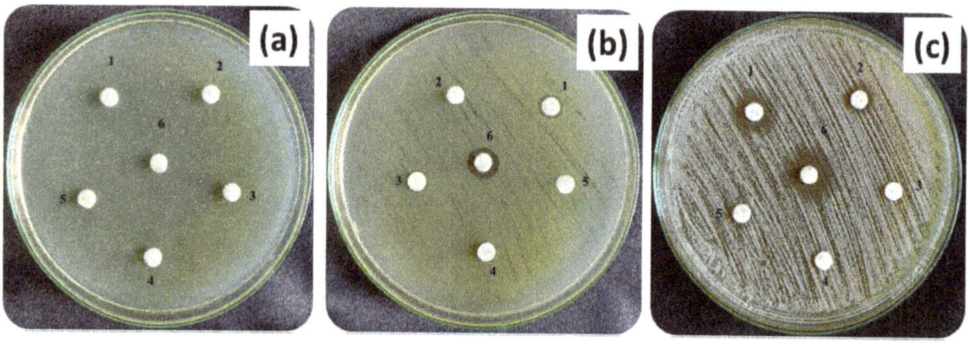

Figure 13. Antimicrobial investigation against (**a**) *S. aureus*, (**b**) *E. coli*, and (**c**) *C. albicans* (where 1. Ag-BCP, 2. HAP, 3. B-TCP, 4. BCP, 5. DMSO, and 6. Standard).

Table 3. Zone of inhibition (mm) data of bioceramic samples treated against *E. coli*, *S. aureus*, and *C. albicans*.

Microorganism	Zone of Inhibition (mm)					
	Ag-BCP	**HAP**	**β-TCP**	**BCP**	**DMSO**	**Std (20 μL)**
E. coli	8 ± 1.15	-	-	-	-	11 ± 1.75
S. aureus	10 ± 2.2	-	-	-	-	17 ± 1.2
C. albicans	10 ± 1.5	-	-	-	-	12 ± 1.0

Figure 14 provides a comparison of in vitro cytocompatibility studies of BCP and Ag-BCP samples when tested on mouse fibroblast L929 cells at various concentrations. We performed an MTT assay carried out with BCP and Ag-BCP samples. Both cells exhibited a significant reduction in the number of cells (detected via their absorption recorded at 450 nm) with an increase in treatment dosage from 31.2 to 1000 μg/mL. When changing the concentration from 31.2 to 62.5 μg/mL, the cell viability % decreased from 80.8 to 50% for the BCP sample and 70 to 32.5% for the Ag-BCP. This determines the IC_{50} value of BCP and Ag-BCP to be 62.5 and 46.8 μg/mL, respectively.

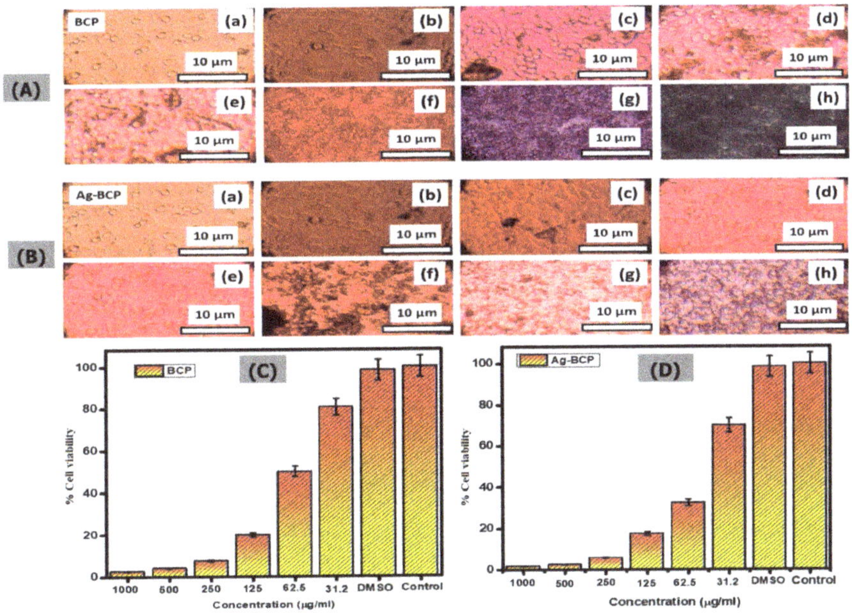

Figure 14. Optical microscopic images of L929 fibroblast cells under various concentrations of BCP (**A**) and Ag-BCP (**B**) ((**a**) for the control, (**b–h**) for the cell concentration in the range of 31.2 to 1000 µg/mL). The corresponding % cell viability changes for BCP (**C**) and Ag-BCP (**D**) in L929 cells.

4. Conclusions

In conclusion, the present study deals with the synthesis, characterization, and testing of BCP and Ag-BCP composites for antimicrobial, drug delivery, and biodegradable characteristics. Bioceramic composites (Venice Mestre, Italy) were characterized for improved interconnectivity, porosity, moderate compressive strength, and biocompatibility where all of these properties are beneficial for bone tissue engineering applications. The Ag-BCP composite was formed by the decoration of Ag NPs with the bioceramic BCP base. The composite maintained its BCP structural framework and at the same time, the porous network structure was formed without compromising its basic characteristics. Nonetheless, the compressive strength and thermal stability increased after Ag loading onto the BCP. Furthermore, the composites showed a difference in pH values in SBF solution, and for the Ag-BCP, the biodegradation rate was reduced. Based on the cumulative results, the Ag-BCP composite would serve as a potential candidate for the efficient growth of damaged or defective bone parts in tissue engineering applications.

Supplementary Materials: The following supporting information can be downloaded at: https://www.mdpi.com/article/10.3390/antibiotics11121780/s1, Table S1: polydispersity index.

Author Contributions: Conceptualization, V.P.P. and P.S.; methodology, V.P.P. and P.S.; validation, B.S., R.G. and M.T.; formal analysis, V.P.P. and P.S.; investigation, R.K. and S.S.; writing—original draft preparation, V.P.P. and P.S.; writing—review and editing, R.G. and M.T.; supervision, R.K. and S.S. All authors have read and agreed to the published version of the manuscript.

Funding: This research received no external funding.

Institutional Review Board Statement: Not applicable.

Informed Consent Statement: Not applicable.

Data Availability Statement: Not applicable.

Acknowledgments: One of the authors, Varun Prasath Padmanabhan, expresses most sincere gratitude to the Council of Scientific and Industrial Research (CSIR), Government of India, for financial support in the form of a Research Associate Fellowships (09/115(0793)/2020 EMR-I.

Conflicts of Interest: The authors declare no conflict of interest.

References

1. Amini, A.R.; Laurencin, C.T.; Nukavarapu, S.P. Bone tissue engineering: Recent advances and challenges. *Crit. Rev. Biomed. Eng.* **2012**, *40*, 363. [CrossRef] [PubMed]
2. Ghassemi, T.; Shahroodi, A.; Ebrahimzadeh, M.H.; Mousavian, A.; Movaffagh, J.; Moradi, A. Current Concepts in Scaffolding for Bone Tissue Engineering. *Arch. Bone Jt. Surg.* **2018**, *6*, 90–99. [PubMed]
3. Zhang, D.; Wu, X.; Chen, J.; Lin, K. The development of collagen based composite scaffolds for bone regeneration. *Bioact. Mater.* **2018**, *3*, 129–138. [CrossRef] [PubMed]
4. Balakrishnan, S.; Padmanabhan, V.P.; Kulandaivelu, R.; Nellaiappan, T.S.N.; Sagadevan, S.; Paiman, S.; Mohammad, F.; Al-Lohedan, H.A.; Obulapuram, P.K.; Oh, W.C. Influence of iron doping towards the physicochemical and biological characteristics of hydroxyapatite. *Ceram. Int.* **2021**, *47*, 5061–5070.
5. O'brien, F.J. Biomaterials & scaffolds for tissue engineering. *Mater. Today* **2011**, *14*, 88–95.
6. Liao, J.; Han, R.; Wu, Y.; Qian, Z. Review of a new bone tumor therapy strategy based on bifunctional biomaterials. *Bone Res.* **2021**, *9*, 1–3. [CrossRef]
7. Canillas, M.; Pena, P.; Antonio, H.; Rodríguez, M.A. Calcium phosphates for biomedical applications. *Boletín Soc. Española Cerámica Y Vidr.* **2017**, *56*, 91–112. [CrossRef]
8. Lobo, S.E.; Arinzeh, T.L. Biphasic calcium phosphate ceramics for bone regeneration and tissue engineering applications. *Materials* **2010**, *3*, 815–826. [CrossRef]
9. Jeong, J.; Kim, J.H.; Shim, J.H.; Hwang, N.S.; Heo, C.Y. Bioactive calcium phosphate materials and applications in bone regeneration. *Biomater. Res.* **2019**, *23*, 1–11. [CrossRef]
10. Chen, F.M.; Liu, X. Advancing biomaterials of human origin for tissue engineering. *Prog. Polym. Sci.* **2016**, *53*, 86–168.
11. Eliaz, N.; Metoki, N. Calcium phosphate bioceramics: A review of their history, structure, properties, coating technologies and biomedical applications. *Materials* **2017**, *10*, 334. [CrossRef] [PubMed]
12. Ebrahimi, M.; Botelho, M.G.; Dorozhkin, S.V. Biphasic calcium phosphates bioceramics (HA/TCP): Concept, physicochemical properties and the impact of standardization of study protocols in biomaterials research. *Mater. Sci. Eng. C* **2017**, *71*, 1293–1312. [CrossRef] [PubMed]
13. de Oliveira Junior, J.M.; Montagner, P.G.; Carrijo, R.C.; Martinez, E.F. Physical characterization of biphasic bioceramic materials with different granulation sizes and their influence on bone repair and inflammation in rat calvaria. *Sci. Rep.* **2021**, *11*, 1–10. [CrossRef] [PubMed]
14. Vallejos Baier, R.; Benjumeda Wijnhoven, I.; Irribarra del Valle, V.; Millán Giovanetti, C.; Vivanco, J.F. Microporosity clustering assessment in calcium phosphate bioceramic particles. *Front. Bioeng. Biotechnol.* **2019**, *7*, 281. [CrossRef] [PubMed]
15. Khang, G.; Kim, S.H.; Kim, M.S.; Lee, H.B. Hybrid, composite, and complex biomaterials for scaffolds. *Princ. Regen. Med.* **2008**, *1*, 636–655.
16. Zhang, F.; Chang, J.; Lu, J.; Lin, K.; Ning, C. Bioinspired structure of bioceramics for bone regeneration in load-bearing sites. *Acta Biomater.* **2007**, *3*, 896–904. [CrossRef]
17. Levengood, S.K.; Polak, S.J.; Wheeler, M.B.; Maki, A.J.; Clark, S.G.; Jamison, R.D.; Johnson, A.J. Multiscale osteointegration as a new paradigm for the design of calcium phosphate scaffolds for bone regeneration. *Biomaterials* **2010**, *31*, 3552–3563. [CrossRef]
18. Miranda, P.; Pajares, A.; Saiz, E.; Tomsia, A.P.; Guiberteau, F. Mechanical properties of calcium phosphate scaffolds fabricated by robocasting. *J. Biomed. Mater. Res. Part A* **2008**, *85*, 218–227. [CrossRef]
19. Abou Neel, E.A.; Aljabo, A.; Strange, A.; Ibrahim, S.; Coathup, M.; Young, A.M.; Bozec, L.; Mudera, V. Demineralization–remineralization dynamics in teeth and bone. *Int. J. Nanomed.* **2016**, *11*, 4743. [CrossRef]
20. Ishikawa, K. Carbonate apatite bone replacement. *Key Eng. Mater.* **2014**, *587*, 17–20.
21. Puttini, I.D.; Poli, P.P.; Maiorana, C.; Vasconcelos, I.R.; Schmidt, L.E.; Colombo, L.T.; Hadad, H.; Santos, G.M.; Carvalho, P.S.; Souza, F.Á. Evaluation of osteoconduction of biphasic calcium phosphate ceramic in the calvaria of rats: Microscopic and histometric analysis. *J. Funct. Biomater.* **2019**, *10*, 7. [CrossRef] [PubMed]
22. Baghbani, F.; Moztarzadeh, F.; Nazari, A.G.; Kamran, A.R.; Tondnevis, F.; Nezafati, N.; Gholipourmalekabadi, M.; Mozafari, M. Biological response of biphasic hydroxyapatite/tricalcium phosphate scaffolds intended for low load-bearing orthopaedic applications. *Adv. Compos. Lett.* **2012**, *21*, 096369351202100102. [CrossRef]
23. Burdușel, A.C.; Gherasim, O.; Grumezescu, A.M.; Mogoantă, L.; Ficai, A.; Andronescu, E. Biomedical applications of silver nanoparticles: An up-to-date overview. *Nanomaterials* **2018**, *8*, 681. [CrossRef] [PubMed]
24. Dikshit, P.K.; Kumar, J.; Das, A.K.; Sadhu, S.; Sharma, S.; Singh, S.; Gupta, P.K.; Kim, B.S. Green synthesis of metallic nanoparticles: Applications and limitations. *Catalysts* **2021**, *11*, 902. [CrossRef]
25. Dakal, T.C.; Kumar, A.; Majumdar, R.S.; Yadav, V. Mechanistic basis of antimicrobial actions of silver nanoparticles. *Front. Microbiol.* **2016**, *7*, 1831. [CrossRef]

26. Nie, L.; Suo, J.; Zou, P.; Feng, S. Preparation and properties of biphasic calcium phosphate scaffolds multiply coated with HA/PLLA nanocomposites for bone tissue engineering applications. *J. Nanomater.* **2012**, *2012*, 213549 . [CrossRef]
27. Xidaki, D.; Agrafioti, P.; Diomatari, D.; Kaminari, A.; Tsalavoutas-Psarras, E.; Alexiou, P.; Psycharis, V.; Tsilibary, E.C.; Silvestros, S.; Sagnou, M. Synthesis of hydroxyapatite, β-tricalcium phosphate and biphasic calcium phosphate particles to act as local delivery carriers of curcumin: Loading, release and in vitro studies. *Materials* **2018**, *11*, 595. [CrossRef]
28. Kim, D.H.; Hwang, K.H.; Lee, J.D.; Park, H.C.; Yoon, S.Y. Long and short range order structural analysis of in-situ formed biphasic calcium phosphates. *Biomater. Res.* **2015**, *19*, 1–5. [CrossRef]
29. Shameli, K.; Ahmad, M.B.; Jazayeri, S.D.; Sedaghat, S.; Shabanzadeh, P.; Jahangirian, H.; Mahdavi, M.; Abdollahi, Y. Synthesis and characterization of polyethylene glycol mediated silver nanoparticles by the green method. *Int. J. Mol. Sci.* **2012**, *13*, 6639–6650. [CrossRef]
30. Ebrahimi, M.; Botelho, M. Biphasic calcium phosphates (BCP) of hydroxyapatite (HA) and tricalcium phosphate (TCP) as bone substitutes: Importance of physicochemical characterizations in biomaterials studies. *Data Brief* **2017**, *10*, 93–97. [CrossRef]
31. Wu, F.; Lin, D.D.; Chang, J.H.; Fischbach, C.; Estroff, L.A.; Gourdon, D. Effect of the materials properties of hydroxyapatite nanoparticles on fibronectin deposition and conformation. *Cryst. Growth Des.* **2015**, *15*, 2452–2460. [CrossRef] [PubMed]
32. van Dijk, L.A.; Duan, R.; Luo, X.; Barbieri, D.; Pelletier, M.; Christou, C.; Rosenberg, A.J.; Yuan, H.; Barrèrre-de Groot, F.; Walsh, W.R.; et al. Biphasic calcium phosphate with submicron surface topography in an Ovine model of instrumented posterolateral spinal fusion. *JOR Spine* **2018**, *1*, e1039. [CrossRef] [PubMed]
33. Lee, Y.H.; Lee, J.W.; Yang, S.Y.; Lee, H.; Koh, Y.H.; Kim, H.E. Dual-scale porous biphasic calcium phosphate gyroid scaffolds using ceramic suspensions containing polymer microsphere porogen for digital light processing. *Ceram. Int.* **2021**, *47*, 11285–11293. [CrossRef]
34. Kanchana, P.; Sekar, C. Effect of magnesium on the mechanical and bioactive properties of biphasic calcium phosphate. *J. Miner. Mater. Charact. Eng.* **2012**, *11*, 982. [CrossRef]
35. Kumar, B.S.; Muthukumar, T.; Deepachitra, R.; Charumathy, R.K.; Hemalatha, T.; Sastry, T.P. In-vitro evaluation of biphasic calcium phosphate/casein incorporated with Myristica fragrans for bone tissue engineering. *Ceram. Int.* **2015**, *41*, 1725–1734. [CrossRef]
36. Jariya, S.I.; Padmanabhan, V.P.; Kulandaivelu, R.; Prakash, N.; Mohammad, F.; Al-Lohedan, H.A.; Paiman, S.; Schirhagl, R.; Hossain, M.M.; Sagadevan, S. Drug delivery and antimicrobial studies of chitosan-alginate based hydroxyapatite bioscaffolds formed by the Casein micelle assisted synthesis. *Mater. Chem. Phys.* **2021**, *272*, 125019. [CrossRef]

antibiotics

MDPI

Article

An Optimization of Oregano, Thyme, and Lemongrass Essential Oil Blend to Simultaneous Inactivation of Relevant Foodborne Pathogens by Simplex–Centroid Mixture Design

Luiz Torres Neto [1,2,3,*], Maria Lúcia Guerra Monteiro [1,2,3,4], Maxsueli Aparecida Moura Machado [1,2,3], Diego Galvan [1,2,5] and Carlos Adam Conte Junior [1,2,3,4,6,*]

1 Center for Food Analysis (NAL), Technological Development Support Laboratory (LADETEC), Cidade Universitária, Rio de Janeiro 21941-598, RJ, Brazil
2 Laboratory of Advanced Analysis in Biochemistry and Molecular Biology (LAABBM), Department of Biochemistry, Federal University of Rio de Janeiro (UFRJ), Cidade Universitária, Rio de Janeiro 21941-909, RJ, Brazil
3 Graduate Program in Food Science (PPGCAL), Institute of Chemistry (IQ), Federal University of Rio de Janeiro (UFRJ), Cidade Universitária, Rio de Janeiro 21941-909, RJ, Brazil
4 Graduate Program in Veterinary Hygiene (PPGHV), Faculty of Veterinary Medicine, Fluminense Federal University (UFF), Vital Brazil Filho, Niterói 24220-000, RJ, Brazil
5 Institute of Chemistry, Federal University of Mato Grosso do Sul (UFMS), Campo Grande 79070-900, MS, Brazil
6 Graduate Program in Sanitary Surveillance (PPGVS), National Institute of Health Quality Control (INCQS), Oswaldo Cruz Foundation (FIOCRUZ), Rio de Janeiro 21040-900, RJ, Brazil
* Correspondence: luiztorresneto@ufrj.br or luiz-torres-neto@hotmail.com (L.T.N.); conte@iq.ufrj.br (C.A.C.J.); Tel.: +55-55-99694-8995 (L.T.N.); +55-21-3938-7825 (C.A.C.J.)

Citation: Torres Neto, L.; Monteiro, M.L.G.; Machado, M.A.M.; Galvan, D.; Conte Junior, C.A. An Optimization of Oregano, Thyme, and Lemongrass Essential Oil Blend to Simultaneous Inactivation of Relevant Foodborne Pathogens by Simplex–Centroid Mixture Design. *Antibiotics* **2022**, *11*, 1572. https://doi.org/10.3390/antibiotics11111572

Academic Editor: Helena P. Felgueiras

Received: 14 October 2022
Accepted: 4 November 2022
Published: 8 November 2022

Abstract: (1) Background: This study aimed to use the simplex–centroid mixture design methodology coupled with a microdilution assay to predict optimal essential oil (EO) formulations against three potential foodborne pathogens simultaneously through the desirability (D) function. (2) Methods: Oregano (ORE; *Origanum vulgare*), thyme (THY; *Thymus vulgaris*), and lemongrass (LG; *Cymbopogon citratus*) and their blends were evaluated concerning minimum inhibitory concentration (MIC) and minimum bactericidal concentration (MBC) for *Salmonella enterica* serotype Enteritidis, *Escherichia coli* and *Staphylococcus aureus*. (3) Results: THY combined with ORE or LG were the most promising EO formulations in inhibiting and killing each bacterium separately. Regarding the simultaneous effect, the optimal proportion for maximum inhibition was composed of 75% ORE, 15% THY, and 10% LG, while for maximum inactivation was 50% ORE, 40% THY, and 10% LG. (4) Conclusion: The multiresponse optimization allowed identifying an EO blend to simultaneously control three potential foodborne pathogens. This first report could be a helpful natural and green alternative for the industry to produce safer food products and mitigate public health risks.

Keywords: natural antimicrobials; volatile oils; bioactive compounds; minimum inhibitory concentration; minimum bactericidal concentration; desirability function

1. Introduction

Foodborne pathogens are a global public health issue with over 600 million cases per year resulting in concern towards morbidity, hospitalizations, and mortality, adding 420,000 deaths, and US$ 110 billion lost each year worldwide, requiring alternatives to produce safer food products [1–4]. In this context, *Salmonella enterica* serotype Enteritidis, *Escherichia coli*, and *Staphylococcus aureus* are responsible for several cases of food outbreaks worldwide [4–6]. They are the most common foodborne pathogens affecting the health of millions of people and adding up to an annual economic loss of billions of dollars [7–10]. Moreover, the infectious diseases caused by them include several harsh symptoms such as stomach cramps, diarrhea, fever, nausea, and/or vomiting, and even life-threatening conditions [11].

Essential oils (EOs) are rich in a broad biological spectrum of compounds such as sesquiterpenes, monoterpenes, aldehydes, alcohols, esters, ketones, polyphenols, and flavonoids, including other classes, which result in a broad biological activity including antimicrobial, antifungal, insecticide, anti-inflammatory, antioxidant, anticarcinogenic, and antiviral [12,13]. Therefore, these oils have been successfully applied in different industrial sectors such as food, cosmetic, pharmaceutical, and agricultural industries [14].

EOs are considered safe and eco-friendly plant-based antimicrobial alternatives to control foodborne pathogens and other microorganisms, including those drug-resistant ones [15,16]. Previous studies have already reported the potential of EOs against *S. enteritidis*, *E. coli*, and *S. aureus*, maintaining food quality and safety [17,18]. However, the EO concentrations needed to achieve antimicrobial effectiveness have generally led to adverse sensory impacts in foods due to their intense aroma. In this way, blended formulations have recently been investigated to reach the antimicrobial goal using lower EO concentrations since the effect of EO mixtures could be boosted by interactions of different functional groups [19–21]. Furthermore, EO blends have been drawing attention as a promising green technology to decrease antibiotic resistance among bacteria, adverse effects on human health, and environmental impacts from synthetic compounds due to their toxicity and slow degradation periods [22,23]. Nevertheless, the wide variety of EOs and foodborne pathogens makes it a stiff challenge.

The Mixture Designs (MDs) can assist in achieving optimized EO blends. In this experimental design, two or more components are combined in different proportions, and the results can be modeled and predicted mathematically and graphically [24,25]. One approach is to apply these models combined with desirability functions to optimize multiple responses simultaneously with a reduced number of experiments but with high quality and low cost [26]. The MDs were already used to maximize or optimize the antimicrobial activity of essential oil blends against different pathogens [16,18,21,27]. However, despite the great potential of this design in evaluating EO blends, this yet is underused. Moreover, these studies evaluate neither the combination of oregano, thyme, and lemongrass nor the simultaneous inhibition and inactivation of foodborne pathogens.

In this context, this study aimed to achieve optimized formulations containing oregano, thyme, and lemongrass EOs for individual and simultaneous inhibition and inactivation of *S. enteritidis*, *E. coli*, and *S. aureus* through an augmented simplex–centroid mixture design, attempting safer food products and public health improvement.

2. Results

2.1. Chemical Composition of Essential oils

The complete profile of compounds in EOs is available in Supplementary Table S1. Oregano EO exhibited a high concentration of carvacrol (70.3%), followed by other compounds such as p-cymene (10.4%), γ-terpinene (4.8%), (E)-β-caryophyllene (4.7%), linalool (2.2%), and myrcene (1.6%). The thyme EO was also predominantly composed of phenolic compounds, which were thymol (31.2%) and carvacrol (25.5%), followed by p-cymene (21.7%), linalool (6%), limonene (3.4%) and borneol (3.4%). Concerning lemongrass EO, its composition was rich in aldehydes (45.5% geranial and 33.7% neral) and other compounds, such as geraniol (4.5%), geranyl acetate (1.5%), citronellal (1.3%), and citronellol (1.1%). Similar chemical composition for oregano [28], thyme [29,30], and lemongrass [31,32] EOs have been previously reported in the literature.

2.2. Single and Combined Antimicrobial Effects through Mixture Design (MD)

Table 1 shows the ANOVA of the regression models for each bacterium concerning MIC and MBC with the corresponding F-values and p-values as well as the corresponding R^2 and adjusted R^2 values demonstrating the quality of the selected quadratic models to data adjustment, except for *S. aureus*. As all least-square regression models were previously evaluated for better data adjustment, this *S. aureus* strain (ATCC 14458) was more complex to model. However, it should be noted that most individual models were significant and did not show a lack of fit in most cases. In addition to these factors, the critical values were experimentally validated, as described in the following sections.

Table 1. The quality of the quadratic model through analysis of variance (ANOVA) for the minimum inhibitory concentration (MIC) and minimum bactericidal concentration (MBC) experiments concerning essential oils (EOs) against *Escherichia coli*, *Staphylococcus aureus*, and *Salmonella enterica* serotype Enteritidis.

E. coli MIC	Sum of Squares	Degrees of freedom	Mean Square	*F*-value	*p*-value	R^2	$R^2_{adj.}$ *
Model	0.026249	5	0.00525	20.83859	0.000995		
Total Error	0.001512	6	0.000252				
Lack of Fit	0.001095	4	0.000274	1.31386	0.475324	0.9456	0.9002
Pure Error	0.000417	2	0.000208				
Total Adjusted	0.02776	11	0.002524				
***S. aureus* MIC**	**Sum of Squares**	**Degrees of freedom**	**Mean Square**	***F*-value**	***p*-value**	R^2	$R^2_{adj.}$ *
Model	0.004934	5	0.000987	2.979	0.108212		
Total Error	0.001988	6	0.000331				
Lack of Fit	0.001987	4	0.000497	1490.163	0.000671	0.7128	0.4735
Pure Error	0.000001	2	0				
Total Adjusted	0.006921	11	0.000629				
***S. enteritidis* MIC**	**Sum of Squares**	**Degrees of freedom**	**Mean Square**	***F*-value**	***p*-value**	R^2	$R^2_{adj.}$ *
Model	0.112971	5	0.022594	14.03543	0.002927		
Total Error	0.009659	6	0.00161				
Lack of Fit	0.009242	4	0.002311	11.09053	0.084416	0.9212	0.8556
Pure Error	0.000417	2	0.000208				
Total Adjusted	0.12263	11	0.011148				
***E. coli* MBC**	**Sum of Squares**	**Degrees of freedom**	**Mean Square**	***F*-value**	***p*-value**	R^2	$R^2_{adj.}$ *
Model	0.026249	5	0.00525	20.83859	0.000995		
Total Error	0.001512	6	0.000252				
Lack of Fit	0.001095	4	0.000274	1.31386	0.475324	0.9456	0.9002
Pure Error	0.000417	2	0.000208				
Total Adjusted	0.02776	11	0.002524				
***S. aureus* MBC**	**Sum of Squares**	**Degrees of freedom**	**Mean Square**	***F*-value**	***p*-value**	R^2	$R^2_{adj.}$ *
Model	0.00622	5	0.001244	1.641339	0.280675		
Total Error	0.004548	6	0.000758				
Lack of Fit	0.002881	4	0.00072	0.864343	0.598648	0.5777	0.2257
Pure Error	0.001667	2	0.000833				
Total Adjusted	0.010768	11	0.000979				
***S. enteritidis* MBC**	**Sum of Squares**	**Degrees of freedom**	**Mean Square**	***F*-value**	***p*-value**	R^2	$R^2_{adj.}$ *
Model	0.106867	5	0.021373	4.81933	0.040813		
Total Error	0.02661	6	0.004435				
Lack of Fit	0.026193	4	0.006548	31.43148	0.031072	0.8006	0.6345
Pure Error	0.000417	2	0.000208				
Total Adjusted	0.133477	11	0.012134				

* R^2_{adj}: R^2 adjusted. MIC: minimum inhibitory concentration; MBC: minimum bactericidal concentration.

The effects obtained by the MD allowed identifying the different bacteriostatic (MIC) and bactericidal (MBC) effects of single and blended EOs (Table 2). Concerning single EOs (100% of ORE, THY, or LG), the individual effects of ORE[1] were not significant in *E. coli*; in contrast, THY[1] and LG[1] demonstrated a significant effect for inhibiting and killing *E. coli*, mainly THY[1], which exhibited a lower coefficient than LG[1], (see Table 2). LG[1] was the only one among the single EOs that showed a significant coefficient regarding bacteriostatic and bactericidal effects for *S. enteritidis*. THY[1] and LG[1] demonstrated the ability to inhibit *S. aureus*, mainly THY[1], due to the lowest coefficient (Table 2). Likewise, for *E. coli*, no significant coefficient effect was observed for ORE[1] on *S. aureus* and *S. enteritidis* inhibition for MIC (Table 2).

Table 2. Coefficients of model fitted for minimum inhibitory concentration (MIC) and minimum bactericidal concentration (MBC) values concerning essential oils (EOs) against *Escherichia coli*, *Staphylococcus aureus*, and *Salmonella enterica* serotype Enteritidis and their level of significance.

	E. coli			*S. aureus*			*S. enteritidis*		
MIC	**Estimation** [€]	**SE** [†]	***p*-Value**	**Estimation**	**SE** [†]	***p*-Value**	**Estimation**	**SE** [†]	***p*-Value**
ORE [1]	0.012896	0.015265	0.430588	0.039735	0.017504	0.063661	0.011105	0.038586	0.783187
THY [1]	0.050495	0.015265	0.016245 **	0.049711	0.017504	0.029566 *	0.033339	0.038586	0.420770
LG [1]	0.192025	0.015265	0.000015 ***	0.106495	0.017504	0.000896 ***	0.376076	0.038586	0.000067 ***
$ORE^{0.5} + THY^{0.5}$	−0.007719	0.068055	0.913390	−0.075322	0.078039	0.371723	0.172999	0.172033	0.353418
$ORE^{0.5} + LG^{0.5}$	−0.225688	0.068055	0.016079 **	−0.060932	0.078039	0.464605	−0.645440	0.172033	0.009489 **
$THY^{0.5} + LG^{0.5}$	−0.300477	0.068055	0.004494 **	−0.039770	0.078039	0.628516	−0.488530	0.172033	0.029577 *
	E. coli			*S. aureus*			*S. enteritidis*		
MBC	**Estimation** [€]	**SE** [†]	***p*-Value**	**Estimation**	**SE** [†]	***p*-Value**	**Estimation**	**SE** [†]	***p*-Value**
ORE [1]	0.012896	0.015265	0.430588	0.086091	0.026477	0.017433 **	0.018346	0.064046	0.784173
THY [1]	0.050495	0.015265	0.016245 **	0.104938	0.026477	0.007424 *	0.058587	0.064046	0.395587
LG [1]	0.192025	0.015265	0.000015 ***	0.104938	0.026477	0.007424 *	0.374725	0.064046	0.001100 **
$ORE^{0.5} + THY^{0.5}$	−0.007719	0.068055	0.913390	−0.188111	0.118046	0.162149	0.018773	0.285542	0.949716
$ORE^{0.5} + LG^{0.5}$	−0.225688	0.068055	0.016079 **	−0.188111	0.118046	0.162149	−0.256087	0.285542	0.404348
$THY^{0.5} + LG^{0.5}$	−0.300477	0.068055	0.004494**	0.051866	0.118046	0.675775	−0.622240	0.285542	0.072145

[†] SE: standard error; * $p < 0.05$; ** $p < 0.02$; *** $p < 0.001$. [1] 100% of EO. [0.5] 50% of each essential EO. [€] Negative coefficient values indicate an increased antibacterial effect, and positive coefficient values suggest an antagonist effect.

Regarding the mixtures of EOs (proportion 50%:50%), the promising blended EOs were those with negative coefficient values, indicating an increased antibacterial effect by acting additively or synergistically (Table 2). Otherwise, a positive coefficient value suggests an antagonist effect [33]. In this way, a significant effect was observed in the $ORE^{0.5} + LG^{0.5}$ and $THY^{0.5} + LG^{0.5}$ for inhibiting and killing *E. coli* (Table 2), especially $THY^{0.5} + LG^{0.5}$, due to the lowest coefficients, revealing an interesting combination against this bacterium. In contrast, although $ORE^{0.5} + THY^{0.5}$ have exhibited a negative coefficient value, its bacteriostatic and bactericidal effects were not significant for *E. coli* (Table 2). The same was observed for *S. enteritidis*, wherein $ORE^{0.5} + THY^{0.5}$ was not effective in inhibiting this bacterium, and none of the three blended EOs (50%:50%) in killing it (MBC) (Table 2). Otherwise, $ORE^{0.5} + LG^{0.5}$ and $THY^{0.5} + LG^{0.5}$ demonstrated a significant effect on *S. enteritidis* inhibition (MIC). Among these two blended EOs, $ORE^{0.5} + LG^{0.5}$ had a lower coefficient than $THY^{0.5} + LG^{0.5}$, indicating that this mixture of oregano and lemongrass has a greater potential to inhibit *S. enteritidis* (Table 2). Regarding *S. aureus*, all blended EOs were not significant for both the MIC and MBC values. Moreover, $THY^{0.5} + LG^{0.5}$ showed a positive coefficient for the MBC values of *S. aureus*, while the lowest ones were against *E. coli* and *S. enteritidis* (Table 2).

The combination of $ORE^{0.5}$ and $THY^{0.5}$ was the only EO blend not significant for the MIC and MBC values of the three bacteria tested. In addition, this EO blend had positive coefficients for the MIC and MBC values of *S. enteritidis*. On the contrary, when combined with LG, ORE, and THY, there is evidence of a synergistic or complementary potential.

The treatments obtained through MD were experimentally evaluated for all bacteria strains. From these data, equations and 3D graphics with the different combinations of the three EOs were generated (Figure 1), which allowed identifying through mathematical models the EO proportions for potentially achieving the lowest MIC and MBC values for all bacteria strains (darkest green regions). The optimal proportions to reach the best inhibitory and bactericidal effect for *E. coli* were 100% ORE (Figure 1A,D). For *S. aureus*, the ideal mixtures observed were 50% ORE and 50% THY to achieve the maximum inhibition (Figure 1B), and 50% ORE and 50% LG or THY (saddle surfaces) to reach the highest bactericidal effect (Figure 1E). For *S. enteritidis*, the optimal proportions to achieve the best inhibitory and bactericidal effect were 75% ORE, 10.7% THY, and 14.3% LG, and 100% ORE, respectively (Figure 1C,F).

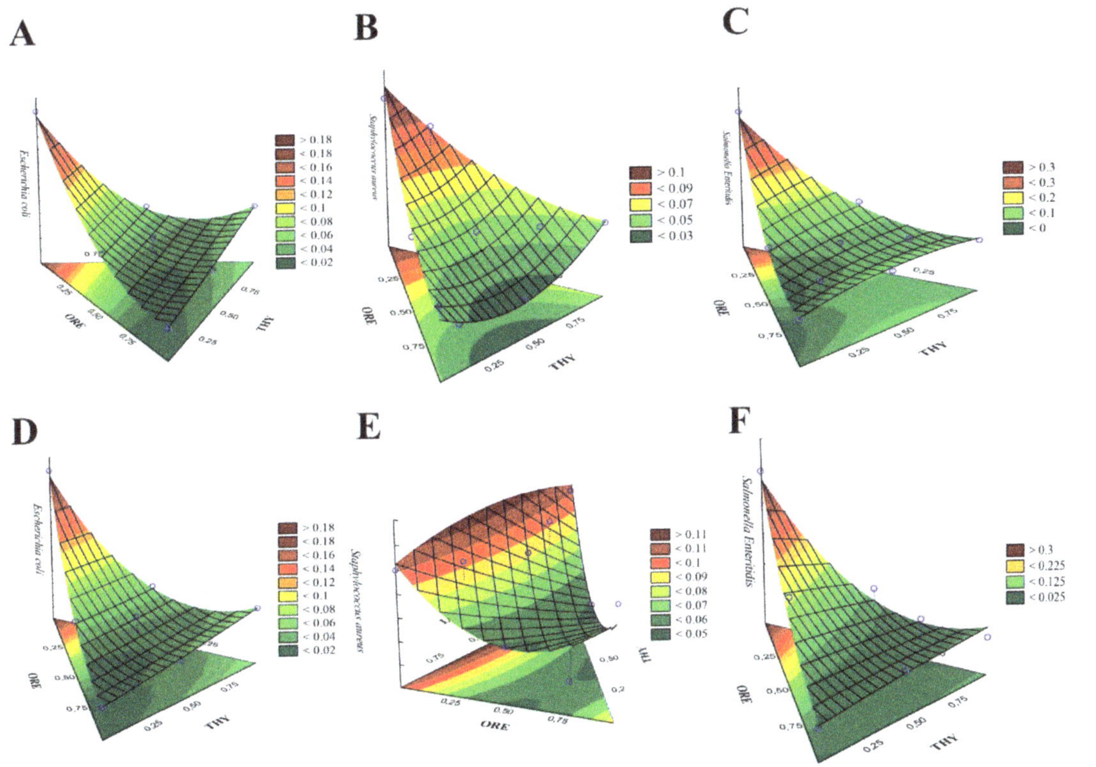

Figure 1. 3D surface plots for the effect of different essential oil blends from oregano (ORE; *Origanum vulgare*), thyme (THY; *Thymus vulgaris*), and lemongrass (LG; *Cymbopogon citratus*) on minimum inhibitory concentration (MIC) value against *Escherichia coli* (**A**), *Staphylococcus aureus* (**B**) and *Salmonella* Enteritidis (**C**) and on minimum bactericidal concentration (MBC) value against *Escherichia coli* (**D**), *Staphylococcus aureus* (**E**) and *Salmonella enterica* serotype Enteritidis (**F**). Results are expressed in percentage (%) and are from twelve experiments, including three central replicates (Section 5.3).

2.3. Mixture Optimization and Validation

The desirability function (D) (Section 5.4.4) allowed the identification of the optimized theoretical mixtures of ORE, THY, and LG to maximize the bacteriostatic (MIC; Figure 2) and bactericidal (MBC; Figure 3) actions against *E. coli*, *S. aureus*, and *S. enteritidis* simultaneously. The optimal EO blend for inhibiting bacteria consisted of 75% ORE, 15% THY, and 10% LG, which minimized the MIC values to 0.014%, 0.034%, and 0.014% for *E. coli*, *S. aureus*, and *S. enteritidis*, respectively (Figure 2). Otherwise, the optimal EO mixture for killing bacteria was composed of 50% ORE, 40% THY, and 10% LG, resulting in the MBC values of 0.021% for *E. coli*, 0.051% for *S. aureus*, and 0.036% for *S. enteritidis* (Figure 3). Therefore, a greater sensitivity was observed in *E. coli*, followed by *S. enteritidis* and *S. aureus* for the MIC and MBC values. The two optimal theoretical proportions were further validated experimentally (Table 3). No difference was observed for the MIC and MBC values of *E. coli*, *S. aureus*, and *S. enteritidis* between theoretical values from desirability function and experimental values (Table 3).

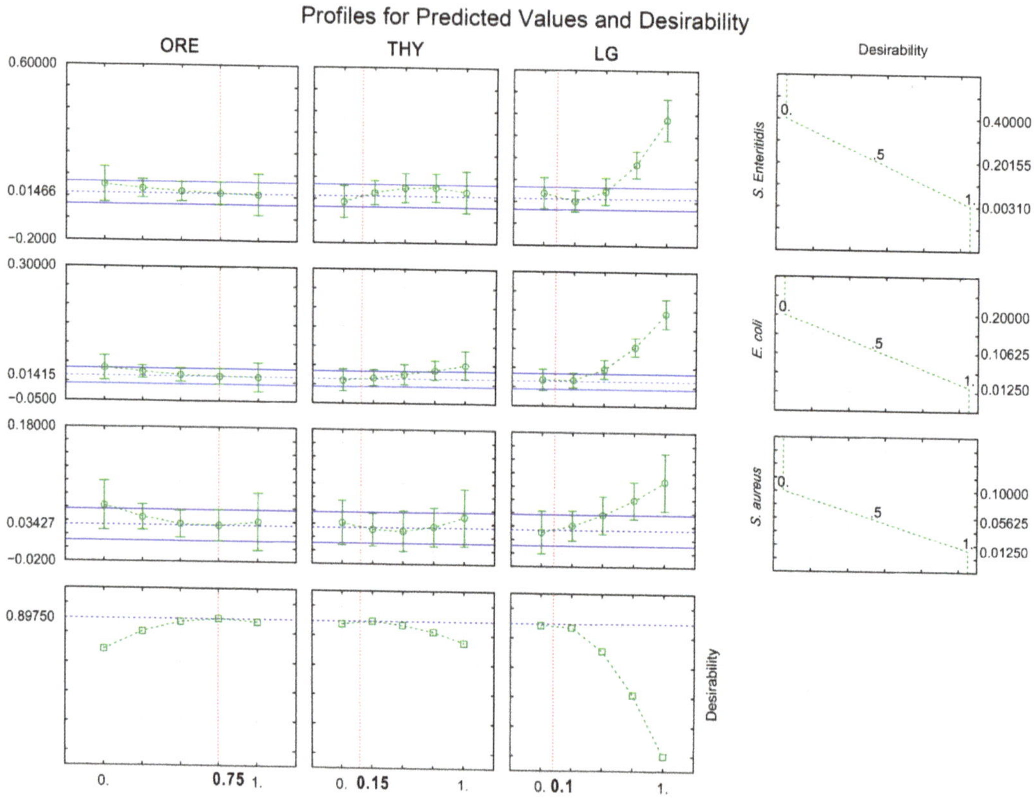

Figure 2. Desirability plot showing the optimal proportions of oregano (ORE; *Origanum vulgare*), thyme (THY; *Thymus vulgaris*), and lemongrass (LG; *Cymbopogon citratus*) to the simultaneous inhibition (MIC) of *Escherichia coli*, *Salmonella enterica* serotype Enteritidis, and *Staphylococcus aureus*. Results are expressed in percentage (%) and are from twelve experiments, including three central replicates (Section 5.3).

Table 3. Observed values and validation for the optimal mixture of oregano (ORE; *Origanum vulgare*), thyme (THY; *Thymus vulgaris*), and lemongrass (LG; *Cymbopogon citratus*) essential oils (EOs) against *Escherichia coli*, *Staphylococcus aureus*, and *Salmonella enterica* serotype Enteritidis considering minimum inhibitory concentration (MIC) and minimum bactericidal concentration (MBC) values.

	MIC (%)$_{(n=6)}$ *				MBC (%)$_{(n=6)}$ *			
	Predicted Value	**Observed Value**	***t*-Test (%)**	**Levene's Test (%)**	**Predicted Value**	**Observed Value**	***t*-Test (%)**	**Levene's Test (%)**
E. coli	0.014	0.013	58.06	48.38	0.021	0.025	5.33	42.27
S. aureus	0.034	0.031	64.94	15.18	0.051	0.075	7.56	38.81
S. enteritidis	0.014	0.021	6.62	6.24	0.036	0.035	86.18	15.18

* The predicted proportions of EOs were 75% (ORE):15% (THY):10% (LG), and 50% (ORE):40% (THY):10% (LG) for MIC and MBC, respectively.

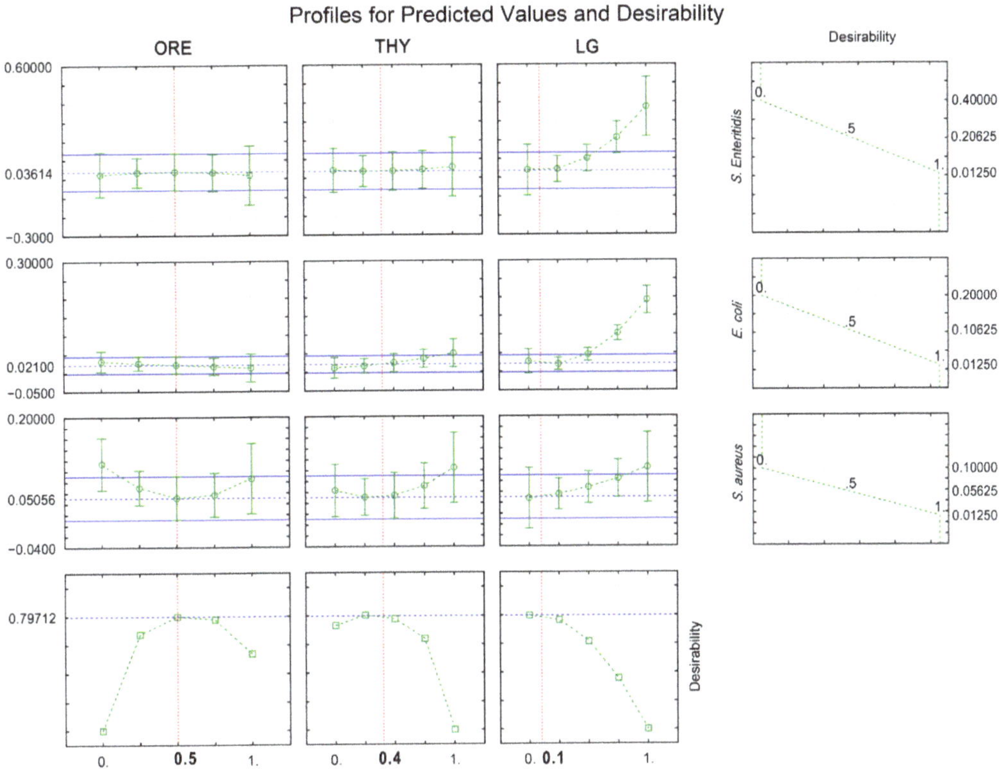

Figure 3. Desirability plot showing the optimal proportions of oregano (ORE; *Origanum vulgare*), thyme (THY; *Thymus vulgaris*), and lemongrass (LG; *Cymbopogon citratus*) to simultaneous inactivation (MBC) of *Escherichia coli*, *Salmonella enterica* serotype Enteritidis, and *Staphylococcus aureus*. Results are expressed in percentage (%) and are from twelve experiments, including three central replicates (Section 5.3).

3. Discussion

It is worth highlighting that this study is the first report investigating the antimicrobial potential of these three combined EOs against *E. coli*, *S. aureus*, and *S. enteritidis* using the mixture design. Moreover, this study aimed to enhance the antibacterial effect through decreased MIC and MBC values (dependent variables), and the lower or negative coefficient values demonstrate that the independent variables (single EO and their mixtures) were effective for increasing the antibacterial activity.

It is well-known that the antimicrobial activity of EOs is attributed to their composition and biochemistry, wherein the lipophilic nature of hydrocarbon skeletons and the hydrophilic nature of the other functional groups play an essential role in this activity [34]. The oregano and thyme EOs showed a high concentration of phenols, followed by hydrocarbons. The oregano EO was also composed of other functional groups, such as alcohol, ether, and ketone, with the thyme of alcohol and ether. Regarding lemongrass, its composition was rich in aldehydes, followed by alcohol, ester, and ketone (Section 2.1).

The single antimicrobial activity of oregano, thyme, and lemongrass EOs has been well-reported in the literature. Previous studies observed bacteriostatic and bactericidal activities of oregano against strains of *E. coli*, *S. enteritidis*, and *S. aureus* [35–37]. However, in our study, single ORE was only effective in killing *S. aureus*. This activity observed in the ORE[1] can be justified by the greater bactericidal effect of the carvacrol against Gram-positive than Gram-negative bacteria [38].

Regarding THY, compounds such as thymol, *p*-cymene, *o*-cymene, γ-terpinene, and linalool are known to be responsible for the broad antibacterial spectrum of this EO [39,40]. The bacteriostatic and bactericidal effects of the thyme EO against *E. coli* [35], *S. enteritidis*, and *S. aureus* [37] have already been reported in the literature. The THY has similar antimicrobial properties to ORE since thymol is analogous to carvacrol [41], and it also possesses the same outer membrane disintegration properties, in addition to affecting a variety of cellular functions [42]. Furthermore, the predominance of carvacrol or thymol associated with the high concentration of hydrocarbons (greater hydrophobicity) may explain our findings concerning the similar activity of both ORE and THY against Gram-negative and Gram-positive bacteria.

The antibacterial activity of LG may be attributed to neral, geranial, and geraniol, among other compounds [43], and was observed by Naik et al. [44] against *E. coli* and *S. aureus*. However, there are no studies to date evaluating the MIC and MBC values of lemongrass against *S. enteritidis*. De Silva et al. [45] reported the bacteriostatic and bactericidal effects of lemongrass on *S. enterica*. Moreover, there is very little information in the literature about the mechanisms of action of neral and geranial compounds, but it is known that aldehydes are more active in Gram-positive bacteria than in Gram-negative ones [46], corroborating our findings (Table 4). In our study, LG[1] showed higher MIC and MBC values (lower antimicrobial activity) than ORE[1] and THY[1] for all evaluated strains (Table 4).

Table 4. Simplex–centroid design experiments, minimum inhibitory concentration (MIC) and minimum bactericidal concentration (MBC) values of single and blended essential oils (EOs) against *Escherichia coli*, *Staphylococcus aureus*, and *Salmonella enterica* serotype Enteritidis.

	Experiments *			MIC (%)			MBC (%)		
	ORE	THY	LG	*S. enteritidis*	*E. coli*	*S. aureus*	*S. enteritidis*	*E. coli*	*S. aureus*
1	1	0	0	0.0031	0.0125	0.05	0.0125	0.0125	0.1
2	0	1	0	0.025	0.05	0.05	0.025	0.05	0.1
3	0	0	1	0.4	0.2	0.1	0.4	0.2	0.1
4	0.5	0.5	0	0.05	0.025	0.025	0.05	0.025	0.05
5	0.5	0	0.5	0.05	0.05	0.05	0.2	0.05	0.05
6	0	0.5	0.5	0.1	0.05	0.05	0.1	0.05	0.1
7	0.33	0.33	0.33	0.05	0.05	0.05	0.05	0.05	0.05
8	0.33	0.33	0.33	0.025	0.025	0.05	0.025	0.025	0.05
9	0.33	0.33	0.33	0.05	0.025	0.025	0.05	0.025	0.1
10	0.67	0.17	0.17	0.025	0.0125	0.0125	0.025	0.0125	0.0125
11	0.17	0.67	0.17	0.05	0.025	0.05	0.1	0.025	0.1
12	0.17	0.17	0.67	0.05	0.05	0.1	0.05	0.05	0.1

The experiment was carried out in triplicate. * Proportion of each EO; ORE: oregano (*Origanum vulgare*) EO; THY: thyme (*Thymus vulgaris*) EO; LG: lemongrass (*Cymbopogon citratus*) EO.

Regarding the EO blends, the interaction between two EOs may be strain-dependent, which could explain our results for *S. aureus*. Furthermore, the similar composition of ORE and THY, with a predominance of phenols followed by hydrocarbons, may have impaired their combined activity since the synergistic or complementary effects are boosted by combining different functional groups [16,47]. According to Gallucci et al. [48], carvacrol and thymol had an antagonistic activity against a Gram-negative (*E. coli*) and a Gram-positive (*S. aureus*) bacteria, corroborating our findings concerning how the combination of $ORE^{0.5}$ and $THY^{0.5}$ has not been significant for the MIC and MBC values of any bacteria tested. On the other hand, according to Kalemba and Kunicka [34], the potential antimicrobial activity from EOs can be ranked in terms of chemical family as follows: phenols > aldehydes > ketones > alcohols > ethers > hydrocarbons, which reinforces our results concerning more antibacterial effectiveness for $ORE^{0.5} + LG^{0.5}$ and $THY^{0.5} + LG^{0.5}$ (Table 2).

The combination of phenols (ORE and THY) and aldehydes (LG) are considered the most active EO functional groups against the bacteria [34,49]. Along with this, the hydrocarbon compounds present in ORE and THY can increase the activity of the aldehydes into LG towards membrane permeability [42,50], allowing better antimicrobial activity. It is worth noting that the antibacterial activity by a combination of LG with ORE or THY has not yet been reported. Nevertheless, an additive and synergistic antibiofilm activity against a Gram-negative bacteria, *Cronobacter sakazakii* (CICC 21544), were observed combining citral with thymol and carvacrol, respectively [51]. The combination observed between aldehydes and phenols corresponds with the major compounds in LG, ORE, and THY (Section 2.1).

Outbreaks with *S. enteritidis*, *E. coli*, and *S. aureus* are very frequent, representing a public health concern in many countries worldwide [52–55], and EOs have been widely studied and reported as being promising and safe antibacterial agents [56]. That provide an alternative against these pathogens in meat, dairy, fruit, and vegetable products and even in drinking water [57,58]. Despite this, there are still neither studies evaluating the action of the combination of oregano, thyme, and lemongrass nor the simultaneous inhibition and inactivation of *S. enteritidis*, *E. coli*, and *S. aureus*. However, some authors also found successful mixtures of other EOs to bacterial and fungal inactivation using a simplex–centroid design. Chraibi et al. [27] observed a synergistic effect between *M. piperita* and *M. pulegium* EOs against *E. coli* (54%/46%), *S. aureus* (56%/44%), and *Candida tropicalis* (55%/45%), and attributed it to a synergy between alcohols and ketones. Likewise, Ouedrhiri et al. [16] reported that a combination of *Origanum compactum* (28%), *Origanum majorana* (30%), and *Thymus serpyllum* (42%) was effective in inactivating *Bacillus subtilis* and *S. aureus*, while for *E. coli* a combination of *O. compactum* (75%) and *O. majorana* (25%) was needed. These authors attributed the effectiveness of the blended EOs to the synergy between alcohols and phenols. The mixture optimization method is still underexplored for the identification and modulation of the antimicrobial activity of EOs. Moreover, the studies with this approach only aim to find the optimal EO ratios for each microorganism without intending simultaneous microbial inactivation.

4. Conclusions

Based on our findings, the antimicrobial efficacy of the studied blends depended on the contribution of each EO in the mixture and the target strains. The most effective EO blends in reducing the minimum inhibitory concentration (MIC), and the minimum bactericidal concentration (MBC) of the bacterial strains tested, were thyme combined with oregano or lemongrass. The ideal mixture for simultaneous inhibition of *S. enteritidis*, *E. coli*, and *S. aureus* was comprised of 75% oregano, 15% thyme, and 10% lemongrass, while inactivation was 50% oregano, 40% thyme, and 10% lemongrass. The EO blends obtained in the present study can be promising alternatives to chemical preservatives against foodborne pathogens besides being used with other technologies, such as modified-release encapsulation systems for food packaging. Furthermore, considering industrial applicability, multiresponse optimization from the desirability function would strongly contribute to ensuring food safety and minimizing public health risks concerning these potential foodborne pathogens simultaneously using natural and green technology.

5. Material and Methods

5.1. Plant Material and Selection of EOs for Study

The EOs of ginger (*Zingiber officinale*), eucalyptus (*Eucalyptus globulus*), oregano (*Origanum vulgare*), thyme (*Thymus vulgaris*), rosemary (*Rosmarinus officinalis*), and lemongrass (*Cymbopogon citratus*) were acquired commercially from Quinari® (Ponta Grossa, PR, Brazil). All EOs were separately submitted to a preliminary test to assess their effectiveness in inhibiting a Gram-positive (*S. aureus*) and a Gram-negative (*E. coli*) bacterium (following Section 2.3). No EO had activity against *S. aureus*; however, oregano, thyme, and lemongrass EOs were the only ones to show inhibition against *E. coli*, and thus these EOs were selected for the present study (data not shown).

5.2. Characterization of the EOs

The composition of EOs was determined using gas chromatography (Agilent 7890A) coupled to mass spectrometry (GC-MS; 5975C mass detector) and an Agilent 7890A gas chromatograph equipped with a flame ionization detector (FID) following the analytical conditions described by Chagas et al. and de Oliveira [59,60]. A 5% diphenyl—95% dimethylpolysiloxane capillary column (DB-5 MS, 30 m × 0.25 mm × 0.25 μm) was used in both chromatography systems. In short, the oven temperature was programmed to rise from 60 °C to 240 °C at 3 °C/minute with helium at 1.0 mL/minute as carrier gas. Furthermore, 1.0 μL (EOs in hexane at 0.1%) was injected at 250 °C. The transfer line was kept at 260 °C, the ion source at 230 °C, and the analyzer at 150 °C. The mass detector was operated in electron ionization mode (70 eV), with 3.15 scans/second, and data were collected in the 40–350 m/z range. For quantification, the samples were injected at 280 °C, using the same column and analytical conditions described above, with hydrogen at 1.5 mL/minute.

5.3. Mixture Design and Statistical Analysis

An augmented simplex–centroid design was used to assess the effect of oregano (ORE), thyme (THY), and lemongrass (LG) EOs in antibacterial activity by Scheffé regression models [25,61,62]. Table 4 shows the experimental design with twelve runs, including three replications (experiments 7, 8, and 9) and additional points (experiments 10, 11, and 12).

The linear, quadratic, and special cubic least-squares regression models were subjected to analysis of variance (ANOVA) and found to understand the best fit of data. Then, the quality of the fitted models was verified based on R^2, $R^2_{adj.}$, and ANOVA. After this preliminary step, it was found that the data fit the quadratic model better. Thus, this model was used to obtain responses of the dependent variables (Y) in the independent ones (X) function, see Equation (1).

$$Y = \alpha_1 X_1 + \alpha_2 X_2 + \alpha_3 X_3 + \alpha_{12} X_{12} + \alpha_{13} X_{13} + \alpha_{23} X_{23} + \varepsilon \quad (1)$$

where Y is the minimum inhibitory concentration (MIC) or minimum bactericidal concentration (MBC) against *E. coli*, *S. enteritidis*, and *S. aureus* expressed in % (w/v); α_1, α_2, α_3 are the estimated parameters of the isolated EOs and α_{12}, α_{13}, α_{23} of the binary mixture; X_1, X_2, and X_3 are the independent variables corresponding to the ratio of ORE, THY, and LG; and ε is an error term. The significance of the estimated coefficients was evaluated through ANOVA with Tukey's post hoc test.

5.4. Antimicrobial Assays

5.4.1. Microorganisms

In order to verify the potential effect of EOs on foodborne pathogens, *E. coli* ATCC 25922, *S. aureus* ATCC 14458, and *S. enteritidis* ATCC 13076 were used in this study. All microorganisms were obtained from the culture bank of the Oswaldo Cruz Foundation (FIOCRUZ, Rio de Janeiro, Brazil), and stored on nutrient agar (Kasvi, Italy) under refrigeration at the Center for Food Analysis (NAL) at the Federal University of Rio de Janeiro, where they were reactivated in 10 mL of brain heart infusion broth (BHI) (Kasvi, Spain) at 37 °C/18–24 h. After, strains were streaked on MacConkey agar (Kasvi, Spain), Baird Parker agar (Kasvi, Spain) supplemented with egg yolk tellurium (Sigma-Aldrich, Germany), and Xylose Lysine Deoxycholate agar (XLD) (Kasvi, Espanha) at 37 °C/18–24 h, respectively. Next, a characteristic colony of *E. coli*, *S. aureus*, and *S. enteritidis* were inoculated in individual tubes containing BHI broth and incubated at 37 °C/18–24 h for subsequent use in the assays described below.

5.4.2. Determination of Minimum Inhibitory Concentration (MIC)

The MIC of EOs and their mixtures were executed through the microdilution method according to the Clinical and Laboratory Standards Institute [63]. For that, an aliquot of the strains in BHI broth was transferred for 5 mL of Mueller Hinton broth (KASVI, Madrid, Spain) and incubated at 37 °C until the turbidity McFarland standard of 0.5 (about 8 log CFU mL^{-1} of each bacterium). The samples were previously diluted in Tween 80 (0.8%; *w*/*v*). In this test, two-fold serial dilutions ranging from 16 to 0.00312% (*w*/*v*) were prepared in Mueller–Hinton broth in a 96-well U-bottom plate. Finally, 10 µL of the suspensions of each bacterium was inoculated into each well. After that, the microplates were then incubated at 37 °C for 24 h. The MIC was determined as the lowest EO concentration to prevent visible growth in each well. The control group was performed with Tween 80 without EO or their mixtures, and as expected, no interferences were observed in the concentration used (0.8% *w*/*v*).

5.4.3. Determination of Minimum Bactericidal Concentration (MBC)

The MBC means the lowest concentration to kill 99.999% of bacteria cells [64]. An aliquot of 50 µL from negative wells (Section 5.4.2) was spread on Plate Count Agar (PCA) (NEOGEN, Heywood, United Kingdom) and then incubated at 37 °C for 24 h. The MBC value corresponded to the lowest concentration of EOs or their mixture when no colony was observed in the culture medium. The MIC and MBC analyses were carried out in triplicate.

5.4.4. Statistical Analysis and Mixture Optimization

The MD approach, regression coefficients, ANOVA, and desirability function (D) were determined using DoE in the Statistica v.9.0 software (Stasoft, Tulsa, OK, USA) [25]. The D function was applied to obtain the optimum EO formulation against *E. coli*, *S. aureus*, and *S. enteritidis* simultaneously [26]. This methodology is based on transforming each response into a dimensionless scale of individual desirability (d_i) where each response ($y_1, y_2, \ldots, y_m$) of the original set is transformed to a range from $0 \le d_i \le 1$. The d_i are then combined using the geometric mean, which gives the overall desirability D (Equation (2)):

$$D = \sqrt[m]{d_1 \times d_2 \times \ldots d_m} \quad (2)$$

where m is the number of responses, and the simultaneous optimization process is reduced to the simple task of the variables' level calculation that maximizes D. A specific response can be maximized, minimized, or assigned a target value. In this study, the minimum function was applied (Equation (3)).

$$d = \begin{cases} 1 \ if \ y_i < L_i \\ \left(\frac{U_i - y_i}{U_i - L_i}\right)^t if \ L_i \le y_i \le U_i \\ 0 \ if \ y_i > U_i \end{cases} \quad (3)$$

where U_i is the maximum acceptable value for a given response, L_i is the lowest allowed value, t is a parameter that expresses the importance of y_i, so that the individual desirability is closer to the minimum in the final result of optimization. The optimal conditions (critical points) were experimentally validated using Tukey and Levene's test.

Supplementary Materials: The following supporting information can be downloaded at: https://www.mdpi.com/article/10.3390/antibiotics11111572/s1, Table S1: Composition of oregano (ORE; *Origanum vulgare*), thyme (THY; *Thymus vulgaris*), and lemongrass (LG; *Cymbopogon citratus*) essential oils through gas chromatography coupled to mass spectrometry (GC-MS) and gas chromatography coupled to flame ionization detector (CG-FID).

Author Contributions: L.T.N.: conceptualization, formal analysis, data curation, writing—original draft. M.L.G.M.: conceptualization, formal analysis, data curation, writing—review and editing. M.A.M.M.: formal analysis, data curation, writing—review and editing. D.G.: formal analysis, data curation, writing—review and editing. C.A.C.J.: funding acquisition, project administration, supervision, and writing—review and editing. All authors have read and agreed to the published version of the manuscript.

Funding: This research was funded by Fundação de Amparo à Pesquisa do Estado do Rio de Janeiro (FAPERJ) Brazil—grant number [E-26/200.891/2021], [E-26/010.000148/2020], and [E-26/201.790/2020]; the Conselho Nacional de Desenvolvimento Científico e Tecnológico (CNPq)—grant number [313119/2020-1], [402215/2022-2] and [200468/2022-7]; and the Coordenação de Aperfeiçoamento Pessoal de Nível Superior (CAPES) Brazil grant number [88887.518752/2020-00].

Institutional Review Board Statement: Not applicable.

Informed Consent Statement: Not applicable.

Data Availability Statement: The data underlying this article will be shared on reasonable request to the corresponding author.

Acknowledgments: The authors are thankful for the financial support provided by the Fundação de Amparo à Pesquisa do Estado do Rio de Janeiro (FAPERJ) Brazil—grant number [E-26/200.891/2021], [E-26/010.000148/2020], and [E-26/201.790/2020]; the Conselho Nacional de Desenvolvimento Científico e Tecnológico (CNPq)—grant number [313119/2020-1], [402215/2022-2] and [200468/2022-7]; and the Coordenação de Aperfeiçoamento Pessoal de Nível Superior (CAPES) Brazil grant number [88887.518752/2020-00].

Conflicts of Interest: The authors declare no conflict of interest and no competing financial interest.

References

1. Pateiro, M.; Munekata, P.E.S.; Sant'Ana, A.S.; Domínguez, R.; Rodríguez-Lázaro, D.; Lorenzo, J.M. Application of essential oils as antimicrobial agents against spoilage and pathogenic microorganisms in meat products. *Int. J. Food Microbiol.* **2021**, *337*, 108966. [CrossRef] [PubMed]
2. WHO. Draft Who Global Strategy for Food Safety 2022–2030. Available online: https://cdn.who.int/media/docs/default-source/food-safety/public-consultation/draft-who-global-strategy-for-food-safety-13may2021.pdf?sfvrsn=ac480bb9_5 (accessed on 30 October 2022).
3. Coimbra, A.; Ferreira, S.; Duarte, A.P. Biological properties of *Thymus zygis* essential oil with emphasis on antimicrobial activity and food application. *Food Chem.* **2022**, *393*, 133370. [CrossRef] [PubMed]
4. WHO. Food Safety. Available online: https://www.who.int/news-room/fact-sheets/detail/food-safety (accessed on 29 October 2022).
5. FDA. Outbreaks of Foodborne Illness. Available online: https://www.fda.gov/food/recalls-outbreaks-emergencies/outbreaks-foodborne-illness#:~:text=Whentwoormorepeople,fromhappeninginthefuture (accessed on 30 October 2022).
6. USDA. Foodborne Illness and Disease. Available online: https://www.fsis.usda.gov/food-safety/foodborne-illness-and-disease#:~:text=What%20Is%20Foodborne%20Illness%3F,comes%20from%20eating%20contaminated%20food (accessed on 4 September 2022).
7. CDC. Staphylococcal (Staph) Food Poisoning. Available online: https://www.cdc.gov/foodsafety/diseases/staphylococcal.html#:~:text=Staph%20food%20poisoning%20is%20characterized,Severe%20illness%20is%20rare (accessed on 4 September 2022).
8. FDA. Get the Facts about Salmonella. Available online: https://www.fda.gov/animal-veterinary/animal-health-literacy/get-facts-about-salmonella (accessed on 30 October 2022).
9. EFSA. Salmonella. Available online: https://www.efsa.europa.eu/en/topics/topic/salmonella (accessed on 30 October 2022).
10. WHO. Estimating the Burden of Foodborne Diseases. Available online: https://www.who.int/activities/estimating-the-burden-of-foodborne-diseases (accessed on 30 October 2022).
11. FDA. *Escherichia coli* (*E. coli*). Available online: https://www.fda.gov/food/foodborne-pathogens/escherichia-coli-e-coli (accessed on 30 October 2022).
12. Galvan, D.; Effting, L.; Torres Neto, L.; Conte-Junior, C.A. An overview of research of essential oils by self-organizing maps: A novel approach for meta-analysis study. *Compr. Rev. Food Sci. Food Saf.* **2021**, *20*, 3136–3163. [CrossRef] [PubMed]
13. Torres Neto, L.; Monteiro, M.L.G.; Galvan, D.; Conte-Junior, C.A. An Evaluation of the Potential of Essential Oils against SARS-CoV-2 from In Silico Studies through the Systematic Review Using a Chemometric Approach. *Pharmaceuticals* **2021**, *14*, 1138. [CrossRef] [PubMed]

14. Tariq, S.; Wani, S.; Rasool, W.; Shafi, K.; Bhat, M.A.; Prabhakar, A.; Shalla, A.H.; Rather, M.A. A comprehensive review of the antibacterial, antifungal and antiviral potential of essential oils and their chemical constituents against drug-resistant microbial pathogens. *Microb. Pathog.* **2019**, *134*, 103580. [CrossRef] [PubMed]
15. Yap, P.S.X.; Yiap, B.C.; Ping, H.C.; Lim, S.H.E. Essential Oils, A New Horizon in Combating Bacterial Antibiotic Resistance. *Open Microbiol. J.* **2014**, *8*, 6–14. [CrossRef]
16. Ouedrhiri, W.; Balouiri, M.; Bouhdid, S.; Moja, S.; Chahdi, F.O.; Taleb, M.; Greche, H. Mixture design of *Origanum compactum*, *Origanum majorana* and *Thymus serpyllum* essential oils: Optimization of their antibacterial effect. *Ind. Crops Prod.* **2016**, *89*, 1–9. [CrossRef]
17. da Silva, B.D.; Bernardes, P.C.; Pinheiro, P.F.; Fantuzzi, E.; Roberto, C.D. Chemical composition, extraction sources and action mechanisms of essential oils: Natural preservative and limitations of use in meat products. *Meat Sci.* **2021**, *176*, 108463. [CrossRef]
18. Falleh, H.; Ben Jemaa, M.; Saada, M.; Ksouri, R. Essential oils: A promising eco-friendly food preservative. *Food Chem.* **2020**, *330*, 127268. [CrossRef]
19. Gutierrez, J.; Barry-Ryan, C.; Bourke, P. Antimicrobial activity of plant essential oils using food model media: Efficacy, synergistic potential and interactions with food components. *Food Microbiol.* **2009**, *26*, 142–150. [CrossRef]
20. Nazer, A.I.; Kobilinsky, A.; Tholozan, J.-L.; Dubois-Brissonnet, F. Combinations of food antimicrobials at low levels to inhibit the growth of Salmonella sv. Typhimurium: A synergistic effect? *Food Microbiol.* **2005**, *22*, 391–398. [CrossRef]
21. Fadil, M.; Fikri-Benbrahim, K.; Rachiq, S.; Ihssane, B.; Lebrazi, S.; Chraibi, M.; Haloui, T.; Farah, A. Combined treatment of *Thymus vulgaris* L., *Rosmarinus officinalis* L. and *Myrtus communis* L. essential oils against Salmonella typhimurium: Optimization of antibacterial activity by mixture design methodology. *Eur. J. Pharm. Biopharm.* **2018**, *126*, 211–220. [CrossRef] [PubMed]
22. Sharma, K.; Guleria, S.; Razdan, V.K.; Babu, V. Synergistic antioxidant and antimicrobial activities of essential oils of some selected medicinal plants in combination and with synthetic compounds. *Ind. Crops Prod.* **2020**, *154*, 112569. [CrossRef]
23. Langeveld, W.T.; Veldhuizen, E.J.A.; Burt, S.A. Synergy between essential oil components and antibiotics: A review. *Crit. Rev. Microbiol.* **2014**, *40*, 76–94. [CrossRef]
24. Maia, E.C.R.; Borsato, D.; Moreira, I.; Spacino, K.R.; Rodrigues, P.R.P.; Gallina, A.L. Study of the biodiesel B100 oxidative stability in mixture with antioxidants. *Fuel Process. Technol.* **2011**, *92*, 1750–1755. [CrossRef]
25. Nunes Filho, R.C.; Galvan, D.; Effting, L.; Terhaag, M.M.; Yamashita, F.; Benassi, M.d.T.; Spinosa, W.A. Effects of adding spices with antioxidants compounds in red ale style craft beer: A simplex-centroid mixture design approach. *Food Chem.* **2021**, *365*, 130478. [CrossRef]
26. Orives, J.R.; Galvan, D.; Coppo, R.L.; Rodrigues, C.H.F.; Angilelli, K.G.; Borsato, D. Multiresponse optimisation on biodiesel obtained through a ternary mixture of vegetable oil and animal fat: Simplex-centroid mixture design application. *Energy Convers. Manag.* **2014**, *79*, 398–404. [CrossRef]
27. Chraibi, M.; Fadil, M.; Farah, A.; Lebrazi, S.; Fikri-Benbrahim, K. Antimicrobial combined action of *Mentha pulegium*, *Ormenis mixta* and *Mentha piperita* essential oils against *S. aureus*, *E. coli* and *C. tropicalis*: Application of mixture design methodology. *LWT* **2021**, *145*, 111352. [CrossRef]
28. Gonçalves, D.C.; Tebaldi de Queiroz, V.; Costa, A.V.; Lima, W.P.; Belan, L.L.; Moraes, W.B.; Pontes Póvoa Iorio, N.L.; Corrêa Póvoa, H.C. Reduction of Fusarium wilt symptoms in tomato seedlings following seed treatment with *Origanum vulgare* L. essential oil and carvacrol. *Crop Prot.* **2021**, *141*, 105487. [CrossRef]
29. Barros, F.A.P.; Radünz, M.; Scariot, M.A.; Camargo, T.M.; Nunes, C.F.P.; de Souza, R.R.; Gilson, I.K.; Hackbart, H.C.S.; Radünz, L.L.; Oliveira, J.V.; et al. Efficacy of encapsulated and non-encapsulated thyme essential oil (*Thymus vulgaris* L.) in the control of Sitophilus zeamais and its effects on the quality of corn grains throughout storage. *Crop Prot.* **2022**, *153*, 105885. [CrossRef]
30. Moazeni, M.; Davari, A.; Shabanzadeh, S.; Akhtari, J.; Saeedi, M.; Mortyeza-Semnani, K.; Abastabar, M.; Nabili, M.; Moghadam, F.H.; Roohi, B.; et al. In Vitro antifungal activity of *Thymus vulgaris* essential oil nanoemulsion. *J. Herb. Med.* **2021**, *28*, 100452. [CrossRef]
31. Ajayi, E.O.; Sadimenko, A.P.; Afolayan, A.J. GC–MS evaluation of *Cymbopogon citratus* (DC) Stapf oil obtained using modified hydrodistillation and microwave extraction methods. *Food Chem.* **2016**, *209*, 262–266. [CrossRef] [PubMed]
32. Aumeeruddy-Elalfi, Z.; Gurib-Fakim, A.; Mahomoodally, M.F. Chemical composition, antimicrobial and antibiotic potentiating activity of essential oils from 10 tropical medicinal plants from Mauritius. *J. Herb. Med.* **2016**, *6*, 88–95. [CrossRef]
33. Kachkoul, R.; Benjelloun Touimi, G.; Bennani, B.; El Habbani, R.; El Mouhri, G.; Mohim, M.; Sqalli Houssaini, T.; Chebaibi, M.; Koulou, A.; Lahrichi, A. The Synergistic Effect of Three Essential Oils against Bacteria Responsible for the Development of Lithiasis Infection: An Optimization by the Mixture Design. *Evid.-Based Complement. Altern. Med.* **2021**, *2021*, 1–17. [CrossRef]
34. Kalemba, D.; Kunicka, A. Antibacterial and Antifungal Properties of Essential Oils. *Curr. Med. Chem.* **2003**, *10*, 813–829. [CrossRef]
35. Lara, V.M.; Carregaro, A.B.; Santurio, D.F.; Sá, M.F.d.; Santurio, J.M.; Alves, S.H. Antimicrobial Susceptibility of *Escherichia coli* Strains Isolated from *Alouatta* spp. Feces to Essential Oils. *Evid.-Based Complement. Altern. Med.* **2016**, *2016*, 1–4. [CrossRef]
36. Boskovic, M.; Djordjevic, J.; Glisic, M.; Ciric, J.; Janjic, J.; Zdravkovic, N.; Krnjaic, D.; Baltic, M.Z. The effect of oregano (*Origanum vulgare*) essential oil on four Salmonella serovars and shelf life of refrigerated pork meat packaged under vacuum and modified atmosphere. *J. Food Process. Preserv.* **2020**, *44*, e14311. [CrossRef]
37. Pesavento, G.; Calonico, C.; Bilia, A.R.; Barnabei, M.; Calesini, F.; Addona, R.; Mencarelli, L.; Carmagnini, L.; Di Martino, M.C.; Lo Nostro, A. Antibacterial activity of Oregano, Rosmarinus and Thymus essential oils against Staphylococcus aureus and Listeria monocytogenes in beef meatballs. *Food Control* **2015**, *54*, 188–199. [CrossRef]

38. Marinelli, L.; Di Stefano, A.; Cacciatore, I. Carvacrol and its derivatives as antibacterial agents. *Phytochem. Rev.* **2018**, *17*, 903–921. [CrossRef]
39. de Carvalho, R.J.; de Souza, G.T.; Honório, V.G.; de Sousa, J.P.; da Conceição, M.L.; Maganani, M.; de Souza, E.L. Comparative inhibitory effects of *Thymus vulgaris* L. essential oil against *Staphylococcus aureus*, *Listeria monocytogenes* and mesophilic starter co-culture in cheese-mimicking models. *Food Microbiol.* **2015**, *52*, 59–65. [CrossRef]
40. Lemos, M.F.; Lemos, M.F.; Pacheco, H.P.; Guimarães, A.C.; Fronza, M.; Endringer, D.C.; Scherer, R. Seasonal variation affects the composition and antibacterial and antioxidant activities of *Thymus vulgaris*. *Ind. Crops Prod.* **2017**, *95*, 543–548. [CrossRef]
41. Dorman, H.J.D.; Deans, S.G. Antimicrobial agents from plants: Antibacterial activity of plant volatile oils. *J. Appl. Microbiol.* **2000**, *88*, 308–316. [CrossRef]
42. Nazzaro, F.; Fratianni, F.; De Martino, L.; Coppola, R.; De Feo, V. Effect of Essential Oils on Pathogenic Bacteria. *Pharmaceuticals* **2013**, *6*, 1451–1474. [CrossRef]
43. Ortega-Ramirez, L.A.; Silva-Espinoza, B.A.; Vargas-Arispuro, I.; Gonzalez-Aguilar, G.A.; Cruz-Valenzuela, M.R.; Nazzaro, F.; Ayala-Zavala, J.F. Combination of Cymbopogon citratus and Allium cepa essential oils increased antibacterial activity in leafy vegetables. *J. Sci. Food Agric.* **2017**, *97*, 2166–2173. [CrossRef]
44. Naik, M.I.; Fomda, B.A.; Jaykumar, E.; Bhat, J.A. Antibacterial activity of lemongrass (*Cymbopogon citratus*) oil against some selected pathogenic bacterias. *Asian Pac. J. Trop. Med.* **2010**, *3*, 535–538. [CrossRef]
45. De Silva, B.C.J.; Jung, W.-G.; Hossain, S.; Wimalasena, S.H.M.P.; Pathirana, H.N.K.S.; Heo, G.-J. Antimicrobial property of lemongrass (*Cymbopogon citratus*) oil against pathogenic bacteria isolated from pet turtles. *Lab. Anim. Res.* **2017**, *33*, 84. [CrossRef]
46. Nakamura, S.; Hatanaka, A. Green-Leaf-Derived C6-Aroma Compounds with Potent Antibacterial Action That Act on Both Gram-Negative and Gram-Positive Bacteria. *J. Agric. Food Chem.* **2002**, *50*, 7639–7644. [CrossRef]
47. Pei, R.; Zhou, F.; Ji, B.; Xu, J. Evaluation of Combined Antibacterial Effects of Eugenol, Cinnamaldehyde, Thymol, and Carvacrol against *E. coli* with an Improved Method. *J. Food Sci.* **2009**, *74*, M379–M383. [CrossRef]
48. Gallucci, M.N.; Oliva, M.; Casero, C.; Dambolena, J.; Luna, A.; Zygadlo, J.; Demo, M. Antimicrobial combined action of terpenes against the food-borne microorganisms Escherichia coli, Staphylococcus aureus and Bacillus cereus. *Flavour Fragr. J.* **2009**, *24*, 348–354. [CrossRef]
49. Nowak, A.; Kalemba, D.; Krala, L.; Piotrowska, M.; Czyzowska, A. The effects of thyme (*Thymus vulgaris*) and rosemary (Rosmarinus officinalis) essential oils on Brochothrix thermosphacta and on the shelf life of beef packaged in high-oxygen modified atmosphere. *Food Microbiol.* **2012**, *32*, 212–216. [CrossRef] [PubMed]
50. Tiwari, B.K.; Valdramidis, V.P.; O' Donnell, C.P.; Muthukumarappan, K.; Bourke, P.; Cullen, P.J. Application of Natural Antimicrobials for Food Preservation. *J. Agric. Food Chem.* **2009**, *57*, 5987–6000. [CrossRef]
51. Cao, Y.; Zhou, D.; Zhang, X.; Xiao, X.; Yu, Y.; Li, X. Synergistic effect of citral and carvacrol and their combination with mild heat against *Cronobacter sakazakii* CICC 21544 in reconstituted infant formula. *LWT* **2021**, *138*, 110617. [CrossRef]
52. CDC. Salmonella Outbreak Linked to Raw Frozen Breaded Stuffed Chicken Products. Available online: https://www.cdc.gov/salmonella/enteritidis-06-21/index.html (accessed on 4 August 2022).
53. CDC. 2019 AR Threats Report. Available online: https://www.cdc.gov/drugresistance/biggest-threats.html (accessed on 4 August 2022).
54. FDA. Outbreak Investigation of *E. coli* O157:H7—Spinach (November 2021). Available online: https://www.fda.gov/food/outbreaks-foodborne-illness/outbreak-investigation-e-coli-o157h7-spinach-november-2021 (accessed on 4 August 2022).
55. Rajakrishnan, S.; Hafiz Ismail, M.Z.; Jamalulail, S.H.; Alias, N.; Ismail, H.; Md Taib, S.; Cheng, L.S.; Zakiman, Z.; Ong, R.; Silverdurai, R.R.; et al. Investigation of a foodborne outbreak at a mass gathering in Petaling District, Selangor, Malaysia. *West. Pacific Surveill. Response J.* **2022**, *13*, 1–5. [CrossRef] [PubMed]
56. Jugreet, B.S.; Suroowan, S.; Rengasamy, R.R.K.; Mahomoodally, M.F. Chemistry, bioactivities, mode of action and industrial applications of essential oils. *Trends Food Sci. Technol.* **2020**, *101*, 89–105. [CrossRef]
57. Sardasht, A.; Tamandani, A.K. Chemical composition of essential oil from *Semenovia suffruticosa* and their antimicrobial's effects in drinking water. *Jordan J. Pharm. Sci.* **2021**, *14*, 37–47.
58. da Silva, B.D.; do Rosário, D.K.A.; Weitz, D.A.; Conte-Junior, C.A. Essential oil nanoemulsions: Properties, development, and application in meat and meat products. *Trends Food Sci. Technol.* **2022**, *121*, 1–13. [CrossRef]
59. Chagas, E.C.; Majolo, C.; Monteiro, P.C.; Oliveira, M.R.d.; Gama, P.E.; Bizzo, H.R.; Chaves, F.C.M. Composition of essential oils of Mentha species and their antimicrobial activity against *Aeromonas* spp. *J. Essent. Oil Res.* **2020**, *32*, 209–215. [CrossRef]
60. De Oliveira, M.I.B.; Brandão, F.R.; Da Silva, M.J.R.; Rosa, M.C.; Farias, C.F.S.; Dos Santos, D.S.; Majolo, C.; Oliveira, M.R.d.; Chaves, F.C.M.; Bizzo, H.R.; et al. In Vitro anthelmintic efficacy of essential oils in the control of *Neoechinorhynchus buttnerae*, an endoparasite of *Colossoma macropomum*. *J. Essent. Oil Res.* **2021**, *33*, 509–522. [CrossRef]
61. Cornell, J.A. *Experiments with Mixtures*; Wiley Series in Probability and Statistics; Wiley: New York, NY, USA, 2002; ISBN 9780471393672.
62. Scheffé, H. The Simplex-Centroid Design for Experiments with Mixtures. *J. R. Stat. Soc. Ser. B* **1963**, *25*, 235–251. [CrossRef]
63. CLSI. *Performance Standards for Antimicrobial Susceptibility Testing*; CLSI: Pennsylvania, PA, USA, 2021.
64. Yang, Z.; He, Q.; Ismail, B.B.; Hu, Y.; Guo, M. Ultrasonication induced nano-emulsification of thyme essential oil: Optimization and antibacterial mechanism against *Escherichia coli*. *Food Control* **2022**, *133*, 108609. [CrossRef]

antibiotics

MDPI

Article

Effect of β-Glucan Supplementation on Growth Performance and Intestinal Epithelium Functions in Weaned Pigs Challenged by Enterotoxigenic *Escherichia coli*

Yuankang Zhou [1,2], Yuheng Luo [1,2], Bing Yu [1,2], Ping Zheng [1,2], Jie Yu [1,2], Zhiqing Huang [1,2], Xiangbing Mao [1,2], Junqiu Luo [1,2], Hui Yan [1,2] and Jun He [1,2,*]

1 Animal Nutrition Research Institute, Sichuan Agricultural University, Chengdu 611130, China; 2020214056@stu.sicau.edu.cn (Y.Z.); luoluo212@126.com (Y.L.); ybingtian@163.com (B.Y.); zpind05@163.com (P.Z.); yujie@sicau.edu.cn (J.Y.); zqhuang@sicau.edu.cn (Z.H.); acatmxb2003@163.com (X.M.); 13910@sicau.edu.cn (J.L.); yan.hui@sicau.edu.cn (H.Y.)

2 Key Laboratory of Animal Disease-Resistance Nutrition, Chengdu 625014, China

* Correspondence: hejun8067@sicau.edu.cn

Abstract: Background: To examine the effect of β-glucan (BGL) supplementation on growth performance and intestinal epithelium functions in weaned pigs upon Enterotoxigenic *Escherichia coli* (ETEC) challenge. Methods: Thirty-two weaned pigs (Duroc × Landrace × Yorkshire) were assigned into four groups. Pigs fed with a basal diet or basal diet containing 500 mg/kg BGL were orally infused with ETEC or culture medium. Results: Results showed BGL tended to increase the average daily gain (ADG) in ETEC-challenged pigs ($0.05 < p < 0.1$). Dietary BGL supplementation had no significant influence on nutrient digestibility ($p > 0.05$). However, BGL improved the serum concentrations of immunoglobulin (Ig) A and IgG, and was beneficial to relieve the increasement of the concentrations of inflammatory cytokines such as the TNF-α and IL-6 upon ETEC-challenge ($p < 0.05$). Interestingly, BGL significantly increased the duodenal, jejunal and ileal villus height, and increased the jejunal ratio of villus height to crypt depth (V/C) upon ETEC challenge ($p < 0.05$). BGL also increased the activities of mucosal, sucrase and maltase in the ETEC-challenged pigs ($p < 0.05$). Moreover, BGL elevated the abundance of *Lactobacillus* and the concentration of propanoic acid in colon in the ETEC-challenged pigs ($p < 0.05$). Importantly, BGL elevated the expression levels of zonula occludins-1 (ZO-1) and mucin-2 (MUC-2) in the small intestinal mucosa upon ETEC challenge ($p < 0.05$). BGL also upregulated the expressions of functional genes such as the claudin-1, cationic amino acid transporter-1 (CAT-1), LAT-1, L amino acid transporter-1 (LAT1), fatty acid transport proteins (FATP1), FATP4, and sodium/glucose cotransporter-1 (SGLT-1) in the duodenum, and the occludin-1 and CAT-1 in the jejunum upon ETEC challenge ($p < 0.05$). Conclusions: These results suggested that BGL can attenuate intestinal damage in weaned pigs upon ETEC challenge, which was connected with the suppressed secretion of inflammatory cytokines and enhanced serum immunoglobulins, as well as improved intestinal epithelium functions and microbiota.

Keywords: β-glucan; intestinal epithelium; inflammation; immunity; weaned pigs

Citation: Zhou, Y.; Luo, Y.; Yu, B.; Zheng, P.; Yu, J.; Huang, Z.; Mao, X.; Luo, J.; Yan, H.; He, J. Effect of β-Glucan Supplementation on Growth Performance and Intestinal Epithelium Functions in Weaned Pigs Challenged by Enterotoxigenic *Escherichia coli*. *Antibiotics* **2022**, *11*, 519. https://doi.org/10.3390/antibiotics11040519

Academic Editor: Helena P. Felgueiras

Received: 15 March 2022
Accepted: 11 April 2022
Published: 13 April 2022

1. Introduction

Weaning is a critical challenge for mammalian animals including pigs. Abrupt changes in diet form and removal of the passive maternal protection in post-weaning piglets increases their susceptibility to diarrhea, resulting in a certain degree of damage to the intestinal structure and mucosal barrier functions [1,2]. ETEC is one of the most critical bacterial causes of intestinal diarrhea [3]. Upon adhering to the intestinal epithelium with their fimbriae, ETEC produces various enterotoxins that can take effect on the small intestinal epithelium and cause the secretion of fluids and electrolytes, which can lead to diarrhea [4]. Over the previous few decades, antibiotics have been extensively used to relieve diarrhea resulting from various bacteria and weaning stress in pig production [5,6]. However, drug residues in the product and the bacterial resistance against classical antibiotics are growing problems all over the world, and it is urgent to develop alternatives to traditionally used antibiotics [7,8].

The most extensive studied alternatives include probiotics, prebiotics, enzymes, acidifiers, and plant extracts [9]. Among the feed additives applied in pigs, prebiotics are regarded as preferable, since they can accelerate competitive exclusion of pathogenic microorganisms and selectively promote the colonization of beneficial microbes [10]. Among the known prebiotics, the mannan-oligosaccharide (MOS) and fructo-oligosaccharide (FOS) have been extensively tested in swine and poultry [11,12]. For instance, both in vivo and in vitro studies showed that in the presence of MOS and FOS, the enteric pathogens do not adhere to the epithelium but to the sugar compounds in the enterocoel, which obviously deceased their colonization and intestinal inflammation [13–15].

β-glucan (BGL) has also been known as an important natural prebiotic that is abundant in lentinan, cereals, and the yeast cell wall [16]. β-(1,3)-glucan can form a triple helix structure, which can facilitate the interaction of molecules and receptors and induce a biological effect. Because of the deficiency of β-glucanase in animals, BGL can escape the enzymatic digestion from the upper gastrointestinal tract and enter the hindgut as a fermentation substrate for microorganisms. Previous studies indicated that BGL can selectively stimulate the growth of beneficial bacteria and help to maintain the intestinal health [17,18]. Moreover, the immunomodulatory effects of BGL have also been well documented [19,20]. For instance, BGL isolated from the yeasts considerably promoted the production of serum cytokines and IgA in hens [21]. It is well known that differences in molecular weight and branching degree of glucan affect the solubility of BGL, which may ultimately affect their immune-modulating properties [22]. As compared to those without branches, BGLs consist of a (1,3)-β-linked backbone with β-(1,6) showed a higher immunomodulatory activity [23]. BGL can also act as a pathogen associated molecular pattern (PAMP), and dectin-1 is the main recognition receptor that is highly expressed in macrophages and dendritic cells [24].

Although numerous studies have indicated a positive role of BGL in regulating the intestinal health and immune functions, data regarding the effect of BGL on the intestinal epithelium functions of the weaned pig exposure to ETEC are scarce. The aim of this study was to examine whether dietary BGL supplementation could relieve ETEC-induced intestinal inflammation and epithelium damage in weaned pigs. This study will also help in the understanding of the mechanisms behind the BGL-regulated intestinal health.

2. Results

2.1. *Effect of BGL on Growth Performance and Nutrients Digestibility in Weaned Pigs upon ETEC Challenge*

As shown in Table 1, no significant differences were observed on ADFI, ADG and F: G among all groups before the ETEC challenge ($p > 0.05$). However, ETEC challenge significant decreased the ADG in the weaned pigs ($p < 0.05$). BGL supplementation relieved the decrement of daily gain (ADG), but there was no significant difference in the ETEC-challenged pigs ($p > 0.05$). In addition, no significant differences were observed on the four treatments for nutrients digestibility (Table 2).

Table 1. Effect of BGL supplementation on ADFI and ADG in weaned pigs upon ETEC challenge.

ITEM	Treatments				SEM	*p*-Value		
	CON	BGL	ECON	EBGL		BGL	ETEC	Interaction
1–25 d								
ADFI (g/d)	427.11	429.16	459.58	459.77	22.47			
ADG (g/d)	280.47	267.2	286.48	302.13	15.61			

Table 1. *Cont.*

ITEM	Treatments				SEM	p-Value		
	CON	BGL	ECON	EBGL		BGL	ETEC	Interaction
F: G	1.58	1.62	1.56	1.55	0.04			
25–28 d								
ADFI (g)	631.68	602.26	537.85	597.22	24.96	0.78	0.35	0.4
ADG (g)	519.9 ^{a}	476.67 ^{ab}	361.33 ^{b}	511.11 ^{ab}	27.9	0.31	0.24	0.08
F: G	1.25	1.3	1.53	1.24	0.07	0.39	0.41	0.22

ADFI, average daily feed intake; ADG, average daily gain; F: G, feed: gain ratio. Mean and total SEM are listed in separate columns ($n = 8$). a, b mean values within a row with unlike superscript letters were significantly different ($p < 0.05$). CON pigs were fed with a basal diet; BGL, pigs were fed with a BGL containing diet (500 mg/kg); ECON pigs were fed with a basal diet and challenged by ETEC; EBGL pigs were fed with a BGL containing diet (500 mg/kg) and challenged by ETEC.

Table 2. Effect of BGL supplementation on nutrients digestibility in weaned pigs.

ITEM	Treatments				SEM	p-Value
	CON	BGL	ECON	EBGL		
DM (%)	88.43	88.69	88.98	89.17	0.40	0.93
CP (%)	86.53	86.49	87.18	87.62	0.68	0.93
EE (%)	84.96	84.89	86.34	86.56	0.58	0.65
Ash (%)	67.65	70.6	69.55	70.91	0.86	0.58
GE (%)	88.56	88.76	89.17	89.27	0.43	0.94

DM, dry matter; CP, crude protein; EE, ether extract; GE, gross energy. Mean and total SEM are list in separate columns ($n = 8$). CON pigs were fed with a basal diet; BGL pigs were fed with a BGL containing diet (500 mg/kg); ECON pigs were fed with a basal diet and challenged by ETEC; EBGL pigs were fed with a BGL containing diet (500 mg/kg) and challenged by ETEC.

2.2. Effect of BGL on Serum Immunoglobulins and Inflammatory Cytokines in Weaned Pigs upon ETEC Challenge

As shown in Figure 1, BGL supplementation increased the concentration of IgA in the non-challenged as well as in the ETEC-challenged pigs ($p < 0.05$). BGL supplementation increased the concentration of IgG in the ETEC-challenged pigs ($p < 0.05$). ETEC enhanced the concentration of IL-1β and IL-6 in the serum ($p < 0.05$). However, BGL supplementation greatly relieved the improvement of the concentration of TNF-α and IL-6 caused by the ETEC challenge ($p < 0.05$).

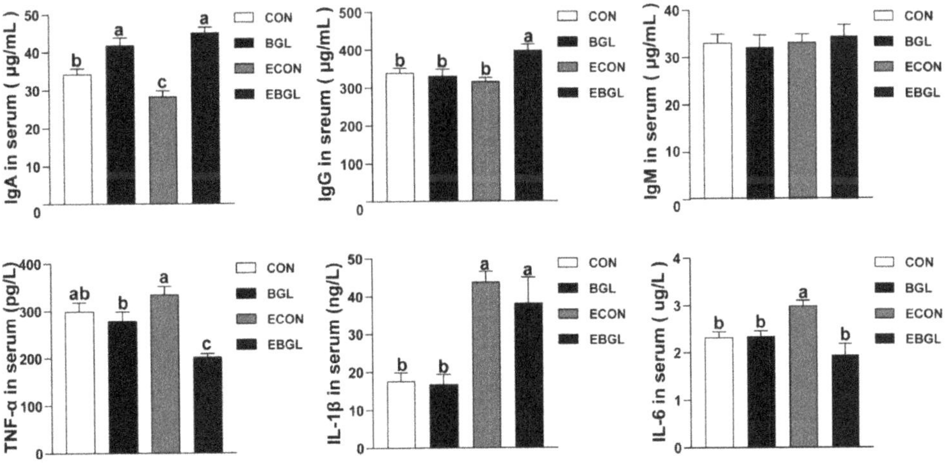

Figure 1. Effect of BGL on serum immunogloblins and inflammatory cytokines in weaned pigs upon ETEC challenge TNF-α, tumor necrosis factor-α; IL-1β, interleukin1-β; IL-6, interleukin-6; IgA,

immunoglobulins A; IgG, immunoglobulins G; IgM, immunoglobulins M. a, b, c mean values within a row with unlike superscript letters were significantly different ($p < 0.05$). CON pigs were fed with a basal diet; BGL pigs were fed with a BGL containing diet (500 mg/kg); ECON pigs were fed with a basal diet and challenged by ETEC; EBGL pigs were fed with a BGL containing diet (500 mg/kg) and challenged by ETEC.

2.3. Effect of BGL Supplementation on Intestinal Morphology and Mucosal Enzyme Activity in Weaned Pigs upon ETEC Challenge

As shown in Table 3 and Figure 2, ETEC challenge acutely reduced the jejunal villus height ($p < 0.01$). Nevertheless, pigs fed with BGL supplementation enormously improved the jejunal villus height and the ratio of V/C upon ETEC challenge ($p < 0.05$). Furthermore, BGL supplementation also enhanced duodenal and ileal villus height upon ETEC challenge ($p < 0.05$). ETEC challenge decreased the activities of maltase, lactase, and sucrase in the jejunum (Table 4). However, BGL supplementation significantly relieved the reduction of their activities upon ETEC challenge ($p < 0.05$). BGL supplementation also enhanced the duodenal and ileal activity of maltase upon ETEC challenge ($p < 0.05$).

Table 3. Effect of BGL supplementation on intestinal morphology in weaned pigs upon ETEC challenge.

ITEM	Treatments				SEM	*p*-Value		
	CON	BGL	ECON	EBGL		BGL	ETEC	Interaction
Duodenum								
Villus height, μm	402.98 ^{ab}	461.37 ^{a}	378.85 ^{b}	454.45 ^{a}	13.79	0.02	0.55	0.74
Crypt depth, μm	135.42	139.42	156.99	154.57	5.4	0.95	0.1	0.78
V:C	3.12 ^{ab}	3.52 ^{a}	2.73 ^{b}	3.19 ^{ab}	0.12	0.07	0.12	0.92
Jejunum								
Villus height, μm	413.89 ^{a}	442.74 ^{a}	299.70 ^{b}	423.62 ^{a}	13.68	<0.01	<0.01	<0.01
Crypt depth, μm	137.96	123.95	144.87	126.52	5.34	0.15	0.67	0.84
V:C	3.11 ^{ab}	3.50 ^{a}	2.62 ^{b}	3.45 ^{a}	0.12	0.01	0.22	0.33
Ileum								
Villus height, μm	318.70 ^{ab}	340.07 ^{a}	265.98 ^{b}	344.75 ^{a}	10.87	0.01	0.21	0.14
Crypt depth, μm	137.96	123.95	144.87	126.52	5.34	0.15	0.67	0.84
V:C	2.42 ^{ab}	3.05 ^{a}	2.23 ^{b}	2.79 ^{ab}	0.14	0.04	0.42	0.9

V:C, Villus height: Crypt depth. Mean and total SEM are listed in separate columns ($n = 8$). a, b mean values within a row with unlike superscript letters were significantly different ($p < 0.05$). CON pigs were fed with a basal diet; BGL pigs were fed with a BGL containing diet (500 mg/kg); ECON pigs were fed with a basal diet and challenged by ETEC; EBGL pigs were fed with a BGL containing diet (500 mg/kg) and challenged by ETEC.

Table 4. Effect of BGL supplementation on mucosal enzyme activity in weaned pigs upon ETEC challenge.

ITEM	Treatments				SEM	*p*-Value		
	CON	BGL	ECON	EBGL		BGL	ETEC	Interaction
Duodenum								
Lactase (U/L)	78.59 ^{a}	66.08 ^{c}	72.92 ^{b}	66.03 ^{c}	1.42	0.001	0.006	0.001
Sucrase (U/L)	255.48 ^{a}	254.47 ^{a}	234.90 ^{b}	264.43 ^{a}	2.76	0.001	0.16	<0.01
Maltase (U/L)	175.56 ^{c}	188.19 ^{b}	166.91 ^{d}	197.63 ^{a}	2.82	0.03	0.88	<0.01
Jejunum								
Lactase (U/L)	107.58 ^{a}	103.48 ^{a}	93.15 ^{b}	102.80 ^{a}	1.77	0.001	0.13	0.14
Sucrase (U/L)	353.32 ^{a}	352.25 ^{a}	328.36 ^{b}	368.95 ^{a}	4.43	0.01	0.56	0.07
Maltase (U/L)	286.45 ^{b}	290.57 ^{ab}	274.63 ^{c}	298.55 ^{a}	2.36	<0.01	0.57	0.007

Table 4. *Cont.*

ITEM	Treatments				SEM	*p*-Value		
	CON	BGL	ECON	EBGL		BGL	ETEC	Interaction
Ileum								
Lactase (U/L)	39.85 ^{a}	40.21 ^{a}	31.15 ^{b}	32.09 ^{b}	1.22	0.72	<0.01	0.87
Sucrase (U/L)	294.08 ^{bc}	315.82 ^{a}	283.07 ^{c}	306.39 ^{ab}	3.27	<0.01	0.03	0.86
Maltase (U/L)	214.18 ^{c}	229.17 ^{b}	202.78 ^{c}	243.26 ^{a}	3.77	<0.01	0.76	0.008

Mean and total SEM are listed in separate columns (n = 8). a, b, c, d mean values within a row with unlike superscript letters were significantly different ($p < 0.05$). CON pigs were fed with a basal diet; BGL pigs were fed with a BGL containing diet (500 mg/kg); ECON pigs were fed with a basal diet and challenged by ETEC; EBGL pigs were fed with a BGL containing diet (500 mg/kg) and challenged by ETEC.

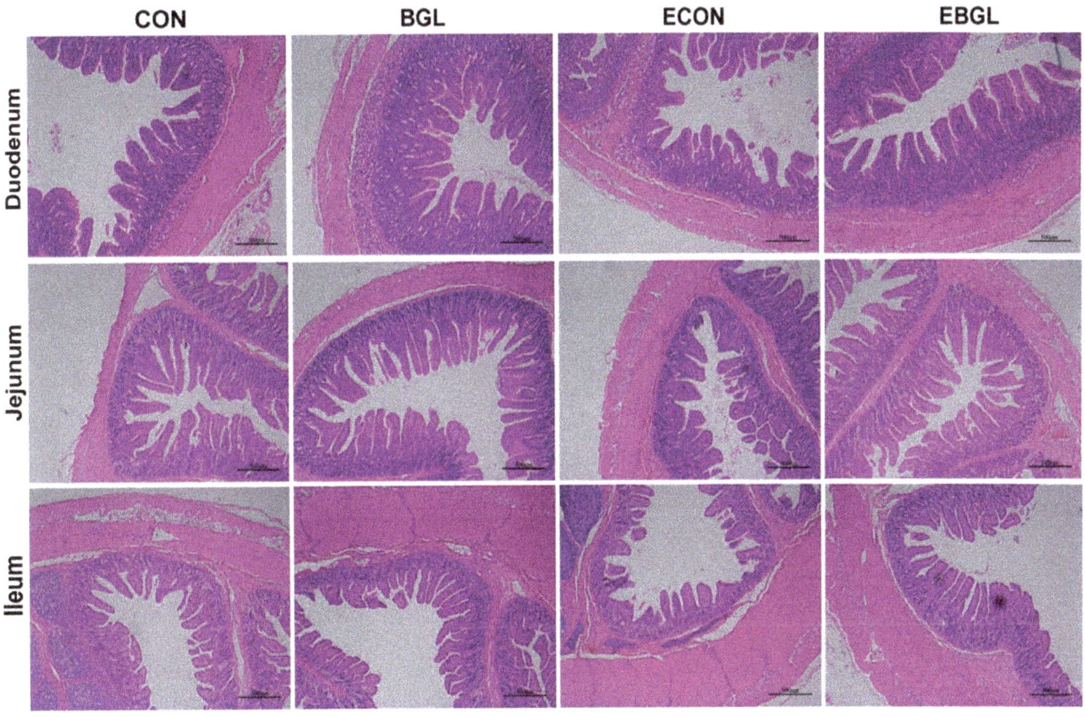

Figure 2. Effect of BGL supplementation on intestinal morphology in weaned pigs upon ETEC challenge (H&E; × 40) CON pigs were fed with a basal diet; BGL pigs were fed with a BGL containing diet, 500 mg/kg; ECON pigs were fed with a basal diet and challenged by ETEC; EBGL pigs were fed with a BGL containing diet (500 mg/kg) and challenged by ETEC. Duodenum, jejunum and ileum in the CON group revealed a normal appearance with regular intestinal villus structure; duodenum, jejunum and ileum in the BGL group, no obvious damages were found. However, in the ECON group, the intestinal lesions were obvious and some intestinal villi were necrotic and shed, or even disappeared, especially in the jejunum; in the duodenum, jejunum and ileum in the EBGL group there were no obvious intestinal lesions and only a few of the intestinal villi were necrotic and shed.

2.4. Effect of BGL Supplementation on Expressions of Critical Genes Involved in Intestinal Epithelium Functions

As shown in Figure 3, the ileal expression levels of ZO-1 obviously decreased upon ETEC challenge ($p < 0.05$). Conversely, BGL supplementation greatly increased the duodenal, jejunal and ileal expression levels in the ETEC-challenged pigs ($p < 0.05$). Pigs fed with BGL improved the duodenal expression levels of claudin-1 and jejunal expression levels of occludin upon ETEC challenge ($p < 0.05$). Moreover, BGL elevated the expression levels of functional genes such as the MUC2, CAT1, LAT1, FATP1, and FATP4 in the duodenum, and increased the jejunal expression levels of MUC2 and CAT1 upon ETEC challenge ($p < 0.05$).

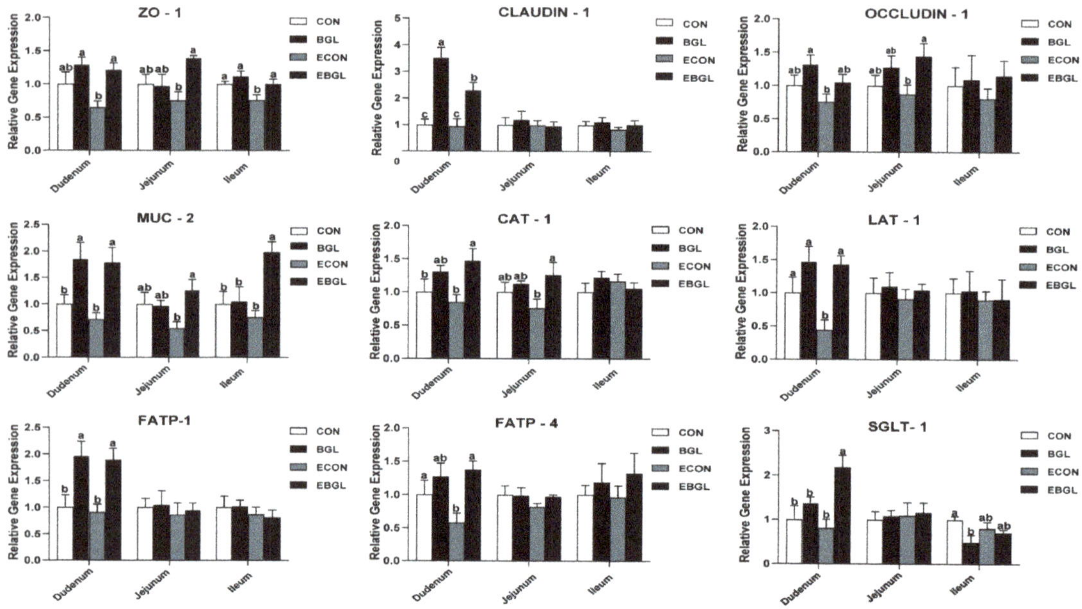

Figure 3. Effect of BGL supplementation on mucosal gene expressions in weaned pigs upon ETEC challenge ZO-1, zonula occludens-1. SGLT-1, sodium/glucose cotransporter-1; CAT-1, cationic amino acid transporter-1; LAT-1, L amino acid transporter-1; FATP, fatty acid transport proteins; a, b, c mean values within a row with unlike superscript letters were significantly different ($p < 0.05$). CON pigs were fed with a basal diet; BGL pigs were fed with a BGL containing diet, 500 mg/kg; ECON pigs were fed with a basal diet and challenged by ETEC; EBGL pigs were fed with a BGL containing diet (500 mg/kg) and challenged by ETEC.

2.5. Effect of BGL Supplementation on Intestinal Microbial Populations in Weaned Pigs upon ETEC Challenge

ETEC challenge enriched the abundance of *Escherichia coli* in cecum ($p < 0.05$). BGL supplementation excellently enhanced the abundance of *Lactobacillus* upon ETEC challenge ($p < 0.05$). BGL supplementation also improved the concentration of propanoic acid in the ETEC-challenged pigs (Table 5).

Table 5. Effect of BGL supplementation on intestinal microbiota and microbial metabolites in weaned pigs upon ETEC challenge.

ITEM	Treatments				SEM	*p*-Value		
	CON	BGL	ECON	EBGL		BGL	ETEC	Interaction
microbial populations (lg(copies/g))								
Total bacteria	11.18	11.03	11.18	11.21	0.05	0.5	0.35	0.34
Escherichia coli	8.35 ^{b}	8.14 ^{b}	9.88 ^{a}	9.50 ^{a}	0.22	0.42	0.001	0.81
Lactobacillus	8.28 ^{ab}	8.75 ^{a}	7.93 ^{b}	8.60 ^{a}	0.1	0.004	0.17	0.58
Bifidobacterium	6.09	6.28	6.2	6.09	0.13	0.88	0.88	0.59
Bacillus	9.16	9.18	9.09	9.03	0.04	0.8	0.18	0.6
VFA (g/g)								
Acetic acid	3.36 ^{ab}	3.65 ^{a}	2.59 ^{b}	3.14 ^{ab}	0.19	0.25	0.09	0.72
Propanoic acid	1.73 ^{ab}	1.74 ^{ab}	1.40 ^{b}	2.00 ^{a}	0.09	0.07	0.83	0.08
Butyric acid	0.78	0.95	0.82	0.78	0.05	0.59	0.6	0.41

VFA, volatile fatty acids. Mean and total SEM are listed in separate columns ($n = 8$). a, b mean values within a row with unlike superscript letters were significantly different ($p < 0.05$). CON pigs were fed with a basal diet; BGL pigs were fed with a BGL containing diet (500 mg/kg); ECON pigs were fed with a basal diet and challenged by ETEC; EBGL pigs were fed with a BGL containing diet (500 mg/kg) and challenged by ETEC.

3. Discussion

As a commercially feasible prebiotic, BGL can escape from enzymatic digestion in the upper intestine, but can be fermented by various microorganisms in the hindgut to produce various short-chain fatty acids that contribute to the maintenance of the intestinal health [25,26]. Importantly, the BGL has long been known as a stimulator of cellular immunity [27]. In this study, the ETEC challenge significantly reduced the ADG in the weaned pigs, but BGL had a tendency to increase the ADG in the ETEC-challenged pigs. The beneficial effects of dietary BGL supplementation on growth performance have also been observed in previous studies using different animal species [28–30].

Immunoglobulins, also known as antibodies, were a class of glycoproteins secreted by plasma cells. They play a key role in the immune response, and they can specifically recognize and bind to particular antigens, such as bacteria or viruses, and assist with their destruction. Currently, five classes (isotypes) of immunoglobulins (IgM, IgG, IgA, IgD, and IgE), were identified according to their heavy chain subunits [31]. Amongst the five classes, IgG is the primary immunoglobulin in serum, which can promote phagocytosis of mononuclear macrophages and neutralize the toxicity of bacterial toxins [32], whereas IgA is the second abundant immunoglobulin in serum [33]. In this study, BGL supplementation maintained a high level of serum IgA and IgG upon ETEC challenge, indicating an immune enhancement in the pigs upon ETEC challenge. As several common pro-inflammatory cytokines, TNF-α, IL-1β and IL-6, are widely considered as the biomarkers of inflammatory response or inflammation [34]. In the present study, the concentrations of them in the serum were elevated upon ETEC challenge, which is consistent with previous studies on pigs [12,15]. However, BGL supplementation greatly relieved the increase of the serum concentration of the TNF-α and IL-6 upon ETEC challenge.

The intestinal tract is the foremost functional site for monogastric animals to digest and absorb nutrients. The intestinal mucosa villus height and crypt depth are popular indexes to characterize intestinal morphology [35]. However, the intestinal morphology can be impaired upon various bacterial or viral infections [36,37]. In this study, the jejunal villus height was significantly reduced upon ETEC challenge, a finding that is in agreement with current studies on pigs [38]. However, BGL supplementation increased the jejunal villus height and the ratio of V/C upon ETEC challenge. Moreover, BGL treatment both elevated the jejunal mucosa liveness of lactase, sucrase, and maltase upon ETEC challenge. The beneficial effect of BGL on gut health may result from its sugar-chain structure, as a wide variety of prebiotics were shown to prevent enteric pathogens (e.g., ETEC) from attaching to the sugar compounds in the intestinal epithelium, which efficaciously cut down their colonization and intestinal inflammation [39–41]. Moreover, the BGL suppressed production of inflammatory factors (for example TNF-α and IL-6) and might also contribute to the improved integrity of the intestinal epithelium upon ETEC challenge, as they were previously reported to induce mucosa atrophy and apoptosis of the intestinal epithelial cells [42].

Tight junctions (TJ) are crucial to retaining the intestinal barrier, and disruption of the intestinal TJ may result in elevated permeability of the intestinal epithelium and entering of harmful substances into the blood [43,44]. In the present study, in pigs challenged with ETEC, the expression levels of the major TJ protein ZO-1 in duodenum and ileum were reduced; however, BGL supplementation significantly increased their expression levels. Moreover, BGL also improved claudin-1 and occludin expression levels in the duodenum and jejunum upon ETEC challenge, respectively. Claudin-1 and occludin are another two critical TJ proteins which are widely expressed in epithelial tissues. A previous study has shown that claudin-1- deficient mice died due to skin wrinkling, severe dehydration and an increase in epidermal permeability [45]. Occludin promotes the formation of reticular TJs and reduces the permeability of ions and macromolecules to enhance the intestinal barrier [46]. The elevation of their expression levels by BGL supplementation indicated an improved integrity of the intestinal barrier upon ETEC challenge. We also explored the expression levels of several key molecules that are associated with the intestinal barrier functions. MUC2 is a major intestinal secretory mucin and acts as a major component of the mucus layer. Importantly, the MUC2 was found to limit the intestinal luminal load by protecting against the accumulation of some less-adherent pathogens (A/E bacteria) [47]. Moreover, the generation of NO is extremely dependent on L-arginine, and the CAT1 is a major L-arginine transmembrane transporter in endothelial cells [48]. In this study, the jejunal expression levels of MUC2 and CAT1 was significantly improved due to BGL treatment after ETEC-challenge. In addition, the duodenal expression levels of LAT1 and FATP4 showed the same result. The LAT1 is one of the major bidirectional transporters of large neutral amino acids, and is responsible for the transportation of Leu into muscle cells [49], whereas the FATP4 is responsible for the uptake of long-chain fatty acids [50]. Elevated expressions of these critical functional genes indicated that BGL treatment is beneficial to enhance intestinal integrity and epithelial functions in weaned piglets under ETEC challenge.

The impact of the intestinal microbiota on health and disease is increasingly emerging. Like with other prebiotics, BGLs are resistant to enzymatic hydrolysis in the small intestine, and thence the majority enters the large intestine in an integral form [51]. In this experiment, BGL supplementation increased the abundance of *Lactobacillus* in the cecum upon ETEC challenge. The *Lactobacillus* is one of the most critical beneficial microorganisms in the gut, which is not only capable of promoting secretion of sIgA from the lamina propria to the surface of the intestinal epithelium, but also capable of activating the macrophages via the TLR2 signaling pathway [52,53]. Prebiotics can be used as substrate for microbial fermentation, causing production of various SCFAs (e.g., acetate, propionate, butyrate) [54]. A previous study has indicated that SCFAs can suppress the ETEC infection in the intestine by reducing virulence gene expression, flagella movement, and colonization [55]. Moreover, the concentration of propanoic acid was elevated by BGL upon ETEC challenge. The increased concentration of propanoic acid was found to increase blood flow and thereby promote the proliferation of intestinal epithelial cells [56]. Moreover, propanoic acid can also enhance the intestinal barrier by increasing the TJs expression levels, such as claudin-1, claudin-8, and occludin [57]. The present experiment was approved a potential mechanism by which the BGL-improved intestinal barrier functions.

4. Materials and Methods

The animal experiment was approved by the Institutional Animal Care and Use Committee of Sichuan Agricultural University (No. 20181105). Glucan product (β-glucan ≥ 60%) was kindly provided by SYNLGHT BIO Co., Ltd. (Guangzhou, China) Pathogenic *Escherichia coli* (ETEC, O149:K91, K88ac) was purchased from the China veterinary culture collection center with a CVCC no. 225 (Beijing, China).

4.1. Animal Diets and Experimental Design

A total of 32 commercial piglets (DLY) weaned at 21 d (with an average body weight of 6.82 ± 0.16 kg) were blocked by weight and allocated as a 2 (BGL) × 2 (ETEC) factorial design to four treatments (n = 8) composed of CON (the basal diet), BGL (500 mg/kg BGL) and ECON (the basal diet and challenged by ETEC), EBGL (500 mg/kg BGL and challenged by ETEC). The basal diet (Table 6) was designed to satisfy the swine nutrient requirements recommended byNRC2012 [58]. Pigs were individually housed in 1.5 × 0.7 m^2 metabolism cages and had free access to feed and water. The temperature was controlled between 27–30 °C and the relative humidity was kept at 65 ± 5%. The experiment lasted for 28 d. The ETEC-challenged pigs were orally infused with 100 mL Luria-Bertani (LB) medium (containing 1 × 10^{10} CFU/mL ETEC) on day 26, and other pigs were infused with an equal volume of LB medium.

Table 6. Experiment basal diet composition and nutrient level.

Ingredients	%	Nutrient Level	Contents
Corn	28.31	Digestible energy (calculated, MJ/kg)	14.78
Extruded corn	24.87	Crude Protein (%)	19.68
Soybean meal	8.5	Calcium (%)	0.81
Extruded full-fat soybean	10.3	Available phosphorus (%)	0.55
Fish meal	4.2	Lysine	1.35
Whey powder	7	Methionine	0.42
Soybean protein concentrate	8	Methionine + cysteine	0.6
Soybean oil	2	Threonine	0.79
Sucrose	4	Tryptophan	0.22
Limestone	0.9		
Dicalcium phosphate	0.5		
NaCl	0.3		
L-Lysine HCl (78%)	0.47		
DL-Methionine	0.15		
L-Threonine (98.5%)	0.13		
Tryptophan (98%)	0.03		
Chloride choline	0.1		
Vitamin premix [1]	0.04		
Mineral premix [2]	0.2		
Total	100		

[1] The vitamin premix provided the following per kg of diet: 9000 IU of VA, 3000 IU of VD 3, 20 IU of VE, 3 mg of VK 3, 1.5 mg of VB1, 4 mg of VB 2, 3 mg of VB6, 0.02 mg of VB12, 30 mg of niacin, 15 mg of pantothenic acid, 0.75 mg of folic acid, and 0.1 mg of biotin. [2] The mineral premix provided the following per kg of diet: 100 mg Fe, 6 mg Cu, 100 mg Zn, 4 mg Mn, 0.30 mg I, 0.3 mg Se. The diet was formulated based on the recommendation of NRC2012.

4.2. Growth Performance Evaluation

Piglets were weighed on day 1, 26, and 29 after 12 h fasting. The average daily gain (ADG), average daily feed intake (ADFI), and feed efficiency (F/G) were calculated by recording the feed intake and body weight gain during the experimental period.

4.3. Sample Collection

Uncontaminated fecal samples were collected straightway after defecation on days 22 to 25, and mixed with 10% H_2SO_4 (10 mL H_2SO_4 per 100 g of fresh fecal) and 1–2 drops of toluene. The four feeds and uncontaminated fecal samples were dried at 65 °C until constant weight, and then stored in powder form of a size that it could pass through a 1-mesh screen. On day 29, blood samples were collected by precaval vein puncture, and serum samples were centrifuged at 3500× *g* for 15 min after standing for 30 min at 4 °C, fecal and serum samples were stored at –20 °C until the next analysis. After blood collection, the pigs were anesthetized by intravenous injection with sodium pentobarbital (200 mg/kg BW), and segments of the duodenum, jejunum and ileum (about 2–4 cm in the middle sections) were isolated and rinsed softly with phosphate buffered saline (PBS), and then fixed in 4% paraformaldehyde solution for intestinal morphological analysis. Moreover, cecal digesta samples were collected into sterile tubes, and the duodenal, jejunal, and ileal mucosa samples were scraped with a glass slide and quick-freeze using liquid N_2, followed by the preservation at −80 °C until further analysis.

4.4. *Apparent Total Tract Nutrient Digestibility Analysis*

Diet and fecal samples were used for the nutrient digestibility analysis, and the acid insoluble ash (AIA) was regarded as an endogenous indicator. The dry matter (DM), crude protein (CP), ether extract (EE), and Ash contents were measured according to AOAC methods [59]. Moreover, the gross energy (GE) was measured by an adiabatic bomb calorimeter (LECO, St. Joseph, MI, USA). All apparent digestibility of nutrients were calculated by the following formula:

$$\text{Apparent digestibility of a nutrient } (\%) = 100 - 100 \times \frac{\text{digesta nutrient } \times \text{ diet AIA}}{\text{diet nutrient } \times \text{ digesta AIA}} \quad (1)$$

4.5. *Serum Proinflammatory Cytokines and Immunoglobulin Detection*

The concentration of proinflammatory cytokines (TNF-α, IL-1β and IL-6) and immunoglobulin (IgG, IgM, and IgA) in serum were determined by Enzyme Linked Immunosorbent Assay (ELISA) kits (Shanghai Meimian Biotechnology Co., Ltd., Shanghai, China). There was less than 10% variation of intra-assay and 12% variation of inter-assay coefficients for each assay.

4.6. *Histomorphology Analysis of Each Intestinal Segment*

The intestinal segment fixed with 4% paraformaldehyde was dewaxed by graded anhydrous ethanol, then stained with hematoxylin and eosin (H&E), dehydrated by graded anhydrous ethanol again, and then sealed with neutral resin. Finally, the crypt depth and villus height of the samples were calculated using an image processing and analysis system (Image-Pro Plus 6.0), and their V/C was calculated. The procedure and statistical method of histomorphology analysis were followed by Wan's. A total of 10 intact, well-oriented villus heights and corresponding crypt depths were obtained per section, and each intestinal segment was analyzed in triplicate [60].

4.7. *Enzyme Activity*

Using the frozen saline as the homogenization medium, the duodenal, jejunal and ileal mucosa were made into 10% homogenate, and then the supernatants were collected after centrifugation at 4000× *g* for 20 min. The lactase, maltase and sucrase were determined by the method of enzyme linked immunosorbent assay (ELISA), and the kits purchased from Shanghai Enzyme-linked Bio-technology Co., Ltd. (Shanghai, China), and kits as follows: lactase (Porcine Lactase ELISA Kit ml712060), sucrase (Porcine Sucrase ELISA Kit ml712026), and maltase (Porcine Maltase ELISA Kit ml712030). All procedures were performed according to the instructions of the kits.

4.8. *Caecal Microbiological Analysis*

About 0.2 g cecum digesta was processed using the Stool DNA Kits (Omega Bio-Tek, Doraville, CA, USA) to total DNA extraction for quantification real-time PCR, which was executed by conventional PCR on the CFX96 Real-Time PCR Detection system (Bio-Rad Laboratories, Hercules, CA, USA). Total bacteria were assessed by the reaction which runs in a total volume of 25 μL with 12.5 μL SYBR Premix Ex Taq (2× concentrated), 1 μL of forward and reverse primers respective (100 nM), 2 μL DNA, and 8.5 μL of RNase-Free ddH_2O. This application program entailed 95 °C for 25 s; followed by 40 cycles of 95 °C for 5 s and 64.5 °C for 25 s; and then a final melting-curve for SYBR Green tests. *Lactobacillus, E. coli, Bacillus and Bifidobacterium* were tested by the SuperReal PreMix (Probe) kit (Tiangen Biotech Co., Ltd., Beijing, China). The reaction run in a total volume of 20 μL with 10 μL 2 × Super Real PreMix (Probe), 0.6 μL of forward and reverse primers (100 nM) respective, 0.4 μL probe (100 nM), 2 μL DNA and 6.4 μL of RNase-Free ddH_2O. All reaction was included in one cycle of pre-denaturation at 95 °C for 15 min; fifty cycles of denaturation at 95 °C for 3 s; annealing and extension at 53 °C for 25 s. The Cycle threshold (Ct) values and baseline settings were determined by automatic analysis settings, and the copy numbers of the target group for each reaction were calculated from the standard curves of plasmid DNA produced by a 10-fold serial dilution of (1×10^1 to 1×10^9 copies/μL).

4.9. Metabolite Concentrations in Cecal Contents

The concentrations of SCFA (acetic acid, propanoic acid, and butyric acid) were determined by a gas chromatograph (VARIAN CP-3800, Varian, Palo Alto, CA, USA; capillary column 30 m × 0.32 mm × 0.25 μm film thickness) according to the previous method [61]. After centrifuging (12,000× *g* for 10 min), the supernatant was mixed with 0.2 mL metaphosphoric acid and 23.3 μL 210 mmol/L crotonic acid in a new tube, and those mixtures were centrifuged in the same conditions again after 30 min incubation at 4 °C. 1 μL of the supernatant were analyzed through the gas chromatograph. The polyethylene glycol column was operated with highly purified N_2 as carrier gas at 1.8 mL/min.

4.10. Isolation and Reverse Transcription of RNA from Intestinal Mucosa and q-PCR

Following the manufacturer's instructions, total RNA was extracted from a bit of duodenal, jejunal and ileal mucosa samples. All mucosa samples were homogenized with 1 mL of RNAiso Plus (Takara Biotechnology Co., Ltd., Dalian, China), and the concentration and fineness of total RNA were measured by a spectrophotometer (NanoDrop 2000, Thermo Fisher Scientific, Inc., Waltham, MA, USA). Approximately 1 μg total RNA was reverse transcribed into cDNA according to the protocol of the PrimeScript™ RT reagent kit with gDNA Eraser (Takara Biotechnology Co., Ltd., Dalian, China). This process was as follows: (1) 37 °C for 15 min, (2): 85 °C for 5 s. The expression level of the target genes was quantified by q-PCR, and the oligonucleotide primer sequences were shown in Table S1. qPCR was performed with the SYBR® Green PCR I PCR reagents (Takara Bio Inc., Dalian, China) using the same PCR system mentioned above. The reaction mixture (total volume of 10 μL) composed of 5 μL SYBR Premix Ex Taq II (Tli RNaseH Plus), 0.4 μL forward primer and reverse primer, 1 μL cDNA and 3.2 μL RNase-Free ddH_2O. The procedure of q-PCR is as follows: 95 °C for 30 s, followed by 40 cycles: at 95 °C for 5 s and 60 °C for 30 s. The mRNA relative expression level of target genes was standardized by the housekeeping gene β-actin, which was calculated based on the $2^{-\Delta\Delta Ct}$ method [62].

4.11. Statistical Analysis

The data was analyzed by two-way ANOVA with the general linear model (GLM) procedure of SPSS as a two (BGL) × 2 (ETEC) factorial design. p-value < 0.05 was considered as significant and the p-value variation from 0.05 to 0.1 was considered as a significant trend. Duncan's multiple range test was used based on the analysis of ANOVA, which showed a significant difference. All data were analyzed by SPSS 27.0 (IBM, Chicago, IL, USA). Results are expressed as means with their standard errors.

5. Conclusions

In conclusion, our results indicated a positive effect of dietary BGL supplementation on the growth performance and intestinal health in the weaned pigs upon ETEC challenge. The mechanisms behind its action may be connected with the suppressed secretion of inflammatory cytokines, improved immunity and intestinal morphology, as well as changes of the microbial fermentation.

Supplementary Materials: The following supporting information can be downloaded at: https://www.mdpi.com/article/10.3390/antibiotics11040519/s1, Table S1: Primers sequences used for quantitative RT-PCR.

Author Contributions: J.H. conceived and designed the experiments. Y.Z. performed animal trial and wrote the manuscript. B.Y., X.M., P.Z., J.Y., H.Y. and Z.H. performed biochemical analysis. J.L., Y.L. and H.Y. gave constructive comments for the results and discussion of the manuscript. All authors have read and agreed to the published version of the manuscript.

Funding: This work was funded by the Key Research and Development Program of Sichuan Province (No. 2020YFN0147) and the National Natural Science Foundation of China (No. 31972599).

Institutional Review Board Statement: The animal experiment was approved by the Institutional Animal Care and Use Committee of Sichuan Agricultural University (No. 20181105).

Informed Consent Statement: Not applicable.

Data Availability Statement: Not applicable.

Acknowledgments: We thank SYNLGHT BIO Co., Ltd. for their generous donation, and Wang Huifen, Wu Fali, and Qiming Duan for their help during the animal trial and sample collections.

Conflicts of Interest: There authors declare that they have no conflict of interest.

References

1. Campbell, J.M.; Crenshaw, J.D.; Polo, J. The biological stress of early weaned piglets. *J. Anim. Sci. Biotechnol.* **2013**, *4*, 19. [CrossRef]
2. Eriksen, E.Ø.; Kudirkiene, E.; Christensen, A.E.; Agerlin, M.V.; Weber, N.R.; Nødtvedt, A.; Nielsen, J.P.; Hartmann, K.T.; Skade, L.; Larsen, L.E.; et al. Post-weaning diarrhea in pigs weaned without medicinal zinc: Risk factors, pathogen dynamics, and association to growth rate. *Porc. Health Manag.* **2021**, *7*, 54. [CrossRef]
3. Fairbrother, J.M.; Nadeau, E.; Gyles, C.L. *Escherichia coli* in postweaning diarrhea in pigs: An update on bacterial types, pathogenesis, and prevention strategies. *Anim. Health Res. Rev.* **2005**, *6*, 17–39. [CrossRef]
4. Madhavan, T.P.V.; Sakellaris, H. Colonization factors of enterotoxigenic *Escherichia coli*. *Adv. Appl. Microbiol.* **2015**, *90*, 155–197. [CrossRef]
5. Jensen, M.L.; Thymann, T.; Cilieborg, M.S.; Lykke, M.; Mølbak, L.; Jensen, B.B.; Schmidt, M.; Kelly, D.; Mulder, I.; Burrin, D.G.; et al. Antibiotics modulate intestinal immunity and prevent necrotizing enterocolitis in preterm neonatal piglets. *Am. J. Physiol. -Gastrointest. Liver Physiol.* **2014**, *306*, G59–G71. [CrossRef]
6. Sato, H.; Zhang, L.S.; Martinez, K.; Chang, E.B.; Yang, Q.; Wang, F.; Howles, P.N.; Hokari, R.; Miura, S.; Tso, P. Antibiotics Suppress Activation of Intestinal Mucosal Mast Cells and Reduce Dietary Lipid Absorption in Sprague-Dawley Rats. *Gastroenterology* **2016**, *151*, 923–932. [CrossRef]
7. Holota, Y.; Dovbynchuk, T.; Kaji, I.; Vareniuk, I.; Dzyubenko, N.; Chervinska, T.; Zakordonets, L.; Stetska, V.; Ostapchenko, L.; Serhiychuk, T.; et al. The long-term consequences of antibiotic therapy: Role of colonic short-chain fatty acids (SCFA) system and intestinal barrier integrity. *PLoS ONE* **2019**, *14*, e0220642. [CrossRef]
8. Oberc, A.M.; Fiebig-Comyn, A.A.; Tsai, C.N.; Elhenawy, W.; Coombes, B.K. Antibiotics Potentiate Adherent-Invasive *E. coli* Infection and Expansion. *Inflamm. Bowel Dis.* **2019**, *25*, 711–721. [CrossRef]
9. Tanaka, T.; Narazaki, M.; Kishimoto, T. Interleukin (IL-6) Immunotherapy. *Cold Spring Harb. Perspect. Biol.* **2018**, *10*, a028456. [CrossRef]
10. Lordan, C.; Thapa, D.; Ross, R.P.; Cotter, P.D. Potential for enriching next-generation health-promoting gut bacteria through prebiotics and other dietary components. *Gut Microbes* **2020**, *11*, 1–20. [CrossRef]
11. Duan, X.; Tian, G.; Chen, D.; Huang, L.; Zhang, D.; Zheng, P.; Mao, X.; Yu, J.; He, J.; Huang, Z.; et al. Mannan oligosaccharide supplementation in diets of sow and (or) their offspring improved immunity and regulated intestinal bacteria in piglet1. *J. Anim. Sci.* **2019**, *97*, 4548–4556. [CrossRef]
12. Liu, L.; Chen, D.; Yu, B.; Yin, H.; Huang, Z.; Luo, Y.; Zheng, P.; Mao, X.; Yu, J.; Luo, J.; et al. Fructooligosaccharides improve growth performance and intestinal epithelium function in weaned pigs exposed to enterotoxigenic *Escherichia coli*. *Food Funct.* **2020**, *11*, 9599–9612. [CrossRef]

13. Froebel, L.K.; Froebel, L.E.; Duong, T. Refined functional carbohydrates reduce adhesion of Salmonella and Campylobacter to poultry epithelial cells in vitro. *Poult. Sci.* **2020**, *99*, 7027–7034. [CrossRef]
14. Adhikari, P.; Cosby, D.E.; Cox, N.A.; Franca, M.S.; Williams, S.M.; Gogal, R.M.; Ritz, C.W.; Kim, W.K. Effect of dietary fructooligosaccharide supplementation on internal organs Salmonella colonization, immune response, ileal morphology, and ileal immunohistochemistry in laying hens challenged with Salmonella enteritidis. *Poult. Sci.* **2018**, *97*, 2525–2533. [CrossRef]
15. Yu, E.; Chen, D.; Yu, B.; Huang, Z.; Mao, X.; Zheng, P.; Luo, Y.; Yin, H.; Yu, J.; Luo, J.; et al. Manno-oligosaccharide attenuates inflammation and intestinal epithelium injury in weaned pigs upon enterotoxigenic *Escherichia coli* K88 challenge. *Br. J. Nutr.* **2021**, *126*, 993–1002. [CrossRef]
16. Du, B.; Meenu, M.; Liu, H.; Xu, B. A Concise Review on the Molecular Structure and Function Relationship of β-Glucan. *Int. J. Mol. Sci.* **2019**, *20*, 4032. [CrossRef]
17. Das, D.; Baruah, R.; Goyal, A. A food additive with prebiotic properties of an α-d-glucan from lactobacillus plantarum DM5. *Int. J. Biol. Macromol.* **2014**, *69*, 20–26. [CrossRef]
18. Wang, H.; Chen, G.; Li, X.; Zheng, F.; Zeng, X. Yeast β-glucan, a potential prebiotic, showed a similar probiotic activity to inulin. *Food Funct.* **2020**, *11*, 10386–10396. [CrossRef]
19. Chae, J.S.; Shin, H.; Song, Y.; Kang, H.; Yeom, C.-H.; Lee, S.; Choi, Y.S. Yeast (1 → 3)-(1 → 6)-β-d-glucan alleviates immunosuppression in gemcitabine-treated mice. *Int. J. Biol. Macromol.* **2019**, *136*, 1169–1175. [CrossRef]
20. Alexander, M.P.; Fiering, S.N.; Ostroff, G.R.; Cramer, R.A.; Mullins, D.W. Beta-glucan-induced inflammatory monocytes mediate antitumor efficacy in the murine lung. *Cancer Immunol. Immunother.* **2018**, *67*, 1731–1742. [CrossRef]
21. Zhen, W.; Shao, Y.; Wu, Y.; Li, L.; van Pham, H.; Abbas, W.; Wan, Z.; Guo, Y.; Wang, Z. Dietary yeast β-glucan supplementation improves eggshell color and fertile eggs hatchability as well as enhances immune functions in breeder laying hens. *Int. J. Biol. Macromol.* **2020**, *159*, 607–621. [CrossRef] [PubMed]
22. Volman, J.J.; Ramakers, J.D.; Plat, J. Dietary modulation of immune function by beta-glucans. *Physiol. Behav.* **2008**, *94*, 276–284. [CrossRef] [PubMed]
23. Lam, K.-L.; Chi-Keung Cheung, P. Non-digestible long chain beta-glucans as novel prebiotics. *Bioact. Carbohydr. Diet. Fibre* **2013**, *2*, 45–64. [CrossRef]
24. Brown, G.D.; Taylor, P.R.; Reid, D.M.; Willment, J.A.; Williams, D.L.; Martinez-Pomares, L.; Wong, S.Y.C.; Gordon, S. Dectin-1 is a major beta-glucan receptor on macrophages. *J. Exp. Med.* **2002**, *196*, 407–412. [CrossRef]
25. El Kaoutari, A.; Armougom, F.; Gordon, J.I.; Raoult, D.; Henrissat, B. The abundance and variety of carbohydrate-active enzymes in the human gut microbiota. *Nat. Rev. Microbiol.* **2013**, *11*, 497–504. [CrossRef] [PubMed]
26. Metzler-Zebeli, B.U.; Zebeli, Q. Cereal β-glucan alters nutrient digestibility and microbial activity in the intestinal tract of pigs, and lower manure ammonia emission: A meta-analysis. *J. Anim. Sci.* **2013**, *91*, 3188–3199. [CrossRef] [PubMed]
27. Vetvicka, V.; Vannucci, L.; Sima, P.; Richter, J. Beta Glucan: Supplement or Drug? From Laboratory to Clinical Trials. *Molecules* **2019**, *24*, 1251. [CrossRef]
28. Abo Ghanima, M.M.; Abd El-Aziz, A.H.; Noreldin, A.E.; Atta, M.S.; Mousa, S.A.; El-Far, A.H. β-glucan administration improves growth performance and gut health in New Zealand White and APRI rabbits with different breed responses. *PLoS ONE* **2020**, *15*, e0234076. [CrossRef]
29. Zhu, M.; Wu, S. The growth performance and nonspecific immunity of loach Paramisgurnus dabryanus as affected by dietary β-1,3-glucan. *Fish Shellfish Immunol.* **2018**, *83*, 368–372. [CrossRef]
30. Teng, P.-Y.; Adhikari, R.; Llamas-Moya, S.; Kim, W.K. Effects of combination of mannan-oligosaccharides and β-glucan on growth performance, intestinal morphology, and immune gene expression in broiler chickens. *Poult. Sci.* **2021**, *100*, 101483. [CrossRef]
31. Markina, Y.V.; Gerasimova, E.V.; Markin, A.M.; Glanz, V.Y.; Wu, W.-K.; Sobenin, I.A.; Orekhov, A.N. Sialylated Immunoglobulins for the Treatment of Immuno-Inflammatory Diseases. *Int. J. Mol. Sci.* **2020**, *21*, 5472. [CrossRef] [PubMed]
32. Bournazos, S.; Ravetch, J.V. Diversification of IgG effector functions. *Int. Immunol.* **2017**, *29*, 303–310. [CrossRef] [PubMed]
33. Schroeder, H.W.; Cavacini, L. Structure and function of immunoglobulins. *J. Allergy Clin. Immunol.* **2010**, *125*, S41–S52. [CrossRef] [PubMed]
34. Ljuca, F.; Gegic, A.; Salkic, N.N.; Pavlovic-Calic, N. Circulating cytokines reflect mucosal inflammatory status in patients with Crohn's disease. *Dig. Dis. Sci.* **2010**, *55*, 2316–2326. [CrossRef] [PubMed]
35. Ji, F.J.; Wang, L.X.; Yang, H.S.; Hu, A.; Yin, Y.L. Review: The roles and functions of glutamine on intestinal health and performance of weaning pigs. *Animal* **2019**, *13*, 2727–2735. [CrossRef]
36. Xia, L.; Yang, Y.; Wang, J.; Jing, Y.; Yang, Q. Impact of TGEV infection on the pig small intestine. *Virol. J.* **2018**, *15*, 102. [CrossRef]
37. Wu, T.; Shi, Y.; Zhang, Y.; Zhang, M.; Zhang, L.; Ma, Z.; Zhao, D.; Wang, L.; Yu, H.; Hou, Y.; et al. Lactobacillus rhamnosus LB1 Alleviates Enterotoxigenic *Escherichia coli*-Induced Adverse Effects in Piglets by Improving Host Immune Response and Anti-Oxidation Stress and Restoring Intestinal Integrity. *Front. Cell. Infect. Microbiol.* **2021**, *11*, 724401. [CrossRef]
38. Kim, S.J.; Kwon, C.H.; Park, B.C.; Lee, C.Y.; Han, J.H. Effects of a lipid-encapsulated zinc oxide dietary supplement, on growth parameters and intestinal morphology in weanling pigs artificially infected with enterotoxigenic *Escherichia coli*. *J. Anim. Sci. Technol.* **2015**, *57*, 4. [CrossRef]
39. Roberts, C.L.; Keita, A.V.; Parsons, B.N.; Prorok-Hamon, M.; Knight, P.; Winstanley, C.; O' Kennedy, N.; Söderholm, J.D.; Rhodes, J.M.; Campbell, B.J. Soluble plantain fibre blocks adhesion and M-cell translocation of intestinal pathogens. *J. Nutr. Biochem.* **2013**, *24*, 97–103. [CrossRef]

40. Shoaf, K.; Mulvey, G.L.; Armstrong, G.D.; Hutkins, R.W. Prebiotic galactooligosaccharides reduce adherence of enteropathogenic *Escherichia coli* to tissue culture cells. *Infect. Immun.* **2006**, *74*, 6920–6928. [CrossRef]
41. Wan, J.; Zhang, J.; Xu, Q.; Yin, H.; Chen, D.; Yu, B.; He, J. Alginate oligosaccharide protects against enterotoxigenic *Escherichia coli*-induced porcine intestinal barrier injury. *Carbohydr. Polym.* **2021**, *270*, 118316. [CrossRef] [PubMed]
42. Stüber, E.; Büschenfeld, A.; von Freier, A.; Arendt, T.; Fölsch, U.R. Intestinal crypt cell apoptosis in murine acute graft versus host disease is mediated by tumour necrosis factor alpha and not by the FasL-Fas interaction: Effect of pentoxifylline on the development of mucosal atrophy. *Gut* **1999**, *45*, 229–235. [CrossRef] [PubMed]
43. Ulluwishewa, D.; Anderson, R.C.; McNabb, W.C.; Moughan, P.J.; Wells, J.M.; Roy, N.C. Regulation of tight junction permeability by intestinal bacteria and dietary components. *J. Nutr.* **2011**, *141*, 769–776. [CrossRef] [PubMed]
44. Suzuki, T. Regulation of intestinal epithelial permeability by tight junctions. *Cell. Mol. Life Sci.* **2013**, *70*, 631–659. [CrossRef] [PubMed]
45. Furuse, M.; Hata, M.; Furuse, K.; Yoshida, Y.; Haratake, A.; Sugitani, Y.; Noda, T.; Kubo, A.; Tsukita, S. Claudin-based tight junctions are crucial for the mammalian epidermal barrier: A lesson from claudin-1-deficient mice. *J. Cell Biol.* **2002**, *156*, 1099–1111. [CrossRef]
46. Saito, A.C.; Higashi, T.; Fukazawa, Y.; Otani, T.; Tauchi, M.; Higashi, A.Y.; Furuse, M.; Chiba, H. Occludin and tricellulin facilitate formation of anastomosing tight-junction strand network to improve barrier function. *Mol. Biol. Cell* **2021**, *32*, 722–738. [CrossRef]
47. Bergstrom, K.S.B.; Kissoon-Singh, V.; Gibson, D.L.; Ma, C.; Montero, M.; Sham, H.P.; Ryz, N.; Huang, T.; Velcich, A.; Finlay, B.B.; et al. Muc2 protects against lethal infectious colitis by disassociating pathogenic and commensal bacteria from the colonic mucosa. *PLoS Pathog.* **2010**, *6*, e1000902. [CrossRef]
48. Rajapakse, N.W.; Nanayakkara, S.; Kaye, D.M. Pathogenesis and treatment of the cardiorenal syndrome: Implications of L-arginine-nitric oxide pathway impairment. *Pharmacol. Ther.* **2015**, *154*, 1–12. [CrossRef]
49. Persaud, A.; Cormerais, Y.; Pouyssegur, J.; Rotin, D. Dynamin inhibitors block activation of mTORC1 by amino acids independently of dynamin. *J. Cell Sci.* **2018**, *131*, jcs211755. [CrossRef]
50. Zhan, T.; Poppelreuther, M.; Ehehalt, R.; Füllekrug, J. Overexpressed FATP1, ACSVL4/FATP4 and ACSL1 increase the cellular fatty acid uptake of 3T3-L1 adipocytes but are localized on intracellular membranes. *PLoS ONE* **2012**, *7*, e45087. [CrossRef]
51. Alessi, A.M.; Gray, V.; Farquharson, F.M.; Flores-López, A.; Shaw, S.; Stead, D.; Wegmann, U.; Shearman, C.; Gasson, M.; Collie-Duguid, E.S.R.; et al. β-Glucan is a major growth substrate for human gut bacteria related to *Coprococcus eutactus*. *Environ. Microbiol.* **2020**, *22*, 2150–2164. [CrossRef] [PubMed]
52. Ya, T.; Zhang, Q.; Chu, F.; Merritt, J.; Bilige, M.; Sun, T.; Du, R.; Zhang, H. Immunological evaluation of *Lactobacillus casei* Zhang: A newly isolated strain from koumiss in Inner Mongolia, China. *BMC Immunol.* **2008**, *9*, 68. [CrossRef] [PubMed]
53. Rocha-Ramírez, L.M.; Hernández-Ochoa, B.; Gómez-Manzo, S.; Marcial-Quino, J.; Cárdenas-Rodríguez, N.; Centeno-Leija, S.; García-Garibay, M. Evaluation of Immunomodulatory Activities of the Heat-Killed Probiotic Strain *Lactobacillus casei* IMAU60214 on Macrophages In Vitro. *Microorganisms* **2020**, *8*, 79. [CrossRef] [PubMed]
54. Ho Do, M.; Seo, Y.S.; Park, H.-Y. Polysaccharides: Bowel health and gut microbiota. *Crit. Rev. Food Sci. Nutr.* **2021**, *61*, 1212–1224. [CrossRef] [PubMed]
55. Lackraj, T.; Kim, J.I.; Tran, S.-L.; Barnett Foster, D.E. Differential modulation of flagella expression in enterohaemorrhagic *Escherichia coli* O157: H7 by intestinal short-chain fatty acid mixes. *Microbiology* **2016**, *162*, 1761–1772. [CrossRef]
56. Mamuad, L.L.; Kim, S.H.; Choi, Y.J.; Soriano, A.P.; Cho, K.K.; Lee, K.; Bae, G.S.; Lee, S.S. Increased propionate concentration in *Lactobacillus mucosae*-fermented wet brewers grains and during *in vitro* rumen fermentation. *J. Appl. Microbiol.* **2017**, *123*, 29–40. [CrossRef]
57. Xia, Z.; Han, Y.; Wang, K.; Guo, S.; Wu, D.; Huang, X.; Li, Z.; Zhu, L. Oral administration of propionic acid during lactation enhances the colonic barrier function. *Lipids Health Dis.* **2017**, *16*, 62. [CrossRef]
58. National Research Council (U.S.). *Nutrient Requirements of Swine*, 11th ed.; National Academies Press: Washington, DC, USA, 2012; ISBN 9780309224239.
59. Horwitz, W. *Official Methods of Analysis of AOAC International*, 18th ed.; Horwitz, W., Latimer, G.W., Eds.; AOAC International: Gaithersburg, MD, USA, 2000.
60. Wan, J.; Zhang, J.; Chen, D.; Yu, B.; Mao, X.; Zheng, P.; Yu, J.; Luo, J.; He, J. Alginate oligosaccharide-induced intestinal morphology, barrier function and epithelium apoptosis modifications have beneficial effects on the growth performance of weaned pigs. *J. Anim. Sci. Biotechnol.* **2018**, *9*, 58. [CrossRef]
61. Franklin, M.A.; Mathew, A.G.; Vickers, J.R.; Clift, R.A. Characterization of microbial populations and volatile fatty acid concentrations in the jejunum, ileum, and cecum of pigs weaned at 17 vs. 24 days of age. *J. Anim. Sci.* **2002**, *80*, 2904–2910. [CrossRef]
62. Fleige, S.; Walf, V.; Huch, S.; Prgomet, C.; Sehm, J.; Pfaffl, M.W. Comparison of relative mRNA quantification models and the impact of RNA integrity in quantitative real-time RT-PCR. *Biotechnol. Lett.* **2006**, *28*, 1601–1613. [CrossRef]

antibiotics

MDPI

Article

Gelidiales Are Not Just Agar—Revealing the Antimicrobial Potential of *Gelidium corneum* for Skin Disorders

Margarida Matias [1], Susete Pinteus [1,*], Alice Martins [1], Joana Silva [1], Celso Alves [1], Teresa Mouga [1], Helena Gaspar [2] and Rui Pedrosa [3,*]

1 MARE—Marine and Environmental Sciences Centre, Polytechnic of Leiria, 2520-630 Peniche, Portugal; maggiemmatias@gmail.com (M.M.); alice.martins@ipleiria.pt (A.M.); joana.m.silva@ipleiria.pt (J.S.); celso.alves@ipleiria.pt (C.A.); mougat@ipleiria.pt (T.M.)
2 BioISI—Biosystems and Integrative Sciences Institute, Faculty of Sciences, University of Lisbon, 1749-016 Lisboa, Portugal; hmgaspar@fc.ul.pt
3 MARE—Marine and Environmental Sciences Centre, ESTM, Polytechnic of Leiria, 2520-630 Peniche, Portugal
* Correspondence: susete.pinteus@ipleiria.pt (S.P.); rui.pedrosa@ipleiria.pt (R.P.); Tel.: +351-262-240-200 (S.P. & R.P.); Fax: +351-262-783-088 (S.P. & R.P.)

Abstract: In recent decades, seaweeds have proven to be an excellent source of bioactive molecules. Presently, the seaweed *Gelidium corneum* is harvested in a small area of the Portuguese coast exclusively for agar extraction. The aim of this work was to fully disclosure *Gelidium corneum* as a sustainable source of antimicrobial ingredients for new dermatological formulations, highlighting its potential to be explored in a circular economy context. For this purpose, after a green sequential extraction, these seaweed fractions (F1–F5) were chemically characterized (^{1}H NMR) and evaluated for their antimicrobial potential against *Staphylococcus aureus*, *Staphylococcus epidermidis* and *Cutibacterium acnes*. The most active fractions were also evaluated for their effects on membrane potential, membrane integrity and DNA damage. Fractions F2 and F3 displayed the best results, with IC_{50} values of 16.1 (7.27–23.02) µg/mL and 51.04 (43.36–59.74) µg/mL against *C. acnes*, respectively, and 53.29 (48.75–57.91) µg/mL and 102.80 (87.15–122.30) µg/mL against *S. epidermidis*, respectively. The antimicrobial effects of both fractions seem to be related to membrane hyperpolarization and DNA damage. This dual mechanism of action may provide therapeutic advantages for the treatment of skin dysbiosis-related diseases.

Keywords: marine natural products; seaweeds; skin microbiota; dermatological applications; antimicrobial activity; Rhodophyta; acne vulgaris; skincare; *Staphylococcus epidermidis*; red algae

Citation: Matias, M.; Pinteus, S.; Martins, A.; Silva, J.; Alves, C.; Mouga, T.; Gaspar, H.; Pedrosa, R. Gelidiales Are Not Just Agar—Revealing the Antimicrobial Potential of *Gelidium corneum* for Skin Disorders. *Antibiotics* **2022**, *11*, 481. https://doi.org/10.3390/antibiotics11040481

Academic Editor: Helena P. Felgueiras

Received: 12 March 2022
Accepted: 1 April 2022
Published: 5 April 2022

Publisher's Note: MDPI stays neutral with regard to jurisdictional claims in published maps and institutional affiliations.

1. Introduction

In recent decades, marine organisms have proven to be an excellent source of bioactive molecules with a wide range of applications. However, few examples have reached the industrial sector, particularly due to the limitations of biomass availability, solvent suitability, low yields of extraction, and specific legal requirements, among others. The seaweed *Gelidium corneum* (former *Gelidium sesquipedale*) is a Rhodophyta belonging to the order Gelidiales, and several species belonging to this order, including *Gelidium corneum*, have the characteristic of being rich in agarans, highly valued in the food industry [1,2]. Presently, this seaweed is harvested in a small area of the Portuguese coast São Martinho do Porto for agar extraction. However, *Gelidium* species are rich in bioactive compounds such as mycosporine-like amino acids, flavonoids, pigments, phycobiliproteins, fatty acids, etc., with relevant biotechnological potential [3–6]. Within a circular economy approach it is highly relevant to understand the full potential of this biomass, to take the maximum advantage of one resource for the development of multiple new products, enhancing economic revenues and, consequently, boosting the local economy.

Within this framework, the present study targeted the red seaweed *Gelidium corneum* as a source of bioactive ingredients with antimicrobial activity with relevance for inclusion in new dermatological formulations.

Skin is the most effective barrier of the human body against external aggressions. Covered by a highly complex microbiome composed of bacteria, fungi and viruses, this living barrier is an effective defence against invading pathogens and heavily contributes to modulating the immune system [7]. Some of the most representative species of skin microbes include *Cutibacterium acnes* (former *Propionibacterium acnes*), *Staphylococcus* spp., *Streptococcus* spp., *Corynebacterium* spp., and *Malassezia* spp., which are distributed according to their specific affinity for a determined microenvironment—moist, dry or sebaceous [7]. While in normal conditions these microorganisms are fundamental for a healthy skin barrier, changes in their normal balance (dysbiosis) may lead to the development of skin pathologies such as acne vulgaris, dermatitis, eczema, and chronic wounds, among others [7]. Even though the skin's microbiome is composed of thousands of different microorganisms, recent studies have focussed on the relationships between two commensal bacteria, *Staphylococcus epidermidis* and *Cutibacterium acnes* [8,9]. The dysbiosis between these two organisms results in a series of complications for the human host. Oily skin or excessive sebum production by the oil glands, may boost the growth of *C. acnes* that, combined with a decreased *S. epidermidis* population, may lead to a skin pathological condition known as acne vulgaris [10,11]. Acne vulgaris is a chronic inflammation of the skin, beginning in the pilosebaceous unit. Not only it is triggered by the increase in sebum production, but also by the hypercornification of the pilosebaceous glands due to the hyperproliferation of keratinocytes on the upper part of the follicle.

The homeostasis of the skin microbiota is considered to be the objective of any acne treatment [10,12]. As a consequence of microbial imbalance, the treatments available are mostly provided by topical antibiotic prescription; however, severe cases require oral treatment. The most common antibiotics for acne topical treatment are tetracycline, clindamycin, and erythromycin, sometimes combined with benzoyl peroxide and zinc acetate [13–15]. A prescription of oral antibiotics is mostly avoided due to the possibility of increased antibiotic resistance. However, *C. acnes* infections can also occur in other organs due to wounds and, in these cases, there is not any other treatment option.

When the balance of these two commensals is disrupted there can also occur an excess of *S. epidermidis* growth, which may result in nosocomial infections [16]. Although *S. epidermidis* rarely lead to severe life-threatening diseases, their infections are extremely common and difficult to treat. This microorganism is frequently involved in vascular graft, prosthetic joint, and cardiac device infections, among others, being the second infectious driver after *Staphylococcus aureus* [16].

The skin microbiome has a vast influence on individuals' wellbeing, thus, finding natural derivatives that can not only deter pathogens but also maintain a microbial balance by targeting key microbes is critical [7].

Knowing that marine seaweeds are a propelling source of new chemical structures with an array of bioactivities, including antimicrobial activity [17,18], this work aims to disclose the antimicrobial potential of *G. corneum*-derived components against microorganisms that are frequently associated with skin disorders, namely *Staphylococcus aureus*, *Staphylococcus epidermidis* and *Cutibacterium acnes*.

2. Results

2.1. Chemical Screening by ^{1}H NMR

The chemical profile of the fractions obtained from *Gelidium corneum* was evaluated by ^{1}H NMR and the corresponding spectra are depicted in Figure 1.

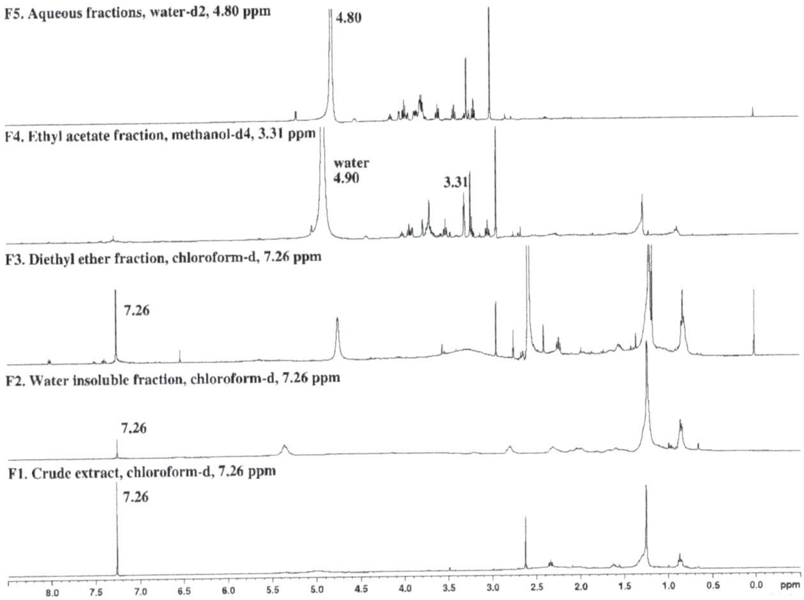

Figure 1. ^{1}H NMR (400 MHz) spectra of *Gelidium corneum* fractions (F1–F5).

Crude extract (F1) evidenced signals in the region of 0.8–2.8 ppm, characteristic of less polar lipophilic compounds such as fatty acids, sterols, terpenes, pigments, and other lipids [19,20]. The presence of these classes of compounds was also evidenced in the insoluble fraction retained in the filter (F2), with the intensity of signals being more pronounced in this fraction due to their higher concentration. More intense proton signals in the region of 0.8–3.1 ppm were also observed in diethyl ether fraction (F3), also suggesting the richness of lipophilic compounds in this fraction. Fraction F3 also evidenced low-intensity signals of aromatic protons (7.5–8.2 ppm), which can be attributed to phenolic compounds. The signals of these compounds are also visible in the spectrum of ethyl acetate fraction (F4), although with less intensity. In this fraction, the most intense signals were observed at 2.7–3.9 ppm, which can be attributed to the presence of amino compounds such as amino acids and proteins [20,21]. However, signals between 3.2 and 4.1 ppm were observed in fractions F4 and F5, which can denote the presence of alcohols, sugars, and esters [20]. Signals between 3.0 and 4.4 ppm are characteristic of the ring hydrogens of polysaccharides [22], supporting the richness of the aqueous fraction (F5) in this group of molecules.

2.2. Antimicrobial Activity of Gelidium corneum Fractions

The antimicrobial potential of *G. corneum* fractions was tested against *C. acnes*, *S. epidermidis* and *S. aureus* growth. The results are shown in Figure 2.

Only the water-insoluble fraction (F2) and diethyl ether fraction (F3) inhibited *C. acnes* and *S. epidermidis* growth by more than 50%. In the case of *S. aureus*, only the F3 fraction promoted inhibition (≈25%). Since fractions F2 and F3 showed high inhibitory activity, a dose–response analysis was conducted and the IC_{50} determined. The results are shown in Table 1.

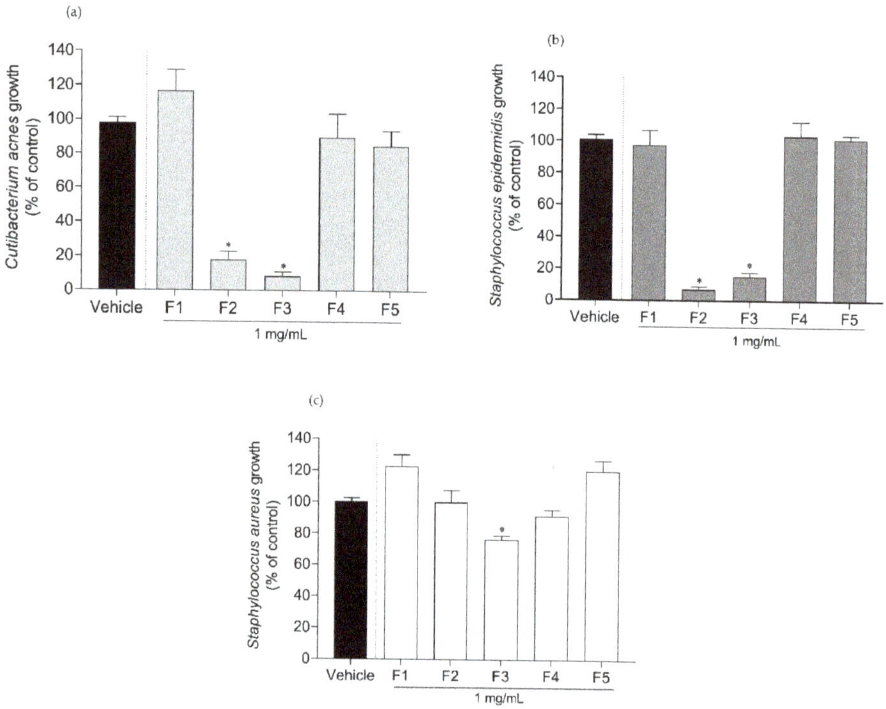

Figure 2. Antimicrobial activity of *Gelidium corneum* fractions (1000 μg/mL) against (**a**) *Cutibacterium acnes*, (**b**) *Staphylococcus aureus*, and (**c**) *Staphylococcus epidermidis*. Values in each column represent the mean ± SEM of three independent experiments carried out in triplicate. Symbol (*) represents significant differences when compared to vehicle (ANOVA, Dunnett's test, $p < 0.05$).

Table 1. Dose–response analysis of *Gelidium corneum* F2 (water insoluble) and F3 (diethyl ether) fractions against *Staphylococcus epidermidis* and *Cutibacterium acnes*.

Fraction	*Staphylococcus epidermidis*	*Cutibacterium acnes*
	IC_{50} (μg/mL)	
F2	53.29 (48.75–57.91)	16.10 (7.27–23.02)
F3	102.80 (87.15–122.30)	51.04 (43.36–59.74)
Oxytetracycline	12.40 (11.22–16.13)	0.07 (0.05–0.09)

Gelidium corneum F2 and F3 fractions exhibited IC_{50} values of 53.29 (48.75–57.91) μg/mL and 102.8 (87.15–122.30) μg/mL, respectively, against *S. epidermidis*. Concerning *C. acnes*, fraction F2 revealed the highest potency with an IC_{50} of 16.10 (7.27–23.02) μg/mL, followed by fraction F3 with an IC_{50} of 51.04 (43.36–59.74) μg/mL.

2.3. Effects of *Gelidium corneum* Fractions on *Cutibacterium acnes* and *Staphylococcus epidermidis* Membrane Potential

Several assays were conducted to understand the possible mechanisms underlying the F2 and F3 fractions' antimicrobial effects. The effects on *C. acnes* and *S. epidermidis* membrane potential are depicted in Figures 3 and 4, respectively.

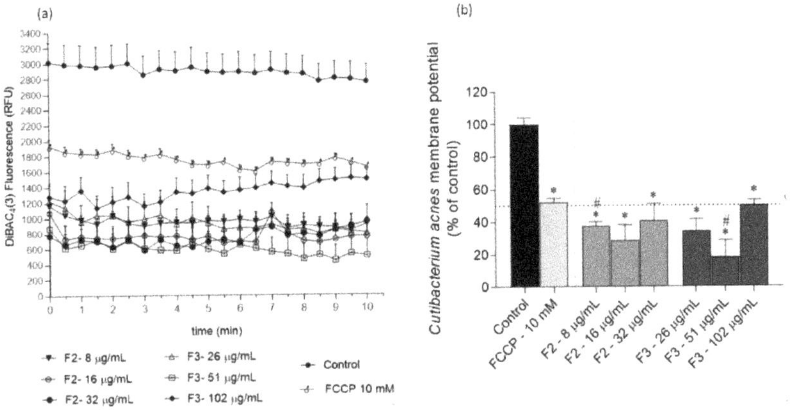

Figure 3. *Cutibacterium acnes* membrane potential when exposed to *Gelidium corneum* F2 (water insoluble) and F3 (diethyl ether) fractions at $\frac{1}{2}$ IC_{50}; IC_{50} and 2 × IC_{50}, labelled with $DiBAC_4(3)$ probe; (**a**) 30 s interval readings; (**b**) after 10 min. FCCP (10 mM) was used as positive control and DMSO as negative control. Each value and bars represent the average of three independent experiments. Vertical lines represent the standard error of the mean (SEM). Symbol (*) represents significant differences when compared to the control. Symbol ($^{\#}$) represents significant differences when compared to FCCP (ANOVA, Dunnett's test, $p < 0.05$).

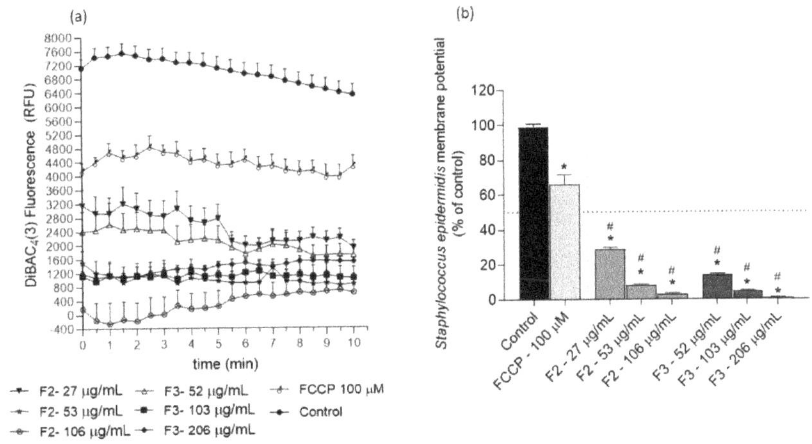

Figure 4. *Staphylococcus epidermidis* membrane potential when exposed to *Gelidium corneum* F2 (water insoluble) and F3 (diethyl ether) fractions at $\frac{1}{2}$ IC_{50}; IC_{50} and 2 × IC_{50}, labelled with $DiBAC_4(3)$ probe; (**a**) 30 s interval readings; (**b**) after 10 min. FCCP (100 µM) was used as positive control and DMSO as negative control. Each value and bars represent the average of three independent experiments. Vertical lines represent the standard error of the mean (SEM). Symbol (*) represents significant differences when compared to the control. Symbol ($^{\#}$) represents significant differences when compared to FCCP (ANOVA, Dunnett's test, $p < 0.05$).

The results obtained with the $DiBAC_4(3)$ method suggest that F2 and F3 fractions promote a membrane hyperpolarization in *C. acnes* and *S. epidermidis*, which is noticeable due to the lower fluorescence emission when compared with the control. Additionally, the fractions exhibited a more marked effect than the positive control FCCP in both bacteria. In *C. acnes*, both fractions decreased the membrane potential in more than 50%, except for F3 at 2 × IC_{50} after 10 min (Figure 3b). This profile was similar against *S. epidermidis*; however, the hyperpolarization was more pronounced, with a reduction in membrane polarization ranging from ≈70% (fraction F2 at $\frac{1}{2}$ IC_{50}) to ≈100% reduction (fraction F3 at 2 × IC_{50}).

2.4. Effects of *Gelidium corneum* Fractions on *Cutibacterium acnes* and *Staphylococcus epidermidis* Membrane Integrity

To further understand if fractions could impair bacterial growth by affecting the membrane integrity, Sytox Green—a green-fluorescent nuclear probe—was used. This probe is impermeant to cells; however, when the membrane is disrupted, this probe links to nucleic acids emitting fluorescence, thus being an extremely useful tool to monitor membrane damage (Figure 5).

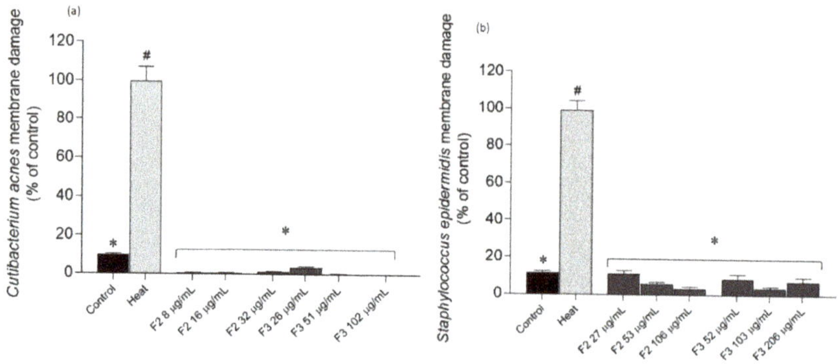

Figure 5. *Cutibacterium acnes* (**a**) and *Staphylococcus epidermidis* (**b**) membrane integrity when exposed to *Gelidium corneum* F2 and F3 fractions at $\frac{1}{2}$ IC_{50}; IC_{50} and 2 × IC_{50}, labelled with Sytox Green probe. Bacterial cells exposed to a heat treatment were used as a positive control. DMSO was used as negative control. Each value represents the average of three independent experiments. Bars represent the standard error of the mean (SEM). Symbols represent significant differences when compared to the heat-treated cells (*) and (#) to control (ANOVA, Dunnett's test, $p < 0.05$).

After an incubation with the most active fractions, it was possible to verify that they did not promote membrane damage, since there were no significant differences in relation to the untreated control.

2.5. Potential of *Gelidium corneum* Fractions to Promote DNA Damage

Since different mechanisms can be behind the antimicrobial effects, the ability of the most active fractions to link and damage DNA was evaluated. The results are shown in Figure 6.

In Figure 6, it is possible to observe a clear smear of the DNA bars in the *G. corneum* F2 and F3 (lanes 7 and 8) when compared to the DMSO controls. Additionally, in lane 8 (F3), both linear and supercoiled DNA have a lower intensity, suggestive of DNA degradation. Ciprofloxacin promoted extensive DNA damage at 30 µg/mL and, at 10 µg/mL, the degradation is particularly evident in the supercoiled DNA (lane 4).

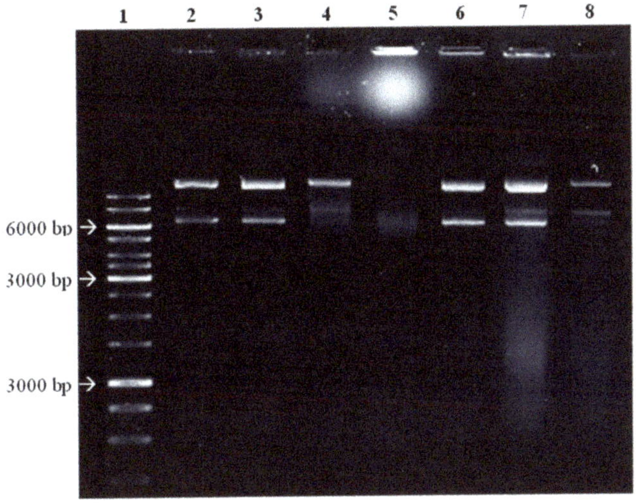

Figure 6. Electrophoresis gel. 1—Marker; 2—DNA; 3—negative control (DMSO); 4—10 µg/mL ciprofloxacin; 5—30 µg/mL ciprofloxacin; 6—negative control (DMSO); 7—F2 (water insoluble) fraction (1 mg/mL); 8—F3 (diethyl ether) fraction (1 mg/mL). Agarose at 0.8%, ran at 85 V for 1 h. Image was obtained through gel imaging system (Gel doc). This figure is representative of three independent experiments.

3. Discussion

In the present work, *Gelidium corneum* was evaluated regarding its potential as a source of antimicrobial compounds through a green extraction approach, aiming to develop new therapeutic alternatives for dysbiosis-related skin disorders. It was verified that two fractions—F2 (water insoluble fraction) and F3 (diethyl ether fraction)—were particularly active against *S. epidermidis* and *C. acnes*.

One of the main causes behind dermatology appointments is acne vulgaris. This disease is the most common skin problem worldwide, occurring in adulthood in about 50% of the population in occidental countries [23–25]. On the other hand, *S. epidermidis* can be an important opportunistic pathogen, with treatment options posing a challenge. Most infections caused by this microorganism start with the introduction of skin bacteria during the insertion of a medical device into the patient [16]. When looking at therapeutical options targeting both microorganisms, the conventionally prescribed topical and oral antibiotics can have some serious side effects, such as disrupting gut health and increasing skin dryness, impacting patients' quality of life. On the other hand, even when the side effects are minimal, bacterial resistance might still occur [26]. This makes seeking new antibiotic options urgent. There are several natural products that could be possible candidates for novel drugs targeting multiple pathogenic factors [27]. Although in the present work the antimicrobial activity against *S. aureus* was low, previous studies with other *Gelidium* species have shown the capacity to deter the growth of *S. aureus* [28–31]. Additionally, a review gathered by Pérez, Falqué and Domínguez [31] cited *G. attenatum*, *G. micropterum*, *G. pulchellum*, *G. pusillum*, *G. robustum*, and *G. spinosum* with antimicrobial activity against different bacteria, including *Escherichia coli*, *Klebsiella pneumoniae*, *Vibrio* spp. and *Enterococcus faecalis*. However, no studies were found with *Gelidium* species against *C. acnes* or *S. epidermidis*. Choi et al. [32] reported 15 seaweed species with antimicrobial activity against *C. acnes*, with the red seaweed *Symphyocladia latiuscula* displaying the highest inhibitory potential (MIC = 160 µg/mL). Comparatively, the results obtained with F2 and F3 fractions from *G. corneum* are quite exciting due to their high potency, particularly against *C. acnes*; thus, they are promising candidates for use in new formulations aiming to treat skin diseases, where *C. acnes* and *S. epidermidis* play a key role. Scientific evidence suggests that *S. epidermidis* has a suppressive effect against *C. acnes*, and thus extracts with a higher antimicrobial effect against *C. acnes* than against *S. epidermidis* may bring additional advantages for acne vulgaris therapeutics [33]. This evidence highlights the relevance of the results obtained with *G. corneum* F2 and F3 fractions.

Antibiotics can work in a synergic way to effectively decrease bacterial resistance. Yet, their mechanism and range of action should be complementary. This is already exploited specifically in acne vulgaris treatment with the combination of different oral antibiotics and benzoyl peroxide [15]. To better understand the antibacterial effect, it is necessary to evaluate the possible mechanisms of action which can be underlying the bacterial growth inhibition. Since fractions F2 and F3 revealed promising results, the antimicrobial effects due to membrane potential disruption, membrane rupture, and/or DNA damage were evaluated.

Membrane potential is central to bacterial development; therefore, disruptions to it might induce an antimicrobial effect [34]. Across the cellular membrane, there is an electrochemical potential that is involved in several functions of bacterial cells, such as intra- and intercellular signalling mediation, which in turn regulates important physiological processes, namely mechano-sensation, spore formation, and biofilm dynamics [35–38]. In addition to its role in bioelectrical signalling, membrane potential is also central to cellular proliferation since it provides the essential driving force for ATP synthesis [39], which is crucial for cell division [40]. Although this potential may fluctuate depending on cells' physiologies, abrupt changes may lead to cellular death. In particular, several studies have shown that a hyperpolarized membrane is associated with bacterial death [34,41]. It was verified that the most active extracts (F2 and F3) promoted a hyperpolarization of bacteria membrane potential (Figures 3 and 4). Although studies surrounding the mechanisms of action of seaweed derivatives are scarce, Patra et al. [42] evaluated the mechanisms underlying the antimicrobial effects of essential oils extracted from *Ulva linza* (former *Enteromorpha linza)* against *Escherichia coli* and verified that the antimicrobial effects were related to changes in the membrane potential. Concerning phytochemical sources, Wu et al. [43] evaluated the antimicrobial effects of a natural compound (2*R*,3*R*-dihydromyricetin) obtained from the pine needles of *Cedrus deodara* against *S. aureus* and verified that this compound promoted a high membrane hyperpolarization, resulting in bactericidal effects. Although the results shown in the present work suggest that *G. corneum* fractions did not promote membrane damage, other seaweed species have shown antimicrobial activity mediated by membrane rupture. Patra and Baek [44] evaluated the antimicrobial mechanisms of oil extracted from the seaweed *U. linza* against *Bacillus cereus* and *S. aureus* and verified that the antimicrobial effects were related to membrane injury. Additionally, El Shafay et al. [45] evaluated the antimicrobial activity of *Sargassum* species against *S. aureus* and *Klebsiella pneumoniae* and concluded that the antimicrobial activity was mediated by cells' membranes rupture and physical distortions. A work conducted with the green seaweed *U. linza* showed that essential oil from this seaweed promotes membrane damage in the foodborne pathogen *Escherichia coli* [44].

It has been suggested that antimicrobial compounds can reach DNA through membranes, with or without membrane rupture [46,47]. Thus, although it was not possible to verify membrane damage, a simple test was conducted to understand if the seaweed fractions could impact DNA. It was shown that both fractions F2 and F3 can promote DNA damage. In accordance with these results, Pinteus et al. [48] also showed that extracts derived from the red seaweed *Asparagopsis armata* had potential to promote DNA damage. Additionally, seaweed polysaccharides have shown antimicrobial activity mediated by DNA damage [49]. Other natural products from various works have also shown the ability to target DNA. Subramanian et al. [50] verified that resveratrol inhibited the growth of *E. coli*, probably by inducing DNA damage. Da et al. [51] evaluated the antimicrobial activity of an extract obtained from *Scutellaria baicalensis* root and verified that it had antimicrobial activity against fungi, possible mediated by DNA damage. Antimicrobial peptides extracted from different sources have also shown potential to target bacteria and fungi

DNA [52]. He et al. [53] isolated a novel polysaccharide from *Streptomyces virginia* and verified that this compound exhibited antimicrobial activity against several microorganisms, possibly mediated by DNA damage.

Fractions 2 and 3 from *Gelidium corneum* are rich in lipophilic compounds, as evidenced by their ^{1}H NMR spectra. Lipophilic compounds from different algae species have shown to display antimicrobial effects [31,49,54–56] and within this group of compounds, different terpenes sourced from red algae show great potential. Rodrigues et al. [57] studied the antimicrobial activity of terpenes isolated from *Sphaerococcus coronopifolius* and one sphaerane bromoditerpene had a great effect on *S. aureus* with an IC_{50} of 6.35 μM. Xu et al. [55] reported that tetracyclic brominated diterpenes from the same algae exhibited a bactericidal effect with MIC values of 16 and 128 μg/mL against multi-resistant *Staphylococcus aureus*. Amongst lipophilic compounds, fatty acids from different species of seaweeds have a reported antimicrobial effect in several microorganisms. For example, *Dunaliella salina* fatty acids are described for their effect over *Listeria monocytogenes* and *B. cereus* with a MIC value of 2.5 mg/mL, and over *Salmonella enteritidis* with a MIC of 1.25 mg/mL [22]. There are no studies regarding compounds isolated from red algae affecting the growth of *S. epidermidis* and *C. acnes*, highlighting the importance of *G. corneum* as a source of compounds with antimicrobial effects against these bacteria. The chemical characterization here presented constitutes the first approach concerning the evaluation of *G. corneum* ingredients for dermatological applications, also validating the effectiveness of the fractionation methodology here reported. Yet, it is important to proceed with a deeper chemical characterization to identify the compound(s) responsible for the observed antimicrobial activities acting individually and/or synergistically.

4. Materials and Methods

4.1. Seaweed Harvest and Sampling

Gelidium corneum (Hudson) J.V.Lamouroux 1813 was collected on October 2020 at Praia dos Barcos (39°22′35.9″ N 9°20′23.7″ W), Peniche, Portugal (identification number: GC.PB.2020-10). It was identified by Professor Teresa Mouga, a botanical expert, and immediately transported to the laboratory facilities (MARE-Polytechnic of Leiria). Samples were washed, firstly with sea water and then with distilled water, to remove invertebrate organisms and debris, and dried at 70 °C with air circulation (Universal laboratory oven UF450, Memmert, Buchenbach, Germany). Dried samples were ground (Moulinex Food processor, Paris, France) and stored in a dry dark place until extraction processing.

4.2. Extraction Procedure

To obtain different fractions from *G. corneum*, a sequential extraction methodology was performed (Figure 1). Solvents were selected according to EU Regulation No. 1223/2009 for cosmetic application and were obtained from VWR-BDH Chemicals (Fontenay-sous-Bois, France). Powdered seaweed (100 g) was extracted with ethanol: water (70:30) under constant stirring (150 rpm), over 17 h, protected from light. The hydroethanolic solution was posteriorly filtered (qualitative filter paper (FP) nr. 4, VWR International, Alcabideche, Portugal) and concentrated under vacuum at low temperature (<40 °C) in a rotary evaporator. In total, 20 g of the dried crude extract (F1) was resuspended in Milli-Q water (Advantage A10 Milli-Q lab, Merck, Darmstadt, Germany) previously warmed to 75 °C and filtered (FP nr. 4), affording a solid insoluble fraction (F2) and an aqueous fraction. After cooling to room temperature (r.t.), this last one was subjected to a liquid–liquid partition, firstly with diethyl ether (F3) and then, with ethyl acetate (F4), which were evaporated until dryness in a rotary evaporator. The remaining aqueous fraction (F5) was also evaporated until a half volume was obtained, and then it was frozen (−20 °C) and lyophilized. The extraction procedure flowchart is depicted in Figure 7.

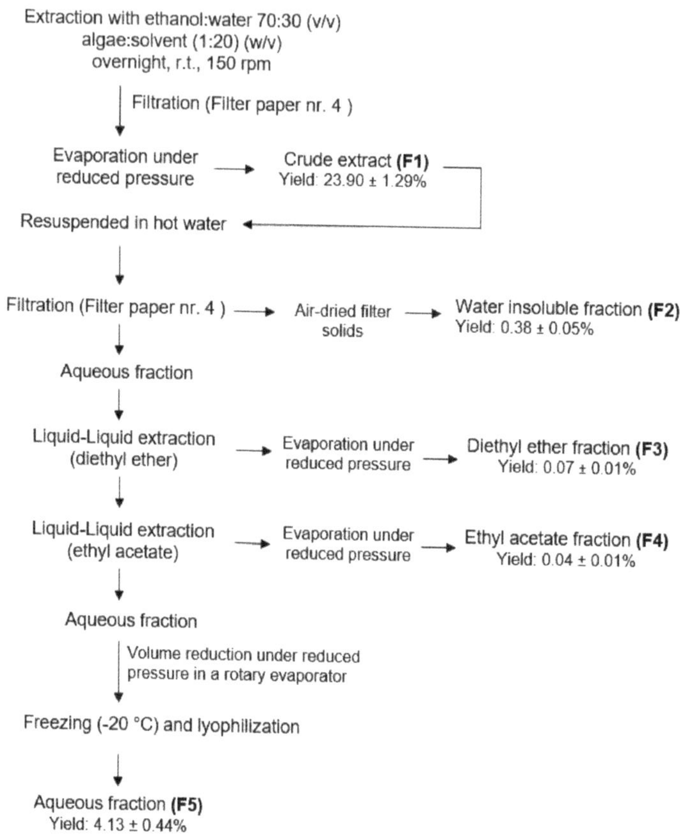

Figure 7. Extraction methodology to obtain *Gelidium corneum* fractions (F1–F5).

4.3. Chemical Screening by ^1H NMR

A preliminary chemical screening of all fractions (F1–F5) obtained from *Gelidium corneum* was attained by proton nuclear magnetic resonance (^1H NMR) spectroscopy. Samples (c.a. 5–6 mg) were dissolved in 0.5 mL of deuterated solvents ($CDCl_3$, MeOD, or D_2O) (Sigma-Aldrich, St. Louis, MO, USA) and the ^1H NMR spectra were recorded at 400.13 MHz on a Bruker AMX400 spectrometer (Bremen, Germany), at 25 °C. Chemical shifts (δ) are expressed in ppm and referenced to the residual solvent signal (δ_H 7.26 $CDCl_3$, δ_H 3.31 MeOD, δ_H 4.79 D_2O).

4.4. Antimicrobial Activity of Gelidium corneum Fractions against Microorganisms Associated with Skin Disorders

The antimicrobial activity of *G. corneum* fractions was evaluated against three microorganisms that belong to the skin's natural microflora, namely, three Gram (+) bacteria, *Staphylococcus epidermidis* (DSM: 1798), *Staphylococcus aureus* (ATCC: 25923) and *Cutibacterium acnes* (DSM: 1897) obtained from the German Collection of Microorganisms and Cell Cultures (DSMZ). The growth conditions and media used for each microorganism were: trypticase soy broth (VWR-BDH Chemicals-Prolabo, Leuven, Belgium) + 0.3 % (*w*/*w*) of yeast extract (VWR-BDH Chemicals-Prolabo, Leuven, Belgium) at 37 °C for *S. epidermidis*; tryptic soy broth (VWR-BDH Chemicals-Prolabo, Leuven, Belgium) in anaerobic conditions at 37 °C for *C. acnes*, and lysogeny broth (LB) medium (VWR-BDH Chemicals-Prolabo, Leuven, Belgium) at 37 °C for *S. aureus*. Fractions were suspended in dimethyl sulfoxide (DMSO) at 100 mg/mL, and the assays were conducted on 96-well microplates with the fractions at a maximum concentration of 1000 µg/mL. Antimicrobial activity was evaluated in the exponential phase of each microorganism. Microorganisms' growth was accompanied spectrophotometrically at 600 nm, and the results were expressed as a percentage of growth inhibition relative to the control (growth medium with microorganism and vehicle (DMSO)). For the fractions that affected the microorganisms' growth by more than 50%, a dose–response was conducted (10–1000 µg/mL) and the IC_{50} was determined. Oxytetracycline was used as a positive control.

4.5. *Mechanisms of Action Underlying the Antimicrobial Activity*

To understand the mechanisms that could be underlying the antimicrobial effects, further tests were conducted with the fractions that exhibited the highest activity (inhibition >50% at 1 mg/mL).

4.5.1. Membrane Potential Analysis

The membrane potential variation assay was adapted from Clementi et al. [58], by combining the potentiometric dye bis-(1,3-dibutylbarbituric acid) trimethine oxonol ($DiBAC_4(3)$) (Thermo Fisher Scientific, Waltham, MA, USA) with the DNA-staining dye propidium iodide (PI) (Sigma Aldrich, Darmstadt, Germany). The extracts with the highest activity were tested at $\frac{1}{2}$ IC_{50}, IC_{50} and 2 × IC_{50}.

Freshly grown bacteria were centrifuged at 1000× *g* and washed twice with 1× PBS (Sigma-Aldrich, Darmstadt, Germany). The pellet was resuspended in 10 mL PBS buffer at 1.5 MacFarland, and 0.25% (*v*/*v*) of a stock glucose solution (1 M) was added to the mix. A blank was prepared with the same proportions but lacking the microbial solution. The microbial solution was then incubated at 37 °C for 30 min. A solution of carbonyl cyanide 4-(trifluoromethoxy) phenylhydrazone (FCCP) (Sigma-Aldrich, Darmstadt, Germany) was used as the positive control for hyperpolarization. A positive control for total membrane permeabilization was conducted with heat-treated (100 °C, 10 min) bacteria.

After incubation, 45 µL of $DiBAC_4(3)$ (25 µM) + 90 µL of PI (1 mg/mL) was added to the microbial solution, distributed through 96-well black plates (99 µL per well), and then incubated in the dark at 37 °C, over one hour. The samples were quickly added (1 µL per well) and the fluorescence read at 490/516 nm excitation/emission (DIBAC) and 535/617 nm (PI) in 30 s intervals for 10 min (Multimodal Synergy H1, BioTek® Instruments, Winooski, VT, USA). The results are presented in $DiBAC_4(3)$ relative fluorescence units (RFU).

4.5.2. Membrane Damage Assay

Membrane damage analysis was performed according to Pinteus et al. [48], as follows: A freshly overnight grown culture was centrifuged (4000 rpm, 5 min) and resuspended in sterile saline solution (0.85%) at an optical density of Abs_{600} = 0.5. Seaweed fractions were added at a concentration of $\frac{1}{2}$ IC_{50}, IC_{50} and 2 × IC_{50}, and incubated over 4 h at 37 °C for *S. epidermidis*, and 6 h for *C. acnes*. DMSO was used as a negative control. Blanks were prepared with samples without microorganism. A positive control was prepared with a thermic treatment (100 °C, 10 min) to induce total membrane permeability. All suspensions were transferred to a black microplate and incubated with 2 µM Sytox Green (Thermo Fisher Scientific, Waltham, MA, USA) for 10 min in the microplate reader. The resultant fluorescence of the DNA-bound dye was quantified on a fluorescence microplate reader (Multimodal Synergy H1, BioTek® Instruments, Winooski, VT, USA). The membrane damage was determined in % of control (DMSO).

4.5.3. DNA Damaging Potential

DNA damaging potential was assessed following the methodology described by Hu et al. [59]. Plasmid (pGADT7—7987 bp) DNA (5 µL; 100 ng) was mixed with seaweed fractions (2 µL; 10 mg/mL) and ultrapure water (13 µL). The reaction mixture was incubated at 37 °C for 1 h before being loaded onto a 0.8% agarose gel containing 1% RedSafe™.

Gene Ruler 1 Kb DNA Ladder (5 µL) was also loaded onto the gel. Electrophoresis was then performed for 60 min under 85 V. DMSO was used in the same conditions as a negative control and ciprofloxacin (10 and 30 µg/mL) was used as positive control.

4.6. Data and Statistical Analysis

To determine possible significant differences relative to the control, a one-way analysis of variance (ANOVA) with Dunnett's multiple comparison test was used. All data were checked for normality (Shapiro–Wilk test) and homoscedasticity (Levene´s test). When requirements for an ANOVA were not met, a non-parametric Kruskal–Wallis and Dunn's multiple comparison test were applied. The IC_{50} values were determined using the GraphPad v9.3.1 software through the equation $y = 100/(1 + 10^{(X - Log\, IC_{50})})$. Calculations were carried out and final graphical representations were made using GraphPad v9.3.1 (GraphPad Software, La Jolla, CA, USA). All data were obtained from at least three independent experiments carried out in triplicate and are presented as standard error of the mean (SEM), with a significance level of 0.05 ($p < 0.05$).

5. Conclusions

With the increasing awareness on the importance of sustainable strategies for the development of new nature-based health and wellness products, this work addressed, for the first time, the potential of *Gelidium corneum* as a source of antimicrobial compounds for skin dysbiosis-related diseases. This seaweed is particularly interesting since it is already explored industrially for agar extraction. Thus, within a circular economy concept, it would be of high economic relevance to define a biorefinery concept to extract both antimicrobial compounds and agar, thus taking the maximum advantage of the same natural resource for maximum environmental and economic sustainability.

The results reveal that *G. corneum* contains compounds with high antimicrobial activity against two important skin opportunistic pathogens, *C. acnes* and *S. epidermidis*. The compounds concentrated in F2 (water insoluble fraction) and F3 (diethyl ether fraction) seem to affect both microorganisms' growth by inducing changes in membrane polarization and by binding to DNA, promoting damage. This dual mechanism of action may provide therapeutic advantages for the treatment of skin dysbiosis-related diseases.

Author Contributions: Conceptualization, S.P., C.A., J.S., A.M. and R.P.; methodology, M.M., S.P., J.S., C.A., A.M. and H.G.; validation, S.P., C.A., A.M., T.M. and R.P.; formal analysis, M.M., S.P., C.A., A.M. and H.G.; investigation, M.M., S.P., J.S., C.A., A.M., T.M. and H.G.; resources, S.P., J.S., C.A., A.M., H.G. and R.P.; writing—original draft preparation, M.M., S.P., A.M. and C.A.; writing—review and editing, all authors; supervision, S.P., C.A., J.S., A.M. and R.P.; project administration, R.P; funding acquisition, S.P., C.A., J.S., A.M. and R.P. All authors have read and agreed to the published version of the manuscript.

Funding: This work was supported by the Portuguese Foundation for Science and Technology (FCT) through the POCI project—01-0247-FEDER-070155—ORCHESTRA—add value to ORChards through thE full valoriSaTion of macroalgae and by FEDER—European Regional Development Fund of the European Union, under the Portugal 2020 Programme, through COMPETE 2020—Competitiveness and Internationalization Operational Program. Additionally, we were also supported through the FCT projects: UIDP/04292/2020 and UIDB/04292/2020 granted to MARE—Marine and Environmental Sciences Centre; UIDP/Multi/04046/2020 and UIDB/04046/2020 granted to BioISI—Biosystems and Integrative Sciences Institute; CROSS-ATLANTIC (PTDC/BIA-OUT/29250/2017), co-financed by COMPETE (POCI-01-0145-FEDER-029250); and Molecules for Health (PTDC/BIA-BQM/28355/2017); National NMR Network (PTNMR), partially supported by Infrastructure Project N° 022161 (co-financed by FEDER through COMPETE 2020, POCI and PORL and FCT through PIDDAC). Additional support was provided by the Project Operational Programme MAR2020 through the project MAR-04.03.01-FEAMP-0101–Limo do Cais.

Institutional Review Board Statement: Not applicable.

Informed Consent Statement: Not applicable.

Data Availability Statement: Not applicable.

Conflicts of Interest: The authors declare no conflict of interest.

References

1. Williams, P.A.; Phillips, G.O. 11—The use of hydrocolloids to improve food texture. In *Texture in Food*; McKenna, B.M., Ed.; Woodhead Publishing: Cambridge, UK, 2003; pp. 251–274. ISBN 185573673X. [CrossRef]
2. Jiao, G.; Yu, G.; Zhang, J.; Ewart, H.S. Chemical Structures and Bioactivities of Sulfated Polysaccharides from Marine Algae. *Mar. Drugs* **2011**, *9*, 196–223. [CrossRef] [PubMed]
3. El-Din, S.M.M.; Alagawany, N.I. Phytochemical Constituents and Anticoagulation Property of Marine Algae *Gelidium crinale*, *Sargassum hornschuchii* and *Ulva linza*. *Thalass. Int. J. Mar. Sci.* **2019**, *35*, 381–397. [CrossRef]
4. Öztürk, B.Y.; Gürsu, B.Y.; Dağ, I. Antibiofilm and antimicrobial activities of green synthesized silver nanoparticles using marine red algae *Gelidium corneum*. *Process Biochem.* **2019**, *89*, 208–219. [CrossRef]
5. Martínez-Sanz, M.; Gómez-Mascaraque, L.G.; Ballester, A.R.; Martinez-Abad, A.; Brodkorb, A.; López-Rubio, A. Production of unpurified agar-based extracts from red seaweed *Gelidium sesquipedale* by means of simplified extraction protocols. *Algal Res.* **2019**, *38*, 101420. [CrossRef]
6. Castejón, N.; Parailloux, M.; Izdebska, A.; Lobinski, R.; Fernandes, S.C.M. Valorization of the Red Algae *Gelidium sesquipedale* by Extracting a Broad Spectrum of Minor Compounds Using Green Approaches. *Mar. Drugs* **2021**, *19*, 574. [CrossRef] [PubMed]
7. Byrd, A.L.; Belkaid, Y.; Segre, J.A. The human skin microbiome. *Nat. Rev. Microbiol.* **2018**, *16*, 143–155. [CrossRef]
8. Dagnelie, M.; Corvec, S.; Timon-David, E.; Khammari, A.; Dréno, B. *Cutibacterium acnes* and *Staphylococcus epidermidis*: The unmissable modulators of skin inflammatory response. *Exp. Dermatol.* **2021**, *31*, 406–412. [CrossRef] [PubMed]
9. Fournière, M.; Latire, T.; Souak, D.; Feuilloley, M.G.J.; Bedoux, G. *Staphylococcus epidermidis* and *Cutibacterium acnes*: Two Major Sentinels of Skin Microbiota and the Influence of Cosmetics. *Microorganisms* **2020**, *8*, 1752. [CrossRef]
10. Claudel, J.-P.; Auffret, N.; Leccia, M.-T.; Poli, F.; Corvec, S.; Dréno, B. Staphylococcus epidermidis: A Potential New Player in the Physiopathology of Acne? *Dermatology* **2019**, *235*, 287–294. [CrossRef] [PubMed]
11. Sfriso, R.; Egert, M.; Gempeler, M.; Voegeli, R.; Campiche, R. Revealing the secret life of skin—With the microbiome you never walk alone. *Int. J. Cosmet. Sci.* **2019**, *42*, 116–126. [CrossRef]
12. Williams, H.C.; Dellavalle, R.; Garner, S. Acne vulgaris. *Lancet* **2011**, *379*, 361–372. [CrossRef]
13. Blaskovich, M.A.T.; Elliott, A.G.; Kavanagh, A.M.; Ramu, S.; Cooper, M.A. In vitro Antimicrobial Activity of Acne Drugs Against Skin-Associated Bacteria. *Sci. Rep.* **2019**, *9*, 14658. [CrossRef] [PubMed]
14. Dréno, B.; Pécastaings, S.; Corvec, S.; Veraldi, S.; Khammari, A.; Roques, C. *Cutibacterium acnes* (*Propionibacterium acnes*) and acne vulgaris: A brief look at the latest updates. *J. Eur. Acad. Dermatol. Venereol.* **2018**, *32* (Suppl. S2), 5–14. [CrossRef] [PubMed]
15. Zaenglein, A.L.; Pathy, A.L.; Schlosser, B.J.; Alikhan, A.; Baldwin, H.E.; Berson, D.S.; Bowe, W.P.; Graber, E.M.; Harper, J.C.; Kang, S.; et al. Guidelines of care for the management of acne vulgaris. *J. Am. Acad. Dermatol.* **2016**, *74*, 945–973.e33. [CrossRef] [PubMed]
16. Otto, M. *Staphylococcus epidermidis*—The 'accidental' pathogen. *Nat. Rev. Genet.* **2009**, *7*, 555–567. [CrossRef] [PubMed]
17. Costantino, V.; Fattorusso, E.; Menna, M.; Taglialatela-Scafati, O. Chemical diversity of bioactive marine natural products: An illustrative case study. *Curr. Med. Chem.* **2004**, *11*, 1671–1692. [CrossRef]
18. Pinteus, S.; Lemos, M.F.; Alves, C.; Neugebauer, A.; Silva, J.; Thomas, O.; Botana, L.M.; Gaspar, H.; Pedrosa, R. Marine invasive macroalgae: Turning a real threat into a major opportunity—The biotechnological potential of *Sargassum muticum* and *Asparagopsis armata*. *Algal Res.* **2018**, *34*, 217–234. [CrossRef]
19. Valentão, P.; Trindade, P.; Gomes, D.; de Pinho, P.G.; Mouga, T.; Andrade, P.B. *Codium tomentosum* and *Plocamium cartilagineum*: Chemistry and antioxidant potential. *Food Chem.* **2009**, *119*, 1359–1368. [CrossRef]
20. Rodrigues, D.; Costa-Pinto, A.R.; Sousa, S.; Vasconcelos, M.W.; Pintado, M.M.; Pereira, L.; Rocha-Santos, T.A.P.; Da Costa, J.P.; Silva, A.M.S.; Duarte, A.C.; et al. *Sargassum muticum* and *Osmundea pinnatifida* Enzymatic Extracts: Chemical, Structural, and Cytotoxic Characterization. *Mar. Drugs* **2019**, *17*, 209. [CrossRef]
21. Date, Y.; Sakata, K.; Kikuchi, J. Chemical profiling of complex biochemical mixtures from various seaweeds. *Polym. J.* **2012**, *44*, 888–894. [CrossRef]
22. Cakmak, Y.S.; Kaya, M.; Ozusaglam, M.A. Biochemical composition and bioactivity screening of various extracts from *Dunaliella salina*, a green microalga. *EXCLI J.* **2014**, *13*, 679–690. [CrossRef] [PubMed]
23. Patel, D.J.; Bhatia, N. Oral Antibiotics for Acne. *Am. J. Clin. Dermatol.* **2020**, *22*, 193–204. [CrossRef] [PubMed]
24. Adamski, Z.; Gornowicz-Porowska, J.; Sobkowska, D.; Kaszuba, K.; Czajkowski, R. Acne—Therapeutic challenges to the cooperation between a dermatologist and a cosmetologist. *Adv. Dermatol. Allergol.* **2021**, *38*, 21–31. [CrossRef] [PubMed]
25. Baldwin, H. Oral Antibiotic Treatment Options for Acne Vulgaris. *J. Clin. Aesthetic Dermatol.* **2020**, *13*, 26–32.
26. Karadag, A.S.; Kayıran, M.A.; Wu, C.; Chen, W.; Parish, L.C. Antibiotic resistance in acne: Changes, consequences and concerns. *J. Eur. Acad. Dermatol. Venereol.* **2020**, *35*, 73–78. [CrossRef] [PubMed]
27. Yang, J.H.; Yoon, J.Y.; Kwon, H.H.; Min, S.; Moon, J.; Suh, D.H. Seeking new acne treatment from natural products, devices and synthetic drug discovery. *Derm. Endocrinol.* **2017**, *9*, e1356520. [CrossRef] [PubMed]

28. Samri, N.; Hsaine, L.; Elkafhi, S.; Khlifi, S.; Etahiri, S. Antimicrobial properties of seven brown algae harvested from the coast of sidi bouzid (El Jadida-Morocco). *Int. J. Pharm. Pharm. Sci.* **2020**, *5*, 57–62. [CrossRef]
29. Metidji, H.; Dob, T.; Mohamed, T.; Krimat, S.; Ksouri, A.; Nouasri, A. In vitro screening of secondary metabolites and evalu-ationof antioxidant, antimicrobial and cytotoxic properties of *Gelidium sesquipedale* Thuret et Bornet red seaweed from Algeria. *J. Mater. Environ. Sci.* **2015**, *6*, 3184–3196.
30. El Wahidi, M.; El Amraoui, B.; El Amraoui, M.; Bamhaoud, T. Screening of antimicrobial activity of macroalgae extracts from the Moroccan Atlantic coast. *Ann. Pharm. Fr.* **2015**, *73*, 190–196. [CrossRef]
31. Pérez, M.J.; Falqué, E.; Domínguez, H. Antimicrobial Action of Compounds from Marine Seaweed. *Mar. Drugs* **2016**, *14*, 52. [CrossRef]
32. Choi, J.-S.; Bae, H.-J.; Kim, S.-J.; Choi, I.S. In vitro antibacterial and anti-inflammatory properties of seaweed extracts against acne inducing bacteria, *Propionibacterium acnes*. *J. Environ. Biol.* **2011**, *32*, 313–318. [PubMed]
33. Januário, A.P.; Félix, R.; Félix, C.; Reboleira, J.; Valentão, P.; Lemos, M.F.L. Red Seaweed-Derived Compounds as a Potential New Approach for Acne Vulgaris Care. *Pharmaceutics* **2021**, *13*, 1930. [CrossRef]
34. Benarroch, J.; Asally, M. The Microbiologist's Guide to Membrane Potential Dynamics. *Trends Microbiol.* **2020**, *28*, 304–314. [CrossRef] [PubMed]
35. Lee, D.-Y.D.; Prindle, A.; Liu, J.; Süel, G.M. SnapShot: Electrochemical Communication in Biofilms. *Cell* **2017**, *170*, 214–214.e1. [CrossRef]
36. Prindle, A.; Liu, J.; Asally, M.; Ly, S.; Garcia-Ojalvo, J.; Süel, G.M. Ion channels enable electrical communication in bacterial communities. *Nature* **2015**, *527*, 59–63. [CrossRef] [PubMed]
37. Bruni, G.N.; Weekley, R.A.; Dodd, B.J.T.; Kralj, J.M. Voltage-gated calcium flux mediates *Escherichia coli* mechanosensation. *Proc. Natl. Acad. Sci. USA* **2017**, *114*, 9445–9450. [CrossRef] [PubMed]
38. Sirec, T.; Benarroch, J.M.; Buffard, P.; Garcia-Ojalvo, J.; Asally, M. Electrical Polarization Enables Integrative Quality Control during Bacterial Differentiation into Spores. *iScience* **2019**, *16*, 378–389. [CrossRef] [PubMed]
39. Kaim, G.; Dimroth, P. ATP synthesis by F-type ATP synthase is obligatorily dependent on the transmembrane voltage. *EMBO J.* **1999**, *18*, 4118–4127. [CrossRef] [PubMed]
40. Strahl, H.; Hamoen, L.W. Membrane potential is important for bacterial cell division. *Proc. Natl. Acad. Sci. USA* **2010**, *107*, 12281–12286. [CrossRef]
41. Lee, D.-Y.D.; Galera-Laporta, L.; Bialecka-Fornal, M.; Moon, E.C.; Shen, Z.; Briggs, S.P.; Garcia-Ojalvo, J.; Süel, G.M. Magnesium Flux Modulates Ribosomes to Increase Bacterial Survival. *Cell* **2019**, *177*, 352–360.e13. [CrossRef]
42. Patra, J.K.; Das, G.; Baek, K.-H. Antibacterial mechanism of the action of *Enteromorpha linza* L. essential oil against Escherichia coli and Salmonella Typhimurium. *Bot. Stud.* **2015**, *56*, 13. [CrossRef] [PubMed]
43. Wu, Y.; Bai, J.; Zhong, K.; Huang, Y.; Gao, H. A dual antibacterial mechanism involved in membrane disruption and DNA binding of 2R,3R-dihydromyricetin from pine needles of *Cedrus deodara* against *Staphylococcus aureus*. *Food Chem.* **2017**, *218*, 463–470. [CrossRef] [PubMed]
44. Patra, J.K.; Baek, K.-H. Antibacterial Activity and Action Mechanism of the Essential Oil from *Enteromorpha linza* L. against Foodborne Pathogenic Bacteria. *Molecules* **2016**, *21*, 388. [CrossRef] [PubMed]
45. El Shafay, S.M.; Ali, S.S.; El-Sheekh, M. Antimicrobial activity of some seaweeds species from Red sea, against multidrug resistant bacteria. *Egypt. J. Aquat. Res.* **2016**, *42*, 65–74. [CrossRef]
46. Tang, Y.-L.; Shi, Y.-H.; Zhao, W.; Hao, G.; Le, G.-W. Interaction of MDpep9, a novel antimicrobial peptide from Chinese tra-ditional edible larvae of housefly, with Escherichia coli genomic DNA. *Food Chem.* **2009**, *115*, 867–872. [CrossRef]
47. Brogden, K.A. Antimicrobial peptides: Pore formers or metabolic inhibitors in bacteria? *Nat. Rev. Microbiol.* **2005**, *3*, 238–250. [CrossRef]
48. Pinteus, S.; Lemos, M.F.; Simões, M.; Alves, C.; Silva, J.; Gaspar, H.; Martins, A.; Rodrigues, A.; Pedrosa, R. Marine invasive species for high-value products' exploration—Unveiling the antimicrobial potential of *Asparagopsis armata* against human pathogens. *Algal Res.* **2020**, *52*, 102091. [CrossRef]
49. Shannon, E.; Abu-Ghannam, N. Antibacterial Derivatives of Marine Algae: An Overview of Pharmacological Mechanisms and Applications. *Mar. Drugs* **2016**, *14*, 81. [CrossRef] [PubMed]
50. Subramanian, M.; Soundar, S.; Mangoli, S. DNA damage is a late event in resveratrol mediated inhibition of *Escherichia coli*. *Free Radic. Res.* **2016**, *50*, 708–719. [CrossRef] [PubMed]
51. Da, X.; Nishiyama, Y.; Tie, D.; Hein, K.Z.; Yamamoto, O.; Morita, E. Antifungal activity and mechanism of action of Ougon (*Scutellaria* root extract) components against pathogenic fungi. *Sci. Rep.* **2019**, *9*, 1683. [CrossRef] [PubMed]
52. Lei, J.; Sun, L.; Huang, S.; Zhu, C.; Li, P.; He, J.; Mackey, V.; Coy, D.H.; He, Q. The antimicrobial peptides and their potential clinical applications. *Am. J. Transl. Res.* **2019**, *11*, 3919–3931. [PubMed]
53. He, F.; Yang, Y.; Yang, G.; Yu, L. Studies on antibacterial activity and antibacterial mechanism of a novel polysaccharide from *Streptomyces virginia* H03. *Food Control* **2010**, *21*, 1257–1262. [CrossRef]
54. Shanmughapriya, S.; Manilal, A.; Sujith, S.; Selvin, J.; Kiran, G.S.; Natarajaseenivasan, K. Antimicrobial activity of seaweeds extracts against multiresistant pathogens. *Ann. Microbiol.* **2008**, *58*, 535–541. [CrossRef]
55. Xu, N.; Fan, X.; Yan, X.; Li, X.; Niu, R.; Tseng, C. Antibacterial bromophenols from the marine red alga *Rhodomela confervoides*. *Phytochemistry* **2003**, *62*, 1221–1224. [CrossRef]

56. Manilal, A.; Sujith, S.; Kiran, G.S.; Selvin, J.; Shakir, C.; Gandhimathi, R.; Lipton, A.P. Antimicrobial potential and seasonality of red algae collected from the southwest coast of India tested against shrimp, human and phytopathogens. *Ann. Microbiol.* **2009**, *59*, 207–219. [CrossRef]
57. Rodrigues, D.; Alves, C.; Horta, A.; Pinteus, S.; Silva, J.; Culioli, G.; Thomas, O.P.; Pedrosa, R. Antitumor and Antimicrobial Potential of Bromoditerpenes Isolated from the Red Alga, *Sphaerococcus coronopifolius*. *Mar. Drugs* **2015**, *13*, 713–726. [CrossRef]
58. Clementi, E.A.; Marks, L.; Roche-Håkansson, H.; Håkansson, A.P. Monitoring Changes in Membrane Polarity, Membrane Integrity, and Intracellular Ion Concentrations in *Streptococcus pneumoniae* Using Fluorescent Dyes. *J. Vis. Exp.* **2014**. [CrossRef] [PubMed]
59. Hu, Q.-P.; Cao, X.-M.; Hao, D.-L.; Zhang, L.-L. Chemical Composition, Antioxidant, DNA Damage Protective, Cytotoxic and Antibacterial Activities of *Cyperus rotundus* Rhizomes Essential Oil against Foodborne Pathogens. *Sci. Rep.* **2017**, *7*, 45231. [CrossRef] [PubMed]

antibiotics

MDPI

Review

Green Antimicrobials as Therapeutic Agents for Diabetic Foot Ulcers

Ines D. Teixeira [1], Eugenia Carvalho [1,2,*] and Ermelindo C. Leal [1,2,*]

[1] Center for Neuroscience and Cell Biology, Center for Innovative Biomedicine and Biotechnology, University of Coimbra, Rua Larga, 3004-504 Coimbra, Portugal
[2] Institute for Interdisciplinary Research, University of Coimbra, 3004-504 Coimbra, Portugal
* Correspondence: ecarvalh@cnc.uc.pt (E.C.); ecleal@cnc.uc.pt (E.C.L.)

Abstract: Diabetic foot ulcers (DFU) are one of the most serious and devastating complications of diabetes and account for a significant decrease in quality of life and costly healthcare expenses worldwide. This condition affects around 15% of diabetic patients and is one of the leading causes of lower limb amputations. DFUs generally present poor clinical outcomes, mainly due to the impaired healing process and the elevated risk of microbial infections which leads to tissue damage. Nowadays, antimicrobial resistance poses a rising threat to global health, thus hampering DFU treatment and care. Faced with this reality, it is pivotal to find greener and less environmentally impactful alternatives for fighting these resistant microbes. Antimicrobial peptides are small molecules that play a crucial role in the innate immune system of the host and can be found in nature. Some of these molecules have shown broad-spectrum antimicrobial properties and wound-healing activity, making them good potential therapeutic compounds to treat DFUs. This review aims to describe antimicrobial peptides derived from green, eco-friendly processes that can be used as potential therapeutic compounds to treat DFUs, thereby granting a better quality of life to patients and their families while protecting our fundamental bio-resources.

Keywords: diabetic foot ulcer; antimicrobials peptides; green synthesis; environmentally friendly

Citation: Teixeira, I.D.; Carvalho, E.; Leal, E.C. Green Antimicrobials as Therapeutic Agents for Diabetic Foot Ulcers. *Antibiotics* **2023**, *12*, 467. https://doi.org/10.3390/antibiotics12030467

Academic Editor: Helena P. Felgueiras

Received: 8 February 2023
Revised: 22 February 2023
Accepted: 23 February 2023
Published: 25 February 2023

1. Introduction

Diabetes mellitus (DM) is a major public health problem with rising prevalence worldwide [1,2]. In 2021, approximately 537 million adults were diagnosed with DM, and this number has been predicted to rise to over 783 million by 2045 [1]. Along with the rising prevalence of diabetes, DM-associated complications are also expected to increase, resulting in high morbidity, mortality, and health expenditure rates, due to the required hospitalizations and specialized care [1,2]. Diabetes can lead to life-threatening and disabling health complications such as retinopathy, neuropathy, cardiovascular diseases, nephropathy, and diabetic foot ulcerations [1,2].

Diabetic foot ulcers (DFUs) are one of the most prevalent complications of diabetes, and are, in part, are associated with peripheral vascular disease and peripheral neuropathy [2]. It is estimated that 19–34% of patients with diabetes will develop DFUs in their lifetime [3]. Approximately 85% of these patients undergo non-traumatic lower limb amputation [2–5].

Wound healing is a dynamic and complex biological process aimed to restore skin function after trauma. Nevertheless, under diabetic conditions, this process is impaired [2,6,7]. Diabetic individuals exhibit multiple risk factors such as hyperglycaemia, prolonged hypoxia, chronic inflammation, peripheral neuropathy, impaired neovascularization, and difficulty fighting infections that compromise and delay the normal wound healing process [2]. This results in the occurrence of chronic non-healing wounds in a persistent pro-inflammatory state [6,7]. Moreover, about 60% of DFUs become infected with bacterial colonies, contributing to chronic wound healing failure [2,6,7].

As a result of diabetic foot infections (DFI), patients are frequently hospitalized and receive multiple antibiotic courses over the sequence of treatments [8]. Today, the emergence and spread of multi-drug resistant bacteria impose a rising threat to global health [8,9]. For this reason, it is pivotal to find green and less environmentally harmful alternatives for fighting these resistant microbials.

Antimicrobial peptides (AMPs) are conserved bioactive molecules of the innate and adaptative immune system which can be found in all forms of life [2,8,9]. Recent studies have demonstrated that these short peptides, composed of 15 to 60 amino acids, have various mechanisms of action that play important roles in the fight against infection by providing broad-spectrum antimicrobial activity against Gram-negative and Gram-positive bacteria, fungi, and viruses, as well as in wound healing [2,8,9]. Nevertheless, these small peptides can present some limitations such as low structure stability, high cytotoxicity, low hydrosolubility, salt sensitivity, and poor selectivity, as well as high production costs [2,8,9].

Direct eradication of microorganisms can be achieved by AMPs through mechanisms such as membrane disruption, interaction with intracellular targets, recruitment, and activation of immune cells [2]. These molecules can also promote wound healing by re-epithelization, support of angiogenesis, and enhancement of extracellular matrix synthesis [2,8]. Thus, AMP-based approaches may be a good solution to fight the emergence of antimicrobial resistance [2].

With the increasing awareness of climate change and high pollution levels, there has been a heightened desire to protect the environment. To address this issue, scientists have been focusing on the development of innovative and environmentally friendly techniques for AMPs extraction and synthesis. The use of plants and microalgae as sources of novel AMPs has been considered to be sustainable and very promising. However, more studies are needed in order to improve the current protocols in use.

Therefore, this review aims to describe AMPs derived from greener, eco-friendlier processes that can be used as potential therapeutic compounds to treat infection, including DFUs, thereby granting a better quality of life to patients and their families while protecting our fundamental bio-resources the environment.

2. Diabetic Foot Ulcers

Wound healing is a dynamic and complex biological process aimed to restore skin function after trauma [6]. This process is achieved through four successive phases with limited overlap: hemostasis, inflammation, proliferation, and remodeling [2,6].

The hemostasis phase initiates immediately after skin injury and is characterized by the constriction of the damaged blood vessels and the activation of platelets. Platelet aggregations will promote the formation of a fibrin clot that seals the damaged endothelium, stopping the bleeding [2,6]. Once the bleeding is controlled, the inflammatory phase begins. Damaged cells release chemokines and cytokines, which recruit inflammatory cells to the site of injury. Neutrophils are the first cells mobilized to the wound site and are responsible for microbial clearance [2,6,7]. These cells are then followed by pro-inflammatory macrophages (M1), which induce the clearing of apoptotic cells, including neutrophils, and secrete growth factors and cytokines that promote the inflammatory response by attracting adaptive immune system cells to the site of injury [2,6,7]. As the inflammatory stage is resolved, these cells undergo a phenotypic alteration to a reparative state (M2) that promotes tissue regeneration [7]. The proliferative phase begins once the inflammation decreases. This phase is characterized by the formation of granulation tissue, contraction of the wound edges, re-epithelization, and neovascularization [2,6,10]. The final step of wound healing is the remodeling phase, which includes collagen fiber reorganization, remodeling, and maturation of the scar tissue, as well as an increase in its tensile strength [2,6,10].

For a wound to heal successfully, all four phases need to be very well coordinated in the right order and within the appropriate time period [2,6]. Nevertheless, under diabetic conditions, this process is impaired resulting in the occurrence of chronic non-healing wounds in a state of persistent pro-inflammation [2,6,7]. This is characterized by an accumulation of immune cells, an increase in the M1/M2 macrophage ratio, increased generation of reactive oxygen species (ROS), and pro-inflammatory cytokines [6]. Moreover, about 60% of DFUs develop bacterial infections which contribute to the failure to heal chronic wounds [2,6].

Patients with DFIs are regularly hospitalized and are frequently exposed to numerous courses of antibiotics [8,9,11]. Today, the emergence and spread of bacteria that are resistant to conventional antibiotics, including those used within the hospital environment, impose a rising threat to global health [9,11]. This problem greatly hampers DFU treatment. For this reason, the development of alternative compounds with the capacity to downregulate the inflammatory response and control pathogen infection is urgently required.

3. Microbiota in Diabetic Foot Ulcers

The skin is the largest organ in the human body, and it has an essential role as a multifunctional barrier, protecting our body against pathogens or toxic substances [12]. The human skin contains a large and diverse composition of living microorganisms known as the skin microbiota. Most of these microorganisms are harmless and even advantageous to their host, protecting against invasion by more pathogenic organisms and contributing to skin homeostasis. However, the disruption in this balanced microbiota system can enhance the susceptibility to skin disorders, including infections [12,13].

The physical and chemical characteristics of the skin surface may vary according to environmental and specific host factors. These features can influence colonization by the skin microbiota and determine unique sets of microorganisms [12]. Under diabetic conditions, skin integrity is compromised by several factors, contributing to impaired tissue regeneration and alterations in the local skin microbiome [14,15]. Persistent hyperglycemia creates an excessive nutrient source for microbes and reduces innate immunity by causing poor chemotaxis, phagocytosis, and cleansing of pathogens by neutrophils [15]. Moreover, peripheral vascular disease hampers the action of the host's immune response, in part due to reduced blood flow [15]. Peripheral neuropathy aggravates minor traumas and increases forefoot pressure, facilitating the entry of microbes [14,15]. All these factors contribute to microbial colonization, biofilm formation, and clinical infections that impair wound healing and contribute to serious complications including osteomyelitis and lower limb amputation [14,15]. Approximately 60% of all DFU cases are estimated to develop DFI [2].

The DFU microbiota has been extensively studied [8,15–18]. A longitudinal study of patients with DFU (n = 100) used shotgun metagenomic sequencing to profile chronic wound microbiota and investigated its role in clinical outcomes and the response to therapy. The majority of bacteria present in diabetic foot ulcers were Gram-positive strains, such as *Staphylococcus aureus*, methicillin-resistant *S.aureus* (MRSA), and *Corynebacterium striatum*, as well as Gram-negative bacteria, including *Pseudomonas aeruginosa* and *Alcaligenes faecalis* [15]. These results were in accordance with other microbial studies that also include *Escherichia coli* and *Proteus* spp. as the most predominant isolated strains from DFU [11,15,16]. Moreover, some anaerobes have also been identified in the deep tissue within diabetic wounds, including *Peptostreptococcus* spp., *Bacteroides* spp., *Prevotella* spp., and *Clostridium* spp. [2,8,15].

The majority of DFI are polymicrobial in nature, and mixed microorganisms, including fungi, are frequently prevalent. The prevalence of pathogenic fungal species and subsequent mycotic infections are responsible for an increased risk of diabetic foot syndrome development and poor clinical outcomes [2,8,15–17]. The most commonly isolated fungi are *Candida* spp., *Trichophyton* spp., *Aspergillus* spp., *Trichosporon* spp., and *Cladosporium herbarum* [2,15–17].

DFUs are considered polymicrobial ecosystems composed of highly dynamic and diverse microbial communities [8,18]. These microorganisms can exist independently or can organize into functionally equivalent pathogroups (FEP), where commensal and pathogenic bacteria co-aggregate symbiotically in a self-produced protective polysaccharide matrix

resistance genes (ARGs) are becoming increasingly prevalent, harming human, animal, and environmental health [32]. Faced with this reality, it is pivotal to find greener and less environmentally harmful alternatives to help fight these resistant pathogens.

Figure 2. Schematic representation of the twelve principles of green chemistry. Figure adapted from [27] and created in Canva.com (accessed on 24 January 2023).

The increasing emphasis on green chemistry has led chemical-pharmaceutical industries and laboratories to concentrate on diminishing their environmental footprint [29,33].

5.1. Green Extraction Methods to Obtain AMPs

To date, several thousand AMPs have been isolated from different natural sources [20,34,35]. However, the techniques used to isolate these small bioactive molecules have not always been the most environmentally friendly. The conventional extraction methods have contained numerous toxic and flammable solvents, used high temperatures and energy demands, and extended extraction times that lead to molecule degradation [36,37]. The implementation of green chemistry in extraction processes is essential to reduce the release of chemicals hazardous to human health and the environment [36]. Extraction methods based on green principles consume less energy, exploit alternative solvents, and utilize sustainable resources while producing high-quality and safe extracts [36,37].

Plants and algae are considered two promising sustainable sources of a wide diversity of natural AMPs [34,35,37]. A variety of methods can be used to extract bioactive molecules from plant-based material. The general approach often includes three main stages: plant material homogenization, extraction, and purification [38]. Some traditional extraction techniques include maceration, hydrodistillation, and Soxhlet extraction [37]. However, many approaches such as ultrasound-assisted extraction (UAE), pressurized liquid extraction (PLE), pulsed electric fields extraction (PEFE), microwave-assisted extraction (MAE), and supercritical fluid extraction (SFE), have been developed and optimized to ensure greater sustainability [36,37].

In ultrasound-assisted extraction, sound waves and frequencies are used to induce the rupture of cell walls and release their content [39–41]. This technique has significant benefits such as shorter extraction time, lower solvent and energy depletion, and greater extraction yield [39–41]. PLE is a solid–liquid method that involves the application of high pressure and temperature. This results in an increase in the solvent's boiling point, promoting its quick infiltration into the sample matrix. Due to its high extraction temperatures, this method is not recommended for compounds that show heat sensitivity [40]. The pulsed electric field extraction (PEFE) method uses electromechanical forces to create irreversible electroporation in the cell membrane, which leads to its increased permeabilization and subsequently enhances the mass transfer rate [41]. Another emerging clean technology is the MAE. The basic principle of this methodology is the transformation of electromagnetic energy into thermal energy. This leads to the heating of the intracellular moisture, causing evaporation, and, consequently, an increase in intracellular pressure that results in both organelle and cell wall disruption [41]. SFE is an advanced separation technique that employs natural chemical components, such as supercritical fluid CO_2, as solvents [36,37,39]. This innovative technique can be applied to both plant and algae-based materials [36]. SFE represents a sustainable alternative to traditional solvent extraction methods, thus granting better efficacy and selectivity, higher diffusivity, with reducing extraction time [36,37,39]. Furthermore, supercritical fluids can be recycled and reused, resulting in a decrease of waste generation [37].

Microalgal AMPs can be obtained by solvent extraction, microbial fermentation, or enzymatic hydrolysis, with the latter being the most used extraction technique [42–44]. Sun et al. isolated the antibacterial peptide SP-1 from *Spirulina platensis* through enzymatic hydrolysis. This peptide showed considerable antibacterial effects against *Escherichia coli* and *Staphylococcus aureus* [45]. Further studies demonstrated antioxidant, antihypertensive, anti-diabetes, and anti-obesity properties of bioactive peptides derived from the same microalgae [46]. In another study, the peptide KLENCNYAVELGK was extracted from pepsin hydrolysates derived from *Limnospira sp.* This bioactive molecule also exhibited antibacterial activity against *Escherichia coli* and *Staphylococcus aureus* [47].

The enzymatic hydrolysis extraction method allows the isolation of bioactive peptides using certain commercial enzymes and physicochemical conditions such as optimal temperatures (below 100 °C) and pHs (close to neutral) [43,46]. In order for an enzyme to perform its functions, it must first bind to the substrate and then proceed with the catalysis. In this process, an enzyme membrane reactor system is used to hydrolyze the extracted proteins [43,44]. The membrane filter retains large particles that are then recycled back into the hydrolysis tank, only allowing hydrolyzed and small fractions to pass [43]. The purification of the obtained peptides can be achieved by various chromatographic methods such as ion exchange, reverse-phase high-performance liquid chromatography (RP-HPLC), and gel chromatography [46]. The purification method is selected based on the physical and chemical characteristics of the bioactive peptides [43]. Finally, to characterize the bioactive peptides at a molecular level, spectrophotometric techniques, such as liquid chromatography-mass spectrophotometry (LC–MS) and mass-mass spectrophotometry (MS–MS) can be used [44,46].

Enzymatic hydrolysis presents some advantages when compared to other techniques. This method has shown an improved reaction rate, higher yields, ease to scale-up, and high specificity, thus allowing the obtention of AMPs with the desired molecular size-weight [43,44,46]. It is considered an environmentally friendly process since it usually does not generate by-products or use any synthetic chemical agents [43]. Moreover, the mild temperatures and pH conditions under which this technique is performed allow for lower operating costs and energy demands [43,44].

In addition to the development of greener extraction methods, other purification techniques have been reviewed and improved in order to reduce their ecological footprint. Recent methodologies include mixed-mode chromatography (MMC) columns, in which the separation is characterized by multiple interactions between the solute and the stationary phase [48]. Another technique is multicolumn counter-current solvent gradient purification (MCSGP). This technology represents a chromatography purification process that allows the separation of different components by countercurrent movements between the stationary and mobile phase. Compared with standard methods, these recent techniques

(MMC and MCSGP) exhibit higher yields and faster separation rates while being able to decrease water consumption and the generation of toxic waste. Moreover, supercritical fluid chromatography (SFC) has shown to be a promising and eco-friendly method compared to HPLC. SFC presents several advantages, such as the use of less organic solvents and toxic modifiers and the possibility of recycling the CO_2 (used as mobile phase) during purification. It is also noteworthy that this process reduces the organic waste to less than one third in comparison with the current methods [48].

5.2. Green Synthesis of AMPs

So far, numerous AMPs have been discovered and isolated from various natural sources. Over the past years, several studies have been focused on unravelling the promising therapeutic properties of these bioactive molecules [2]. However, most of these small peptides possess some inherent limiting characteristics, such as low hydrosolubility, low stability, salt sensitivity, poor selectivity, and sometimes high cytotoxicity to the host [2,8,9,49–51].

To solve the hurdles presented by some natural AMPs, scientists began the development of new and improved synthetic antimicrobial peptides (SAMPs). These new molecules use the natural AMPs amino acid (aa) sequences as templates but undergo some specific chemical alterations, new formulations, or fusion with various synthetic elements in order to enhance their properties [49–51]. In comparison with natural AMPs, SAMPs exhibit improved effectiveness, reduced cytotoxicity, and greater resistance to protease degradation [49,50]. These characteristics improve their therapeutic potential in clinical applications [2,50].

In the development of synthetic peptides, a variety of approaches have been used [2,50,51]. One of the main strategies consists of chemical modifications. These approaches are characterized by punctual alterations (substitution, deletion, or addition of new aa) in the amino acid sequence of the active sites of naturally occurring AMPs, resulting in semi-synthetic antimicrobial peptides [49,51]. Different modification approaches revealed significant advantages in AMP manipulation: glycosylation (glycan is covalently attached to the peptide) [52]; lipidation (lipid group is covalently attached to the peptide) [2,53,54]; hydrazidation (attachment of a hydrazide to the peptide) [2,53,54].; guanidination (lysine residues are converted into homoarginine residues) [2,54]; and small molecule conjugation (small molecules such as antibiotics, peptides, and others, are incorporated into the AMP structure) [2,54].

Chemical synthesis has quickly become the most used method for obtaining synthetic AMPs [51,55]. The solid-phase peptide synthesis (SPPS) is widely used as the standard technique to obtain small or medium-sized peptides (30 to 50 residues). In fact, it is demonstrated that this method improved potency, reliability, and speed [50,51,56]. Zapotoczna et al. have demonstrated the antibacterial effectiveness of seven synthetic AMPs (D-Bac8c, WMR, HB43, P18, Omiganan, Polyphemusin, and Ranalexin) against *Staphylococcus aureus* biofilm infections. Among these SAMPs, D-Bac8c has shown higher potency in microbial eradication of *S. aureus* biofilm infections in both in vitro and in vivo studies [57]. In the SPPS process, successive and protected amino acids are added to a growing peptide chain, anchored to a solid support (usually resin). In addition, it involves several deprotection and washing steps to eliminate solvent waste [55,58]. The generation and accumulation of by-products are directly affected by the sequence and length of the desired peptide as well as the choice of resin used in this process, thereby affecting the efficiency of the peptide synthesis [59].

Despite the significant achievements of chemical synthesis, this process has a huge negative environmental impact. It is particularly responsible for enormous amounts of hazardous waste [58]. Moreover, this technique requires high consumption of toxic solvents, such as N,N-Dimethylformamide (DMF), dichloromethane (DCM), and *N*-methyl-2-pyrrolidone (NMP) [58–60]. In fact, some studies associated exposure to these compounds with hepatotoxicity and chronic toxicity [58]. The rising health and environmental concerns prompted the scientific community to develop greener and environmentally friendly alternative approaches for peptide synthesis [58–60]. Overall, several attempts have provided some promising results while granting high yields of peptides [58].

Most efforts have been made regarding greener and safer solvents. In 2009, Declerck et al. reported the first successful solvent-free peptide synthesis by the ball-milling technique [61]. Later, an in-depth study by Lopez et al. found that the best suitable replacement for the toxic compound DMF in SPPS was the green solvent *N*-butylpyrrolidinone (NBP) [60]. More recently, Jadhav et al. revealed that varying the composition of green binary solvent mixtures resulted in the mitigation of SPPS side reactions [59].

The use of greener resins is also another important research field currently in development [48,62,63]. New eco-friendly resins, such as polyethylene glycol-based, are able to swell to a significant extent in the majority of solvents and show great compatibility with new green solvents [48]. Recently, another greener resin derived from renewable resources, poly-ε-lysine resin SpheriTide, has proven to be biodegradable and to have a high loading capacity [48,63].

Over the last decades, green analytical methodologies have been introduced in order to make experimental protocols even safer and more environmentally benign [64–66]. In this context, green chromatographic methods have been given increasing importance [64]. The primary goal of green chromatography is to make the process of analysis greener at every stage by replacing existing solvents with greener alternatives, reducing solvent use, and decreasing waste generation [64,65]. In liquid chromatography (LC), the solvent reduction can be achieved by several strategies such as decreasing the column length with small particle size, using ultra-high performance liquid chromatography systems (short columns with reduced internal diameters), microflow and capillary HPLC columns, or even through the use of elevated temperatures [64]. Moreover, several green LC methodologies already implement the use of greener solvents (Brij 35, sodium dodecyl sulfate, and propylene carbonate) as a substitute for toxic solvents [64,65]. For example, replacing acetonitrile with ethanol (a nontoxic alternative) is a good strategy to improve LC techniques [64]. Additionally, as referred before, several processes can be optimized by the implementation of greener solvents and materials/instruments similar to resins.

5.3. Green and Sustainable Sources of AMPs

5.3.1. Plants

As plants have evolved, they have developed refined defense mechanisms that enable them to protect themselves from potentially harmful organisms [34]. These include chemical barriers, in which plants produce a high number of toxic defense molecules, including AMPs [34,67]. In response to pathogen stimulation, multiple AMPs can be produced from different parts of the plant such as roots, seeds, flowers, fruits, stems, and leaves, or even from the whole plant [67]. A single plant species can contain numerous AMPs with various functional characteristics, structures, different expression patterns, and particular targets [34,67]. Based on their structure, plant AMPs can be classified into eight different families: thionins, defensins, hevein-like peptides, knotting-type peptides, α-hairpinin, lipid transfer proteins, cyclotides, and snakins [34,67].

Further research has revealed that plant AMPs have many other physiological functions in addition to their antimicrobial roles. These include the regulation of plant growth and development, as well as the ability to treat numerous diseases effectively [34]. Therefore, natural peptide extraction is very important since these molecules represent a promising alternative to conventional antibiotics with broad potential applications [34,38,67].

In 1970, Okada and Yoshizumi isolated the first AMP of plant origin, the purothionins, from barley (*Hordeum vulgare*) endosperm. These peptides showed antimicrobial effects against *Pseudomonas solanacearum*, *Erwinia amylovora*, *Xanthomonas phaseoli*, and *X. campestris*, *Corynebacterium poinsettiae*, *C. fascians*, *C. flaccumfaciens*, and *C. sepedonicum* [68]. Since then, many other AMPs have been isolated from different plant species. Vilas Boas et al. identified the AMP kalata B1 (cyclotide) from the leaves of the African plant *Oldenlandia affinis*. This peptide showed strong anti-viral activity against human immunodeficiency virus (HIV), having the ability to destroy viral particles and prevent their fusion to the host cell membrane [69]. An antifungal peptide PvD1 (defensin) was successfully purified from *Phaseolus vulgaris* seeds and demonstrated inhibitory growth effects against several yeasts (*Candida albicans, C. parapsilosis, C. guilliermondii, C. tropicalis, Saccharomyces cerevisiae,* and *Kluyveromyces marxiannus*) and fungi (*Rhizoctonia solani, Fusarium solani, F. oxysporum,* and *F. lateritium*) [70]. Next, the AMP snakin-Z was isolated from Ziziphus jujube fruits, showing a great antifungal and antibacterial potential against *Staphylococcus aureus, Escherichia coli, Bacillus subtili,* and *Klebsiella pneumoniae* [71]. The cyclotides Cycloviolacin O2 and Cycloviolacin O8 were isolated from *Viola odorata* [72]. The two compounds demonstrated strong antibacterial activity against various pathogenic bacteria.

It is noteworthy that some parts of the plant do not require extraction. However, in some of these cases, the purification and isolation of these peptides can require more stages [38]. A study conducted by Mandal et al. was able to successfully purify and identify three AMPs (Cn-AMP1, Cn-AMP2, Cn-AMP3) from coconut water (Cocos nucifera L.) [73]. Immediately following the selection of the material, the authors performed dialysis against distilled water with acetic acid addition up to pH 2.0. In order to isolate and purify the AMPs, the resuspended samples were fractionated onto reverse phase chromatography (HPLC). Moreover, these peptides showed antimicrobial activity against *Escherichia coli, Bacillus subtilis, Pseudomonas aeruginosa,* and *Staphylococcus aureus.*

Over the last decades, many efforts have been made in order to use environmentally friendly approaches to extract AMPs from natural sources. Despite all efforts, more studies are needed concerning the development of complete green extraction protocols. Recently, many methodologies of extraction implemented greener steps involving eco-friendly solvents and materials, as previously described. Certain AMPs have already been obtained using some of these techniques. For example, Song et al. identified nine novel AMPs (CHQQEQRP, DENFRKF, EWPEEGQRR, KPPIMPIGKG, KDFPGRR, LGLRSGIILCNV, PRNFQQQLR, QNLNALQPK, and SQEATSPR) from cottonseed protein by enzymatic hydrolysis. These peptides demonstrate a successful inhibitory effect against *Staphylococcus aureus, Escherichia coli, Salmonella* sp., and *Streptococcus* sp. [74]. A recent study conducted by Farhadpour et al. reported the isolation of five vigno cyclotides (vigno 1–5) from *Viola ignobilis* by the microwave-assisted extraction method [75]. Further data have shown that vigno 5 has chemotherapeutic effects on cervical cancer [76].

Over the last three decades, the use of plants and plant cells as system platforms to produce AMPs, also known as molecular farming, has been extensively studied by the biopharmaceutical industry in order to increase the accessibility of recombinant peptides for clinical use [77]. Some of their advantages include a more affordable and rapid process, feasible large-scale production with high yields, the use of simpler manufacturing procedures that reduce extensive purification, and enhanced growth conditions free from toxic contaminants and pathogens. In addition, this expression system offers the major benefit of allowing post-translational modifications to occur, which may be crucial for protein folding and AMPs biological functions. However, regulatory compliance is still a big disadvantage that must be overcome [77]. LFchimera, a chimerical antimicrobial peptide, was recombinantly expressed in the hairy roots of *Nicotiana tabacum* [78]. This peptide exhibited strong antibacterial activity against *Escherichia coli.* Lojewska et al. expressed the recombinant Colicin M peptide in *Nicotiana tabacum* plants and verified its action against *Escherichia coli* and *Klebsiella pneumoniae* [79]. Another study conducted by Patiño-Rodríguez used the same tobacco species as previously reported to express the recombinant broad-spectrum AMP, Protegrin-1 [80]. The authors demonstrated its effectiveness against *Klebsiella pneumoniae, Staphylococcus aureus, Escherichia coli, Mycobacterium bovis,* and the fungal pathogen *Candida albicans.*

5.3.2. Algae

The search for new natural sources of AMPs, turned the scientist's attention to the marine environment. Throughout evolution, marine organisms have developed a variety of bioactive molecules and strategies to defend themselves from prokaryotic and viral infections [81]. In this context, microalgae and cyanobacteria represent promising resources of molecules with antimicrobial properties [35]. As a result of their flexible metabolism, they are able to adapt to a variety of environmental conditions, including highly competitive environments, and respond to different environmental stresses and nutrient sources. Furthermore, they are exposed to a wide range of predators and microbial pathogens [35]. In response to these stimuli, these microorganisms have developed a wide diversity of antimicrobial compounds that can serve a broad spectrum of applications in the fields of biotechnology, medicine, agriculture, and aquaculture [82].

Cyanobacteria, also known as blue-green algae, are an ancient group of photosynthetic microbes with great ecological importance [83]. These organisms exist as single cells, in pluricellular forms, or as symbiotic partners of other plants and animals [82,83]. Antibacterial peptides isolated from cyanobacteria have been extensively studied and reviewed in the literature over the last decades [35,84,85]. These peptides can be classified into six different categories: cyclic peptides (41.6%), lipopeptides (20.8%), cyclic lipopeptides (16,6%), cyclic depsipeptides (12.5%), linear (4.2%), and depsipeptides (4.2%) [35]. Zainuddin et al. identified four cyclic undecapeptides named lyngbyazothrins A, B, C, and D from the freshwater strain *Lyngbya* sp. as binary mixtures (A/B and C/D). In this study, the authors showed that lyngbyazothrins C/D had antimicrobial activity against both Gram-positive (*Bacillus subtilis*) and Gram-negative (*Escherichia coli, Pseudomonas aeruginosa, Serratia marcesens)* bacteria [84]. Montaser et al., isolated pitipeptolides C-F from the marine cyanobacterium *Lyngbya majuscula* in Piti Bomb Holes, Guam. This study showed that pitipeptolides F was the most potent compound in a disc diffusion assay against *Mycobacterium tuberculosis* [86]. The cyclic undecapeptide, Kawaguchipeptin B, was extracted from the cultured cyanobacterium *Microcystis aeruginosa* (NIES-88) and exhibited antibacterial effects against *Staphylococcus aureus* [87]. Another study conducted by Dussault et al. investigated the effects of ten cyanobacterial isolates on the growth of foodborne pathogens, and concluded that several cyanobacterial peptides (laxaphycins A, B and B3) had antibacterial activity against Gram-positive bacteria [88].

Marine microalgae are unicellular photosynthetic eukaryotic microorganisms that can be found in a variety of aquatic habitats [89]. In the last few decades, these microorganisms have been studied in the search for new high-value molecules with antimicrobial activity that can be used to develop future environmentally friendly antibiotics to combat microbial antibiotic resistance [35,42,82,89,90]. There are a number of advantages to using marine microalgae in drug discovery that include the ease of culture (both on small and large scales), the limited generation time (5–8 h), and the ability to conduct an eco-friendly approach to drug discovery [42,90]. Enzymatic hydrolysis is the most used extraction technique to obtain microalgal AMPs [42]. AMPs derived from microalgae extracts have been predominantly obtained from protein hydrolyzates of different species such as *Chlorella vulgaris, Chlorella ellipsoidea, Tetradesmus obliquus, Navicula incerta*, and *Nannochloropsis oculate* [89]. Several studies have demonstrated that these bioactive microalgal peptides can exert several biological functions, including antioxidant, anticancer, antihypertensive, and antimicrobial properties, with beneficial health effects and potential therapeutic applications [42,89,90]. Guzmán et al. reported three antibacterial peptides from the marine microalgae *Tetraselmis suecica* [89]. In this study, one of the peptides, AQ-1766, which exhibited high activity against Gram-negative bacteria (*E. coli, S. typhimurium*, and *P. aeruginosa*) and Gram-positive bacterial strains (*B. cereus*, methicillin-resistant *S. aureus* (MRSA), *L. monocytogenes* and *M. luteus)*, was subjected to a lysine replacement of some residues obtaining six peptides with improved antibacterial activity (AQ-3001, AQ-3002, AQ3369, AQ-3370, AQ-3371, and AQ-3372).

6. Therapeutic Use of Green AMPs in DFU

AMPs are a promising alternative for infected wounds since they are active against a wide range of Gram-positive and Gram-negative bacteria. In addition, AMPs can exhibit immunomodulatory and angiogenic properties, stimulate cell proliferation and migration, and accelerate wound healing [2]. Although the AMPs referred above were obtained using green processes and can be relevant as therapeutic option for DFUs (Table 1), some of their inherent steps still need to be improved to develop a completely green protocol. Only a few AMPs are being extracted using more environmentally friendly approaches: Cn-AMPs, SP-1, KLENCNYAVELGK, Lyngbyazothrins mixture C/D, and Laxaphycin A, B, and B3. This may be explained by the recent increase in the awareness for the need of more sustainable and safer processes. AMPs extracted by greener methodologies include the Cn-AMPs: three antimicrobial peptides extracted from coconut water [73]. The extraction of these AMPs has been achieved through simple methodologies that did not use toxic solvents. Nevertheless, reverse phase chromatography (HPLC) was used in their purification stage. In order to reduce the generation of organic waste and accomplish a complete green extraction protocol, this technique could be replaced by new promising and eco-friendly approaches such as MMC, MCSGP, or even SFC [49]. These peptides showed antimicrobial activity against multiple Gram-positive and Gram-negative bacteria and fungi [73]. Interestingly, further studies investigated the promiscuity of these peptides and demonstrated that in addition to their antimicrobial activities, they also present activity against cancerous cells and immunostimulatory effects [91–93]. The immunostimulatory effect is a very important property, particularly in diabetes due to the low-grade inflammation, and it is difficult to have a controlled inflammatory phase of wound healing. This is of utmost importance as it will dictate the progression of wound closure [94,95].

Another example is SP-1, an AMP extracted by enzymatic hydrolysis from the cyanobacteria *Spirulina platensis*. The peptide KLENCNYAVELGK was successfully extracted from pepsin hydrolysates derived from *Limnospira sp*. This peptide showed antibacterial activity against *Escherichia coli* and *Staphylococcus aureus* [47]. It also presents antioxidant, anti-hypertensive, anti-diabetes, and anti-obesity activities [45,46]. Most of these peptides have antioxidant properties which are important to promote cell function in wounds under diabetic conditions, particularly, cell migration and proliferation, as well as angiogenic properties [96,97]. Moreover, for a proper wound healing, the overall condition of the patient is very important. The anti-hypertensive and anti-diabetic properties of the peptides will contribute to the control and improvement of the patient's general condition, and consequently better wound healing. In this sense, peptides that improve the general condition of diabetes could make a better contribution to the treatment of DFUs (Figure 3). In addition, some peptides such as PvD1, cycloviolacin O2, pitipeptolides C-F, and laxaphycins A, B, and B3 have shown to have high cytotoxicity against tumor cells [98–100]. This is an important property since diabetic patients present an elevated risk of developing different types of cancers [101,102].

Furthermore, the peptides in the Lyngbyazothrins mixture C/D and Laxaphycin A, B, and B3 obtained by greener methodologies also showed activity against common pathogens found in chronic DFUs [84,88]. The cyclic undecapeptides Lyngbyazothrins C/D showed antimicrobial activity against *Escherichia coli, Pseudomonas aeruginosa, Serratia marcesens,* and *Bacillus subtilis* [84]. Cyanobacterial peptides laxaphycins A, B, and B3 demonstrated antibacterial activity against *Listeria monocytogenes, Bacillus cereus,* and *Staphylococcus aureus* [88].

In addition to the aforementioned peptides, there are others obtained by less eco-friendly methodologies that present interesting properties concerning DFUs treatment. A study reported a peptide known as AQ-1766 was extracted from the marine microalgae *Tetraselmis suecica,* with high activity against *E. coli, P. aeruginosa,* MRSA, *L. monocytogenes,* and *M. luteus)* [89]. The fact that this particular AMP possesses high activity against MRSA, an antibiotic-resistant and prevalent pathogen in infected chronic DFUs, makes them an interesting and therapeutic approach for DFUs. Moreover, further research demonstrated that the peptide Snakin-Z also exhibited high antioxidant activity in addition to its antimicrobial properties [103].

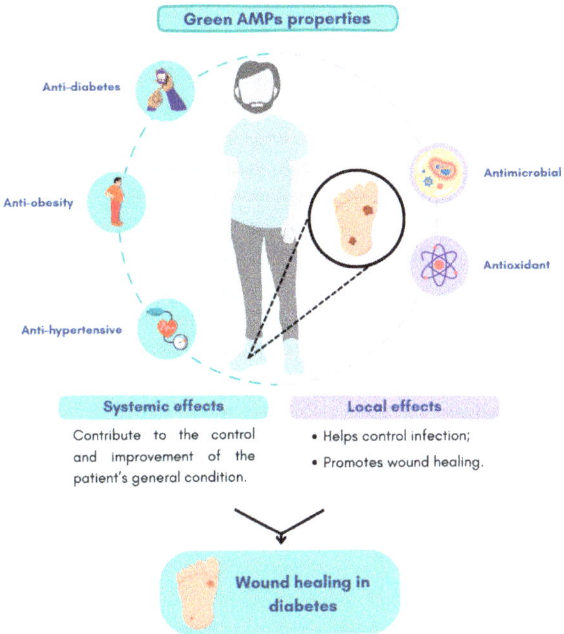

Figure 3. Effects of more environmentally friendly AMPs as potential therapeutic agents in chronic DFUs. Figure created in Canva.com (accessed on 21 February 2023).

Plants and microalgae have been considered sustainable and attractive sources of novel AMPs. Nevertheless, many of the extraction and production processes currently in use still employ hazardous substances [87,88]. The industrial scale-up production of AMPs is still a significant challenge due to highly expensive chemical methodologies [93].

Similarly, the current synthesis methodologies in use are still not completely green, in part due to their use of or the creation of toxic by-products. However, efforts are being made in order to develop more sustainable methodologies. This can be achieved by the replacement of hazardous substances and techniques by more sustainable options. Thus, it is imperative that the current synthesis of synthetic AMPs already identified as potential therapeutic agents for DFUs can be produced using greener techniques.

The potential therapeutic role of most of these peptides for DFU have not yet been investigated. However, due to their multiple properties, it would be important to further study the effects of novel greener AMPs as promising therapeutic agents for non-healing chronic infected wounds.

Table 1. Antimicrobial peptides extracted from sustainable sources as therapeutic options for DFU.

Source	AMPs/Sequences	Susceptible Species	Other Effects	Ref.
Plants				
Phaseolus vulgaris seeds	PvD1	*Candida albicans* *Candida parapsilosis* *Candida guilliermondii* *Candida tropicalis* *Saccharomyces cerevisiae*	Activity against tumor cells	[70,98]
Ziziphus jujuba fruits	Snakin-Z	*Staphylococcus aureus* *Escherichia coli* *Bacillus subtili* *Klebsiella pneumoniae*	Antioxidant activity	[71,103]

Table 1. *Cont.*

Source	AMPs/Sequences	Susceptible Species	Other Effects	Ref.
Viola odorata	Cycloviolacin O2	*S. enterica serovar* Typhimurium LT2 *Escherichia coli* *Klebsiella pneumoniae* *Pseudomonas aeruginosa*	Activity against tumor cells	[72,99]
Cocos nucifera L.	Cn-AMP1 Cn-AMP2 Cn-AMP3	*Escherichia coli* *Bacillus subtilis* *Pseudomonas aeruginosa* *Staphylococcus aureus*	Activity against tumor cells Immunostimulatory activity	[73]
Cottonseed defatted protein powder	CHQQEQRP DENFRKF EWPEEGQRR KPPIMPIGKG KDFPGRR LGLRSGIILCNV PRNFQQQLR QNLNALQPK SQEATSPR	*Staphylococcus aureus* (ATCC27068) *Escherichia coli* (ATCC25922) *Steptococcus* (CMCC35668) *Salmonella* (CMCC50013)	-	[74]
Nicotiana tabacum	LFchimera	*Escherichia coli*	-	[78]
Nicotiana tabacum	Colicin M	*Escherichia coli* *Klebsiella pneumoniae*	-	[79]
Nicotiana tabacum	Protegrin-1	*Klebsiella pneumoniae* *Staphylococcus aureus* *Escherichia coli* *Mycobacterium bovis* *Candida albicans*	-	[80]
Microalgae				
Spirulina platensis	SP-1	*Escherichia coli* *Staphylococcus aureus*	Antioxidant, antihypertensive, anti-diabetes, and anti-obesity	[45,46]
Limnospira maxima	KLENCNYAVELGK	*Escherichia coli* *Staphylococcus aureus*	-	[47]
Lyngbya sp.	Lyngbyazothrins mixture C/D	*Bacillus subtilis* *Escherichia coli* *Pseudomonas aeruginosa* *Serratia marcesens*	-	[84]
Lyngbya majuscula	Pitipeptolides C-F	*Mycobacterium tuberculosis*	Activity against tumor cells	[86,100]
Microcystis aeruginosa (NIES-88)	Kawaguchipeptin B	*Staphylococcus aureus*	-	[87]
Hawaii and Caribbean collection of cyanobacteria	Laxaphycin A	*Listeria monocytogenes* *Bacillus cereus* *Staphylococcus aureus*	Activity against tumor cells	[88,100]
	Laxaphycin B	*Listeria monocytogenes* *Bacillus cereus* *Staphylococcus aureus*		
	Laxaphycin B3	*Bacillus cereus*		
Tetraselmis suecica	AQ-1766 AQ-3001 AQ-3002 AQ3369 AQ-3370 AQ-3371 AQ-3372	*Escherichia coli* *Salmonella typhimurium* *Pseudomonas aeruginosa* *Bacillus cereus* Methicillin-resistant *S. aureus* (MRSA) *Listeria monocytogenes*	-	[89]

7. Conclusions and Future Perspectives

Wound healing is a complex biological process which is impaired under diabetes conditions, often leading to chronic non-healing wound infections. The advances in therapeutic approaches to diabetic wound care recognizes the potential of AMPs against infectious pathogens, and also provides other regenerative functions to improve wound healing. Green AMPs represent a promising alternative to conventional antibiotics and could be the answer to help fight antimicrobial resistance. Over the last decades, the raising awareness of climate change and high levels of pollution has increased the consciousness of the need to protect the environment. In this context, the scientific community has been working on the development of novel, greener, and more environmentally friendly methodologies for AMP extraction and synthesis. Plants and microalgae have been considered sustainable, and attractive sources of novel AMPs. Nevertheless, these green sources still have some limitations that need to be overcome. This review underlines the need for further research on new eco-friendly methodologies that allow complete green protocols for AMPs extraction, purification, and synthesis. Moreover, future research on the potential therapeutic role of these green peptides for treating non-healing infected wounds may contribute to the improving the treatment of DFUs.

Author Contributions: Investigation (literature review): I.D.T.; writing—original draft preparation: I.D.T.; writing—review and editing: I.D.T., E.C.L. and E.C.; supervision, E.C.L. and E.C.; funding acquisition: E.C.L. and E.C. All authors have read and agreed to the published version of the manuscript.

Funding: This work was financed by the European Regional Development Fund (ERDF) through the Centro 2020 Regional Operational Programme and through the COMPETE 2020—Operational Programme for Competitiveness and Internationalization and Portuguese national funds via FCT—Fundação para a Ciência e a Tecnologia, under projects POCI-01-0145-FEDER-007440, DL57/2016/CP1448/CT0024, UIDB/04539/2020, UIDP/04539/2020 and LA/P/0058/2020.

Institutional Review Board Statement: Not applicable.

Informed Consent Statement: Not applicable.

Data Availability Statement: Not applicable.

Acknowledgments: The figures were created with Canva.com (available online: Canva.com, accessed on 24 January 2023).

Conflicts of Interest: The authors declare no conflict of interest.

References

1. International Diabetes Federation. *IDF Diabetes Atlas*, 10th ed.; International Diabetes Federation: Brussels, Belgium, 2021; ISBN 978-2-930229-98-0.
2. Da Silva, J.; Leal, E.C.; Carvalho, E. Bioactive antimicrobial peptides as therapeutic agents for infected diabetic foot ulcers. *Biomolecules* **2021**, *11*, 1894. [CrossRef] [PubMed]
3. Armstrong, D.G.; Boulton, A.J.; Bus, S.A. Diabetic Foot Ulcers and Their Recurrence. *New Engl J. Med.* **2017**, *376*, 2367–2375. [CrossRef]
4. Vibha, S.P.; Kulkarni, M.M.; Kirthinath Ballala, A.B.; Kamath, A.; Maiya, G.A. Community based study to assess the prevalence of diabetic foot syndrome and associated risk factors among people with diabetes mellitus. *BMC Endocr. Disord.* **2018**, *18*, 43. [CrossRef] [PubMed]
5. Edmonds, M.; Manu, C.; Vas, P. The current burden of diabetic foot disease. *J. Clin. Orthop. Trauma* **2021**, *17*, 88–93. [CrossRef] [PubMed]
6. Petkovic, M.; Sørensen, A.E.; Leal, E.C.; Carvalho, E.; Dalgaard, L.T. Mechanistic actions of microRNAs in diabetic wound healing. *Cells* **2020**, *9*, 2228. [CrossRef]
7. Guo, S.A.; DiPietro, L.A. Factors affecting wound healing. *J. Dent. Res.* **2010**, *89*, 219–229. [CrossRef]
8. Ramirez-Acuña, J.M.; Cardenas-Cadena, S.A.; Marquez-Salas, P.A.; Garza-Veloz, I.; Perez-Favila, A.; Cid-Baez, M.A.; Flores-Morales, V.; Martinez-Fierro, M.L. Diabetic foot ulcers: Current advances in antimicrobial therapies and emerging treatments. *Antibiotics* **2019**, *8*, 193. [CrossRef]
9. Browne, K.; Chakraborty, S.; Chen, R.; Willcox, M.D.; Black, D.S.; Walsh, W.R.; Kumar, N. A new era of antibiotics: The clinical potential of antimicrobial peptides. *Int. J. Mol. Sci.* **2020**, *21*, 7047. [CrossRef]

10. Singer, A.J.; Clark, R.A.F. Cutaneous Wound Healing. *New Engl. J. Med.* **1999**, *341*, 738–746. [CrossRef]
11. Uivaraseanu, B.; Bungau, S.; Tit, D.M.; Fratila, O.; Rus, M.; Maghiar, T.A.; Pantis, C.; Vesa, C.M.; Zaha, D.C. Clinical, pathological and microbiological evaluation of diabetic foot syndrome. *Medicina* **2020**, *56*, 380. [CrossRef]
12. Grice, E.A.; Segre, J.A. The skin microbiome. *Nat. Rev. Microbiol.* **2011**, *9*, 244–253. [CrossRef]
13. Rademacher, F.; Gläser, R.; Harder, J. Antimicrobial peptides and proteins: Interaction with the skin microbiota. *Exp. Dermatol.* **2021**, *30*, 1496–1508. [CrossRef]
14. Kalan, L.R.; Meisel, J.S.; Loesche, M.A.; Horwinski, J.; Soaita, I.; Chen, X.; Uberoi, A.; Gardner, S.E.; Grice, E.A. Strain-and species-level variation in the microbiome of diabetic wounds is associated with clinical outcomes and therapeutic efficacy. *Cell Host Microbe* **2019**, *25*, 641–655. [CrossRef] [PubMed]
15. Kalan, L.R.; Brennan, M.B. The role of the microbiome in nonhealing diabetic wounds. *Ann. N. Y. Acad. Sci.* **2019**, *1435*, 79–92. [CrossRef] [PubMed]
16. Kareliya, H.; Bichile, L.; Bal, A.; Varaiya, A.; Bhalekar, P. Fungal Infection in Diabetic Foot a Clinicomicrobiological Study. *Acta Sci. Microbiol.* **2019**, *2*, 49–55.
17. Kalan, L.; Loesche, M.; Hodkinson, B.P.; Heilmann, K.; Ruthel, G.; Gardner, S.E.; Grice, E.A. Redefining the Chronic-Wound Microbiome: Fungal Communities Are Prevalent, Dynamic, and Associated with Delayed Healing. *MBio* **2016**, *7*, e01058-16. [CrossRef] [PubMed]
18. Pouget, C.; Dunyach-Remy, C.; Pantel, A.; Schuldiner, S.; Sotto, A.; Lavigne, J.P. Biofilms in diabetic foot ulcers: Significance and clinical relevance. *Microorganisms* **2020**, *8*, 1580. [CrossRef]
19. Banu, A.; Noorul Hassan, M.M.; Rajkumar, J.; Srinivasa, S. Spectrum of bacteria associated with diabetic foot ulcer and biofilm formation: A prospective study. *Australas. Med. J.* **2015**, *8*, 280–285. [CrossRef]
20. Fry, D.E. Antimicrobial peptides. *Surg. Infect.* **2018**, *19*, 804–811. [CrossRef]
21. Depta, J.; Małkowska, P.; Wysokińska, M.; Todorska, K.; Sierawska, O.; Hrynkiewicz, R.; Bebnowska, D.; Niedźwiedzka-Rystwej, P. Therapeutic Role of Antimicrobial Peptides in Diabetes Mellitus. *Biologics* **2022**, *2*, 92–106. [CrossRef]
22. Gaiser, R.A. Antimicrobial Peptides and the Interplay between Microbes and Host: Towards Preventing Porcine Infections with Streptococcus Suis. Ph.D. Thesis, Wageningen University and Research, Wageningen, The Netherlands, 2016.
23. Gallo, R.L.; Nizet, V. Endogenous production of antimicrobial peptides in innate immunity and human disease. *Curr. Allergy Asthma Rep.* **2003**, *3*, 402–409. [CrossRef]
24. Zhang, L.J.; Gallo, R.L. Antimicrobial peptides. *Curr. Biol.* **2016**, *26*, R14–R19. [CrossRef] [PubMed]
25. Gonzalez-Curiel, I.; Trujillo, V.; Montoya-Rosales, A.; Rincon, K.; Rivas-Calderon, B.; De Haro-Acosta, J.; Marin-Luevano, P.; Lozano-Lopez, D.; Enciso-Moreno, J.A.; Rivas-Santiago, B. 1,25-dihydroxyvitamin D3 induces LL-37 and HBD-2 production in keratinocytes from diabetic foot ulcers promoting wound healing: An in vitro model. *PLoS ONE* **2014**, *9*, e111355. [CrossRef] [PubMed]
26. Al-Shibly, I.K.; Alhamdany, M.H.; Al-Kaif, R.A.I.; Al-Kaif, L.A. Immunological Base Behind the Increased Susceptibility of Diabetic Patients for Infections. *Indian J. Public Health* **2019**, *10*, 3047–3051. [CrossRef]
27. De Marco, B.A.; Rechelo, B.S.; Tótoli, E.G.; Kogawa, A.C.; Salgado, H.R.N. Evolution of green chemistry and its multidimensional impacts: A review. *Saudi Pharmaceutical Journal* **2019**, *27*, 1–8. [CrossRef]
28. Anastas, P.T.; Warner, J.C. *Green Chemistry: Theory and Practice*; Oxford University Press: New York, NY, USA, 1998; p. 30.
29. Abdussalam-Mohammed, W.; Ali, A.Q.; Errayes, A.O. Green chemistry: Principles, applications, and disadvantages. *Chem. Methodol.* **2020**, *4*, 408–423.
30. Rodgers, K.; McLellan, I.; Peshkur, T.; Williams, R.; Tonner, R.; Hursthouse, A.S.; Knapp, C.W.; Henriquez, F.L. Can the legacy of industrial pollution influence antimicrobial resistance in estuarine sediments? *Environ. Chem. Lett.* **2019**, *17*, 595–607. [CrossRef]
31. Wilkinson, J.L.; Boxall, A.B.; Kolpin, D.W.; Leung, K.M.; Lai, R.W.; Galbán-Malagón, C.; Teta, C. Pharmaceutical pollution of the world's rivers. *Proc. Natl. Acad. Sci. USA* **2022**, *119*, e2113947119. [CrossRef]
32. Buelow, E.; Ploy, M.C.; Dagot, C. Role of pollution on the selection of antibiotic resistance and bacterial pathogens in the environment. *Curr. Opin. Microbiol.* **2021**, *64*, 117–124. [CrossRef]
33. Cue, B.W.; Zhang, J. Green process chemistry in the pharmaceutical industry. *Green Chem. Lett. Rev.* **2009**, *2*, 193–211. [CrossRef]
34. Li, J.; Hu, S.; Jian, W.; Xie, C.; Yang, X. Plant antimicrobial peptides: Structures, functions, and applications. *Bot. Stud.* **2021**, *62*, 5. [CrossRef] [PubMed]
35. Rojas, V.; Rivas, L.; Cárdenas, C.; Guzmán, F. Cyanobacteria and eukaryotic microalgae as emerging sources of antibacterial peptides. *Molecules* **2020**, *25*, 5804. [CrossRef]
36. Singh, S.; Verma, D.K.; Thakur, M.; Tripathy, S.; Patel, A.R.; Shah, N.; Aguilar, C.N. Supercritical fluid extraction (SCFE) as green extraction technology for high-value metabolites of algae, its potential trends in food and human health. *Food Res. Int.* **2021**, *150*, 110746. [CrossRef] [PubMed]
37. Abou Elmaaty, T.; Sayed-Ahmed, K.; Elsisi, H.; Magdi, M. Optimization of extraction of natural antimicrobial pigments using supercritical fluids: A review. *Processes* **2022**, *10*, 2111. [CrossRef]
38. Barashkova, A.S.; Rogozhin, E.A. Isolation of antimicrobial peptides from different plant sources: Does a general extraction method exist? *Plant Methods* **2020**, *16*, 1–10. [CrossRef]
39. Uwineza, P.A.; Waśkiewicz, A. Recent advances in supercritical fluid extraction of natural bioactive compounds from natural plant materials. *Molecules* **2020**, *25*, 3847. [CrossRef]

40. Chemat, F.; Rombaut, N.; Sicaire, A.G.; Meullemiestre, A.; Fabiano-Tixier, A.S.; Abert-Vian, M. Ultrasound assisted extraction of food and natural products. Mechanisms, techniques, combinations, protocols and applications. A review. *Ultrason. Sonochemistry* **2017**, *34*, 540–560. [CrossRef]
41. Kumari, B.; Tiwari, B.K.; Hossain, M.B.; Brunton, N.P.; Rai, D.K. Recent advances on application of ultrasound and pulsed electric field technologies in the extraction of bioactives from agro-industrial by-products. *Food Bioprocess Technol.* **2018**, *11*, 223–241. [CrossRef]
42. Giordano, D.; Costantini, M.; Coppola, D.; Lauritano, C.; Pons, L.N.; Ruocco, N.; di Prisco, G.; Ianora, A.; Verde, C. Biotechnological Applications of Bioactive Peptides from Marine Sources. *Adv. Microb. Physiol.* **2018**, *73*, 171–220. [PubMed]
43. Cruz-Casas, D.E.; Aguilar, C.N.; Ascacio-Valdés, J.A.; Rodríguez-Herrera, R.; Chávez-González, M.L.; Flores-Gallegos, A.C. Enzymatic hydrolysis and microbial fermentation: The most favorable biotechnological methods for the release of bioactive peptides. *Food Chem. Mol. Sci.* **2021**, *3*, 100047. [CrossRef]
44. Fan, X.; Bai, L.; Zhu, L.; Yang, L.; Zhang, X. Marine algae-derived bioactive peptides for human nutrition and health. *J. Agric. Food Chem.* **2014**, *62*, 9211–9222. [CrossRef]
45. Sun, Y.; Chang, R.; Li, Q.; Li, B. Isolation and characterization of an antibacterial peptide from protein hydrolysates of Spirulina platensis. *Eur. Food Res. Technol.* **2016**, *242*, 685–692. [CrossRef]
46. Yücetepe, A.; Özçelik, B. Bioactive peptides isolated from microalgae Spirulina platensis and their biofunctional activities. *Akad. Gıda* **2016**, *14*, 412–417.
47. Sathya, R.; MubarakAli, D.; Mehboob Nousheen, M.G.; Vasimalai, N.; Thajuddin, N.; Kim, J.-W. An Investigation of Pepsin Hydrolysate of Short Antibacterial Peptides Derived from *Limnospira* Sp. *Appl. Biochem. Biotechnol.* **2022**, *194*, 5580–5593. [CrossRef]
48. Isidro-Llobet, A.; Kenworthy, M.N.; Mukherjee, S.; Kopach, M.E.; Wegner, K.; Gallou, F.; Smith, A.G.; Roschangar, F. Sustainability challenges in peptide synthesis and purification: From R&D to production. *J. Org. Chem.* **2019**, *84*, 4615–4628.
49. Lima, P.G.; Oliveira, J.T.; Amaral, J.L.; Freitas, C.D.; Souza, P.F. Synthetic antimicrobial peptides: Characteristics, design, and potential as alternative molecules to overcome microbial resistance. *Life Sci.* **2021**, *278*, 119647. [CrossRef] [PubMed]
50. Sarkar, T.; Chetia, M.; Chatterjee, S. Antimicrobial peptides and proteins: From nature's reservoir to the laboratory and beyond. *Front. Chem.* **2021**, *9*, 432. [CrossRef] [PubMed]
51. Vanzolini, T.; Bruschi, M.; Rinaldi, A.C.; Magnani, M.; Fraternale, A. Multitalented synthetic antimicrobial peptides and their antibacterial, antifungal and antiviral mechanisms. *Int. J. Mol. Sci.* **2022**, *23*, 545. [CrossRef] [PubMed]
52. Bednarska, N.G.; Wren, B.W.; Willcocks, S.J. The importance of the glycosylation of antimicrobial peptides: Natural and synthetic approaches. *Drug Discov. Today* **2017**, *22*, 919–926. [CrossRef]
53. Rounds, T.; Straus, S.K. Lipidation of antimicrobial peptides as a design strategy for future alternatives to antibiotics. *Int. J. Mol. Sci.* **2020**, *21*, 9692. [CrossRef]
54. Li, W.; Separovic, F.; O'Brien-Simpson, N.M.; Wade, J.D. Chemically modified and conjugated antimicrobial peptides against superbugs. *Chem. Soc. Rev.* **2021**, *50*, 4932–4973. [CrossRef] [PubMed]
55. Li, F.; Brimble, M. Using chemical synthesis to optimise antimicrobial peptides in the fight against antimicrobial resistance. *Pure Appl. Chem.* **2019**, *91*, 181–198. [CrossRef]
56. Gan, B.H.; Gaynord, J.; Rowe, S.M.; Deingruber, T.; Spring, D.R. The multifaceted nature of antimicrobial peptides: Current synthetic chemistry approaches and future directions. *Chem. Soc. Rev.* **2021**, *50*, 7820–7880. [CrossRef] [PubMed]
57. Zapotoczna, M.; Forde, É.; Hogan, S.; Humphreys, H.; O'Gara, J.P.; Fitzgerald-Hughes, D.; Devolcelle, M.; O'Neill, E. Eradication of Staphylococcus aureus biofilm infections using synthetic antimicrobial peptides. *J. Infect. Dis.* **2017**, *215*, 975–983. [CrossRef]
58. Varnava, K.G.; Sarojini, V. Making solid-phase peptide synthesis greener: A review of the literature. *Chem. Asian J.* **2019**, *14*, 1088–1097. [CrossRef]
59. Jadhav, S.; Martin, V.; Egelund, P.H.; Castro, H.J.; Krüger, T.; Richner, F.; Quement, S.T.; Albericio, F.; Dettner, F.; Lechner, C.; et al. Replacing DMF in solid-phase peptide synthesis: Varying the composition of green binary solvent mixtures as a tool to mitigate common side-reactions. *Green Chem.* **2021**, *23*, 3312–3321. [CrossRef]
60. Lopez, J.; Pletscher, S.; Aemissegger, A.; Bucher, C.; Gallou, F. N-butylpyrrolidinone as alternative solvent for solid-phase peptide synthesis. *Org. Process Res. Dev.* **2018**, *22*, 494–503. [CrossRef]
61. Declerck, V.; Nun, P.; Martinez, J.; Lamaty, F. Solvent-free synthesis of peptides. *Angew. Chem. Int. Ed.* **2009**, *48*, 9318–9321. [CrossRef]
62. Lawrenson, S.; North, M.; Peigneguy, F.; Routledge, A. Greener solvents for solid-phase synthesis. *Green Chem.* **2017**, *19*, 952–962. [CrossRef]
63. Wegner, K.; Barnes, D.; Manzor, K.; Jardine, A.; Moran, D. Evaluation of greener solvents for solid-phase peptide synthesis. *Green Chem. Lett. Rev.* **2021**, *14*, 153–164. [CrossRef]
64. Korany, M.A.; Mahgoub, H.; Haggag, R.S.; Ragab, M.A.; Elmallah, O.A. Green chemistry: Analytical and chromatography. *J. Liq. Chromatogr. Relat. Technol.* **2017**, *40*, 839–852. [CrossRef]
65. Kannaiah, K.P.; Sugumaran, A.; Chanduluru, H.K.; Rathinam, S. Environmental impact of greenness assessment tools in liquid chromatography–A review. *Microchem. J.* **2021**, *170*, 106685. [CrossRef]
66. Sajid, M.; Płotka-Wasylka, J. Green analytical chemistry metrics: A review. *Talanta* **2022**, *238*, 123046. [CrossRef] [PubMed]

67. Nawrot, R.; Barylski, J.; Nowicki, G.; Broniarczyk, J.; Buchwald, W.; Goździcka-Józefiak, A. Plant antimicrobial peptides. *Folia Microbiol.* **2014**, *59*, 181–196. [CrossRef] [PubMed]
68. Okada, T.; Yoshizumi, H.; Terashima, Y. A Lethal Toxic Substance for Brewing Yeast in Wheat and Barley: Part I. Assay of Toxicity on Various Grains, and Sensitivity of Various Yeast Strains Part II. Isolation and Some Properties of Toxic Principle. *Agric. Biol. Chem.* **1970**, *34*, 1084–1094. [CrossRef]
69. Vilas Boas, L.C.P.; Campos, M.L.; Berlanda, R.L.A.; de Carvalho Neves, N.; Franco, O.L. Antiviral peptides as promising therapeutic drugs. *Cell. Mol. Life Sci.* **2019**, *76*, 3525–3542. [CrossRef]
70. Games, P.D.; Dos Santos, I.S.; Mello, É.O.; Diz, M.S.; Carvalho, A.O.; de Souza-Filho, G.A.; Cunha, M.; Vasconcelos, I.M.; Ferreira, B.D.S.; Gomes, V.M. Isolation, characterization and cloning of a cDNA encoding a new antifungal defensin from Phaseolus vulgaris L. seeds. *Peptides* **2008**, *29*, 2090–2100. [CrossRef]
71. Daneshmand, F.; Zare-Zardini, H.; Ebrahimi, L. Investigation of the antimicrobial activities of Snakin-Z, a new cationic peptide derived from Zizyphus jujuba fruits. *Nat. Prod. Res.* **2013**, *27*, 2292–2296. [CrossRef]
72. Pränting, M.; Lööv, C.; Burman, R.; Göransson, U.L.F.; Andersson, D.I. The cyclotide cycloviolacin O2 from Viola odorata has potent bactericidal activity against Gram-negative bacteria. *J. Antimicrob. Chemother.* **2010**, *65*, 1964–1971. [CrossRef]
73. Mandal, S.M.; Dey, S.; Mandal, M.; Sarkar, S.; Maria-Neto, S.; Franco, O.L. Identification and structural insights of three novel antimicrobial peptides isolated from green coconut water. *Peptides* **2009**, *30*, 633–637. [CrossRef]
74. Song, W.; Kong, X.; Hua, Y.; Chen, Y.; Zhang, C.; Chen, Y. Identification of antibacterial peptides generated from enzymatic hydrolysis of cottonseed proteins. *LWT* **2020**, *125*, 109199. [CrossRef]
75. Farhadpour, M.; Hashempour, H.; Talebpour, Z.; A-Bagheri, N.; Shushtarian, M.S.; Gruber, C.W.; Ghassempour, A. Microwave-assisted extraction of cyclotides from Viola ignobilis. *Anal. Biochem.* **2016**, *497*, 83–89. [CrossRef] [PubMed]
76. Esmaeili, M.A.; Abagheri-Mahabadi, N.; Hashempour, H.; Farhadpour, M.; Gruber, C.W.; Ghassempour, A. Viola plant cyclotide vigno 5 induces mitochondria-mediated apoptosis via cytochrome C release and caspases activation in cervical cancer cells. *Fitoterapia* **2016**, *109*, 162–168. [CrossRef]
77. Shanmugaraj, B.K.; Bulaon, C.J.I.; Malla, A.; Phoolcharoen, W. Biotechnological insights on the expression and production of antimicrobial peptides in plants. *Molecules* **2021**, *26*, 4032. [CrossRef] [PubMed]
78. Chahardoli, M.; Fazeli, A.; Ghabooli, M. Recombinant production of bovine Lactoferrin-derived antimicrobial peptide in tobacco hairy roots expression system. *Plant Physiol. Biochem.* **2018**, *123*, 414–421. [CrossRef]
79. Łojewska, E.; Sakowicz, T.; Kowalczyk, A.; Konieczka, M.; Grzegorczyk, J.; Sitarek, P.; Skała, E.; Czarny, P.; Sliwinski, T.; Kowalczyk, T. Production of recombinant colicin M in *Nicotiana tabacum* plants and its antimicrobial activity. *Plant Biotechnol. Rep.* **2020**, *14*, 33–43. [CrossRef]
80. Patiño-Rodríguez, O.; Ortega-Berlanga, B.; Llamas-González, Y.Y.; Flores-Valdez, M.A.; Herrera-Díaz, A.; Montes-de-Oca-Luna, R.; Korban, S.S.; Alpuche-Solís, A.G. Transient expression and characterization of the antimicrobial peptide protegrin-1 in *Nicotiana tabacum* for control of bacterial and fungal mammalian pathogens. *Plant Cell Tissue Organ Cult. PCTOC* **2013**, *115*, 99–106. [CrossRef]
81. Smith, V.J.; Desbois, A.P.; Dyrynda, E.A. Conventional and unconventional antimicrobials from fish, marine invertebrates and micro-algae. *Mar. Drugs* **2010**, *8*, 1213–1262. [CrossRef]
82. Falaise, C.; François, C.; Travers, M.A.; Morga, B.; Haure, J.; Tremblay, R.; Turcotte, F.; Pasetto, P.; Gastineau, R.; Hardivillier, Y.; et al. Antimicrobial compounds from eukaryotic microalgae against human pathogens and diseases in aquaculture. *Marine drugs* **2016**, *14*, 159. [CrossRef]
83. Vincent, W.F. Cyanobacteria. In *Encyclopedia of Inland Waters*, 1st ed.; Elsevier Academic Press: Amsterdam, The Netherlands; pp. 226–232.
84. Zainuddin, E.N.; Jansen, R.; Nimtz, M.; Wray, V.; Preisitsch, M.; Lalk, M.; Mundt, S. LyngbyazothrinsA−D, Antimicrobial CycliUndecapeptides from the Cultured Cyanobacterium Lyngbya sp. *J. Nat. Prod.* **2009**, *72*, 2080. [CrossRef]
85. Xue, Y.; Zhao, P.; Quan, C.; Zhao, Z.; Gao, W.; Li, J.; Zu, X.; Fu, D.; Feng, S.; Bai, X. Cyanobacteria-derived peptide antibiotics discovered since 2000. *Peptides* **2018**, *107*, 17–24. [CrossRef]
86. Montaser, R.; Paul, V.J.; Luesch, H. Pitipeptolides C–F, antimycobacterial cyclodepsipeptides from the marine cyanobacterium Lyngbya majuscula from Guam. *Phytochemistry* **2011**, *72*, 2068–2074. [CrossRef]
87. Ishida, K.; Matsuda, H.; Murakami, M.; Yamaguchi, K. Kawaguchipeptin B, an antibacterial cyclic undecapeptide from the cyanobacterium Microcystis aeruginosa. *J. Nat. Prod.* **1997**, *60*, 724–726. [CrossRef]
88. Dussault, D.; Vu, K.D.; Vansach, T.; Horgen, F.D.; Lacroix, M. Antimicrobial effects of marine algal extracts and cyanobacterial pure compounds against five foodborne pathogens. *Food Chem.* **2016**, *199*, 114–118. [CrossRef]
89. Guzmán, F.; Wong, G.; Román, T.; Cárdenas, C.; Alvárez, C.; Schmitt, P.; Albericio, F.; Rojas, V. Identification of antimicrobial peptides from the microalgae Tetraselmis suecica (Kylin) Butcher and bactericidal activity improvement. *Mar. Drugs* **2019**, *17*, 453. [CrossRef]
90. Lauritano, C.; Andersen, J.H.; Hansen, E.; Albrigtsen, M.; Escalera, L.; Esposito, F. Bioactivity screening of microalgae for antioxidant, anti-inflammatory, anticancer, anti-diabetes and antibacterial activities. *Front. Mar. Sci.* **2016**, *3*, 68. [CrossRef]
91. Silva, O.N.; Porto, W.F.; Migliolo, L.; Mandal, S.M.; Gomes, D.G.; Holanda, H.H.; Silva, R.S.P.; Dias, S.C.; Costa, M.P.; Costa, C.R.; et al. Cn-AMP1: A new promiscuous peptide with potential for microbial infections treatment. *Pept. Sci.* **2012**, *98*, 322–331. [CrossRef] [PubMed]

92. Santana, M.J.; de Oliveira, A.L.; Queiroz Júnior, L.H.; Mandal, S.M.; Matos, C.O.; Dias, R.D.O.; Franco, O.L.; Lião, L.M. Structural insights into Cn-AMP1, a short disulfide-free multifunctional peptide from green coconut water. *FEBS Lett.* **2015**, *589*, 639–644. [CrossRef] [PubMed]
93. Sathya, R.; MubarakAli, D.; MohamedSaalis, J.; Kim, J.W. A systemic review on microalgal peptides: Bioprocess and sustainable applications. *Sustainability* **2021**, *13*, 3262. [CrossRef]
94. Leal, E.C.; Emanuelli, T.; Santos, D.; Moura, J.; Fonseca, A.C.R.; Burgeiro, A.; Carvalho, E. Dysregulation of endoplasmic reticulum stress response in skin wounds in a streptozotocin-induced diabetes mouse model. *J. Mol. Endocrinol.* **2023**. [CrossRef]
95. Leal, E.C.; Carvalho, E.; Tellechea, A.; Kafanas, A.; Tecilazich, F.; Kearney, C.; Kuchibhotla, S.; Auster, M.E.; Kokkotou, E.; Mooney, D.J.; et al. Substance P promotes wound healing in diabetes by modulating inflammation and macrophage phenotype. *Am. J. Pathol.* **2015**, *185*, 1638–1648. [CrossRef] [PubMed]
96. Figueiredo, A.; Leal, E.C.; Carvalho, E. Protein tyrosine phosphatase 1B inhibition as a potential therapeutic target for chronic wounds in diabetes. *Pharmacol. Res.* **2020**, *159*, 104977. [CrossRef] [PubMed]
97. Leal, E.C.; Carvalho, E. Heme Oxygenase-1 as Therapeutic Target for Diabetic Foot Ulcers. *Int. J. Mol. Sci.* **2022**, *23*, 12043. [CrossRef] [PubMed]
98. Figueira, T.N.; Oliveira, F.D.; Almeida, I.; Mello, É.O.; Gomes, V.M.; Castanho, M.A.; Gaspar, D. Challenging metastatic breast cancer with the natural defensin PvD1. *Nanoscale* **2017**, *9*, 16887–16899. [CrossRef]
99. Gerlach, S.L.; Rathinakumar, R.; Chakravarty, G.; Göransson, U.; Wimley, W.C.; Darwin, S.P.; Mondal, D. Anticancer and chemosensitizing abilities of cycloviolacin O2 from Viola odorata and psyle cyclotides from Psychotria leptothyrsa. *Pept. Sci.* **2010**, *94*, 617–625. [CrossRef] [PubMed]
100. Qamar, H.; Hussain, K.; Soni, A.; Khan, A.; Hussain, T.; Chénais, B. Cyanobacteria as Natural Therapeutics and Pharmaceutical Potential: Role in Antitumor Activity and as Nanovectors. *Molecules* **2021**, *26*, 247. [CrossRef] [PubMed]
101. Kim, D.S.; Scherer, D.S.P.E. Obesity, Diabetes, and Increased Cancer Progression. *Diabetes Metab. J.* **2021**, *45*, 799–812. [CrossRef]
102. Wang, M.; Yang, Y.; Liao, Z. Diabetes and cancer: Epidemiological and biological links. *World J. Diabetes* **2020**, *11*, 227–238. [CrossRef]
103. Zare-Zardini, H.; Tolueinia, B.; Hashemi, A.; Ebrahimi, L.; Fesahat, F. Antioxidant and cholinesterase inhibitory activity of a new peptide from Ziziphus jujuba fruits. *Am. J. Alzheimer's Dis. Other Dement.®* **2013**, *28*, 702–709. [CrossRef]

antibiotics

MDPI

Review

Antiseptic Agents for Chronic Wounds: A Systematic Review

Koko Barrigah-Benissan [1,†], Jérôme Ory [1,†], Albert Sotto [2], Florian Salipante [3], Jean-Philippe Lavigne [1,*] and Paul Loubet [2]

1 Bacterial Virulence and Chronic Infections, INSERM U1047, Department of Microbiology and Hospital Hygiene, University of Montpellier, CHU Nîmes, 30029 Nîmes, France; koko.barrigahbenissan@chu-nimes.fr (K.B.-B.); jerome.ory@chu-nimes.fr (J.O.)
2 Bacterial Virulence and Chronic Infections, INSERM U1047, Department of Infectious Diseases, University of Montpellier, CHU Nîmes, 30029 Nîmes, France; albert.sotto@chu-nimes.fr (A.S.); paul.loubet@chu-nimes.fr (P.L.)
3 Department of Biostatistics, Epidemiology, Public Health and Innovation in Methodology, University of Montpellier, CHU Nîmes, 30029 Nîmes, France; florian.salipante@chu-nimes.fr
* Correspondence: jean.philippe.lavigne@chu-nimes.fr; Tel.: +33-466683202
† These authors contributed equally to this work.

Abstract: In many parts of the world, antiseptic agents remain non-indicated in chronic wound care. In the current context of bacterial resistance to antibiotics and the development of new-generation antiseptic agents, wound antisepsis represents an asset for the prevention of wound infection. We aimed to evaluate four common antiseptic agents in chronic wound care complete healing. The review protocol was based on the Cochrane Handbook for Systematic Reviews of Intervention and devised in accordance with the Preferred Reporting Items for Systematic Review and Meta-Analyses (PRISMA) statement guidelines. Five databases and three clinical trials registries were searched from inception to 30 June 2021 without language restrictions. We included randomised trials evaluating the efficacy of antiseptic agents in chronic wound care in adults. Interventions considered were those using antiseptics for cleansing or within a dressing. Risk of bias was assessed using the bias excel tool provided by the Bristol Academy. Evidence quality was assessed using Grading of Recommendation Assessment, Development and Evaluation (GRADE) criteria. Of 838 studies, 6 were finally included, with a total of 725 patients. The included studies assessed iodine (cadexomer or povidone iodine) (n = 3), polyhexanide (n = 2), and octenidine (n = 1). Limited evidence suggested a better wound healing completion with iodine compared to saline (two randomised controlled trials (RCT), 195 patients, pooled RR 1.85 (95%CI (1.27 to 2.69)), moderate-quality evidence). There was not enough evidence to suggest a difference in wound healing using octenidine or polyhexamide. None of the antiseptic agents influenced adverse event occurrence compared to saline.

Keywords: antiseptic agents; efficiency; iodine; systematic review; wound healing; wound infection

Citation: Barrigah-Benissan, K.; Ory, J.; Sotto, A.; Salipante, F.; Lavigne, J.-P.; Loubet, P. Antiseptic Agents for Chronic Wounds: A Systematic Review. *Antibiotics* **2022**, *11*, 350. https://doi.org/10.3390/antibiotics11030350

Academic Editor: Helena P. Felgueiras

Received: 18 January 2022
Accepted: 3 March 2022
Published: 6 March 2022

1. Introduction

Chronic wounds are wounds failing to proceed through the normal phases of healing in an orderly and timely manner. The definition of time without complete or partial healing differs across countries, ranging from 4 weeks to 3 months [1]. The Wound Healing Society defines four types of chronic wounds: diabetic foot ulcers (DFU), vascular ulcers (containing venous and arterial ulcers), and pressure ulcers (PU) [2].

Infection is a common complication of chronic wounds. Historically, the research on wound infection control and improvement was focused on reducing the "pathogen burden". However, quantitative consideration of microbial load is insufficient for assessing wound improvement or wound risk of infection [3,4]. Microorganisms in a chronic wound are co-aggregated together within a protective extracellular matrix, constituting a biofilm. This biofilm conformation induces a dramatically increased tolerance to host immune defences and a greater resistance to antimicrobials [5]. Biofilm delays wound healing by inducing an ineffective host inflammatory response and damaging host tissues [6]. For this reason, a management of the biofilm is more relevant for the treatment of chronic wound [7]. In 2011, Dissemond et al. classified wounds into four categories, depending on wound bed clinical and microbiological situation: (1) contaminated or colonised wounds without risk of infection; (2) colonised wounds at risk (WAR) or critically colonised wounds; (3) wounds with local infection; and (4) systemic infections and infected wounds. The authors suggested the use of antiseptic agents for wounds from the second category alongside other treatments [8].

Chronic wound care commences with wound bed preparation via (i) wound cleansing to create a wet or moist environment, favourable to healing. (ii) Wound debridement via removal of devitalised, contaminated tissue from within or adjacent to a wound, until surrounding healthy tissue is exposed [9,10]. Debridement can be mechanical (sharp debridement, surgical), enzymatical, or bio-surgical (e.g., maggot therapy) [11–15]. Negative pressure wound therapy has also been used for bacterial decontamination and wound bed preparation [16]. (iii) Application of a suitable dressing, according to the type of wound. (iv) Antibiotic treatment, exclusively for infected wounds. Other therapies are beneficial to specific wounds: compression therapy is required for venous leg ulcers (VLU) [17]; arterial revascularisation, offloading foot ulcers, and diabetes control are essential in DFU [18,19]; and skin assessment and care, offloading and pressure redistribution, dressings [19,20], and structured educational program are useful for all types of chronic wounds.

In some countries, tap water or saline remain the only recommended agents for wound cleansing. Antisepsis is a common, yet controversial, wound cleansing method. Some studies consider debridement alone insufficient to reduce the biofilm that delays wound healing and suggest antiseptics to delay biofilm reformation and reduce the risk of infection [7,20]. Antiseptic agents may complement the debridement process and control infection.

The primary mode of action of antiseptics can be pharmacological, metabolic, and/or immunological [21]. For the purpose of this review, antiseptic agents are defined as medication that can prevent the growth or destroy microorganisms in or on a living tissue. Following this definition of medication, antiseptic agents must pass through a drug authorisation procedure with a medicine agency [22]. The main antiseptic agents used in chronic wound care are halogenated compounds, alcohol-based agents, biguanides (e.g., polyhexanide also called polyhexamethylenebiguanide or PHMB, chlorhexidine), and quaternary ammoniums (e.g., octenidine). Halogenated compounds include subfamilies such as the iodine/iodophor agents (e.g., povidone iodine, cadexomer iodine) and chlorous agents (hypochlorite, hypochlorous acid) [21,22]. Alternative therapeutics (e.g., honey, silver), while been antimicrobial agents are not antiseptic agents as they did not go through an authorisation procedure for this purpose [23]. They therefore are not part of the antiseptic agents' classification. International guidelines recommend against the routine use of topical antiseptics to manage infected chronic wounds [24–27].

The emergence and diffusion of multidrug resistant bacteria and a better formulation of antiseptics with less side effects has renewed consideration of antisepsis. No resistance or adaptation has yet been observed in antiseptic agents with unspecific effects such as iodine agents, polyhexanide, octenidine, or oxidising agents (e.g., hypochlorous acid) [21]. In countries where antiseptic agents are a part of chronic wound care management protocol, there is no consensus on the best antiseptic agents for chronic wound care. Our study aimed to assess the evidence of four common antiseptic agents on chronic wound healing.

2. Results

2.1. Characteristics of the Included Studies

We identified 838 studies including 61 registered clinical trials through database searching and 31 through manual searching. Of these studies, 149 duplicates were removed, and a further 613 studies were excluded by title or abstract. Three studies could not be retrieved and they were withdrawn from the database. We fully screened 73 studies, leaving 6 studies eligible for inclusion (Figure 1) [28–33]. A total of 17 studies were excluded after full screening: 23 studies met a population exclusion criterion (10 studies were not clinical trials and 13 studies included non-chronic or infected wounds); 21 out of 23 were identified through databases and registers searching and the last 2 were identified via other methods. A total of 18 studies met an intervention exclusion criterion (non-antiseptic agents or unbalanced methodology; 16 were identified by databases or register searching and the last 2 by other methods). Nine studies met a comparison exclusion criterion (no control group), 11 studies were uncompleted or prematurely ended, and six studies were ongoing. The six included studies were published from 1989 to 2020 and included 725 patients [28–33]. No unpublished studies were included.

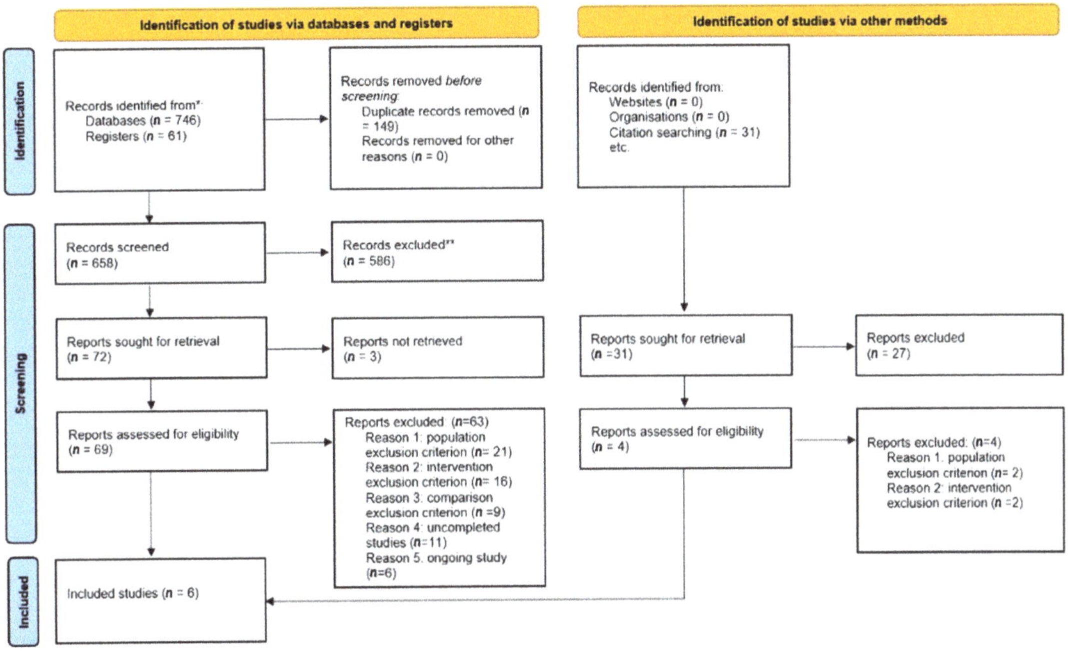

Figure 1. Study flow diagram.

The included studies are listed in Table S1. No studies assessing chlorous agents met our eligibility criteria. Saline was the comparator in all studies. None of the studies compared two or more antiseptic agents. Three studies evaluated the efficacy of iodine (cadexomer iodine or polyvidone iodine) [28,31,32], two studies evaluated polyhexanide [30,33], and one last study evaluated octenidine [29]. Five studies had two comparative groups of patients, and one had three comparative groups [32]. All studies were multicentre, taking place in 55 centres, both inpatient and outpatient settings across South Korea, India, France, Hungary, the UK, the USA, Germany, Canada, and Italy. Different types of chronic wounds were studied: one including DFU only [28]; one including VLU only [29]; two including PU only [30,31]; one including DFU, VLU, and PU [32]; and the last including chronic wounds without precision of the type of the wound [33]. The minimum duration used to define chronic wounds ranged from 4 weeks [28,29] to 3 months [31], although others did not state the definition, but gave the mean duration of the wounds [32], or merely described wounds as "chronic" [33], or only gave the duration of the comorbidity [30]. Study duration ranged from 4 weeks [30,33] to 24 weeks [31].

The seven domains of risk of bias were assessed following the Cochrane recommendation of 2011 [34] and are presented in Table S2. The most common source of bias was outcome assessment blinding (see Table S2). The Risk of Bias tool (2019) released by the

Cochrane collaboration was also used for the risk of bias assessment and presented in Table S3. The overall risk of bias among included studies was rated with some concerns for five studies [28–30,32,33] and high for one [31] (Table S3). A plot of the percentage of risk of bias assessments per domain is presented in Figure S1, as recommended by the Cochrane collaboration tool for risk of bias assessment in randomised trials (2019 version).

Three studies reported the primary outcome (wound healing) as the proportion of patients with complete wound healing at 4 and 8 weeks [29], at 12 weeks [32], and as healing time from wound size [29]. All studies assessed wound healing rate as wound size reduction by planimetry measurement. Three studies assessed pain as a secondary outcome with various pain scales (e.g., verbal rating scales, visual analogue scales) [30,31,33]. Two studies evaluated bacterial bioburden reduction [31,33]. Four studies reported adverse events (AEs) [28,29,31,32]. No AEs were noted in the last two studies [30,33]. Table 1 reports the summary of findings for the primary outcome and Table S4 the different outcomes reported among the studies.

2.2. *Iodine vs. Saline (3 Randomised Controlled Trials (RCT), 260 Patients)*

Three studies, with a total of 260 patients, compared iodine to saline [28,31,32]. In 1984, Holloway et al. studied the efficacy of cadexomer iodine in 3 months venous stasis ulcers over 24 weeks [31]. The study was rated at high risk of bias. In 2018, Raju et al. also studied the efficacy of cadexomer iodine in the treatment of various 1 month chronic ulcers over 12 weeks [32]. In 2019, Gwak et al. studied the efficacy of an 8 weeks treatment of povidone iodine in patients with DFU with a mean duration of 7 weeks [28]. The risk of bias was unclear for these studies.

Two studies assessed the primary outcome as the proportion of patients with complete wound healing (Table S1) [28,32]. Gwak et al. observed no significant difference in the two groups at 8 weeks (44.4% vs. 44.1%; $p = 0.978$) [28], whereas Raju et al. found a significant complete wound healing for patients treated with two different formulations of cadexomer iodine compared to saline at 12 weeks (61.9% vs. 20%; $p < 0.001$) [32]. The pooled data showed that iodine has a higher percentage of patients achieving complete wound healing (RR 1.85 (95%CI (1.27 to 2.69), $n = 2$ studies, moderate quality evidence) (Table 2). In the last study, the endpoint was the time to complete healing (Table S1) [31]. The authors found no significant difference between the two groups (31.0 days $\pm$ 14.1 for povidone iodine vs. 33.3 $\pm$ 12.6 for saline; $p = 6.54$).

Table 1. Description of included studies.

Study	Country	Date	Setting	Number of Participants	Diagnosis	Main Outcome (Scale Details)	Duration of the Study	Intervention	Comparator
Gwak et al., 2020 [28]	Korea	March 2016–September 2017	Multicentre 10 sites	71	DFU ≥ 1 cm^2 post debridement and no clinical signs of infection	Proportion of patients with complete wound healing	8 weeks	PVP-I 44.4% (16/36)	Saline 44.1% (15/35)
Raju et al., 2019 [32]	India	March 2016–March 2017	Multicentre 15 sites	124	Chronic ulcers (single VLU, DFU, PU with adequate arterial blood supply)	Proportion of patients with complete wound healing	12 weeks	Cadexomer iodine ointment: 65.9% (27/41) Cadexomer iodine powder: 58.1% (25/43)	Saline 20% (8/40)
Bellingeri et al., 2016 [30]	Italy	June 2010–December 2013	Multicentre 6 sites	289	PU less than 80 cm^2	Wound improvement measured by BWAT scale tool	4 weeks	PHMB	Saline
Vanscheidt et al., 2011 [29]	Germany, France, Hungary, UK	November 2007–December 2009	Multicentre 15 sites	126	Chronic venous ulcer locally infected	Time to complete wound healing Proportion of patients with complete wound healing	12 weeks	Octenidine 92 days 30.6% (15/49)	Saline 87 days 32.0% (16/50)
Sibbald et al., 2011 [33]	Canada	February 2008–April 2009	Multicentre 2 sites	40	Chronic wounds > 1 cm^2	Healing rate	4 weeks	PHMB 35% reduction in wound surface	Saline 28% reduction in wound surface
Holloway et al., 1989 [31]	USA	NG	Multicentre 3 sites	75	At least a venous stasis ulcer present for a minimum of 3 months	Healing rate	24 weeks	Cadexomer iodine 0.95 cm^2 per week	Saline 0.41 cm^2 per week

BWAT, Bates Jensen Wound Assessment Tool; DFU, diabetic foot ulcer; NG, not given; PHMB, polyhexamethylenebiguanide; PU, pressure ulcer; PVP-I, povidone-iodine; VLU, venous leg ulcer.

Table 2. Summary of findings table with comparison between iodine and saline solution in chronic wound care.

Comparison One: Iodine Compared to Normal Saline for Chronic Wound Care					
Patient or Population: Chronic Wound Care Intervention: Iodine Comparison: Normal Saline					
Outcomes	**Anticipated Absolute Effects * (95% CI)**		**Relative Effect (95% CI)**	**№ of Participants (Studies)**	**Certainty of the Evidence (GRADE)**
	Risk with Normal Saline	**Risk with Iodine**			
Proportion of patients with complete wound healing assessed with: visual assessment follow up: range 8 weeks to 12 weeks	307 per 1000	567 per 1000 (390 to 824)	RR 1.8478 (1.2706 to 2.6874)	195 (2 RCTs)	⊕⊕⊕○ MODERATE [a]
Adverse events assessed with: report follow-up: range 8 weeks to 24 weeks	115 per 1000	166 per 1000 (89 to 308)	RR 1.440 (0.774 to 2.676)	270 (3 RCTs)	⊕⊕⊕○ MODERATE [a]
Ulcer healing rate (healing rate) assessed with: planimetry follow-up: range 8 weeks to 24 weeks	Raju et al. [32] found a significantly ($p < 0.001$) higher percentage of reduction for both formulations of iodine (94.3% and 90.4%) compared to saline (67.8%). Holloway et al. [31] found a rate reported to baseline of 0.04 ± 0.01 $cm^2/week/cm^2$ for cadexomer iodine and 0.03 ± 0.01 $cm^2/week/cm^2$ for saline. There was no significant difference ($p = 0.079$). Gwak et al. [28] presented the healing rate with three different visual displays showing the percentage change rate for the length, the width, and the area. They found no difference.			270 (3 RCTs)	⊕○○○ VERY LOW [b,c,d]
Pain evaluation (Pain) assessed with: mean rate of change follow up: mean 24 weeks	The mean rate of change in pain scores were -2.44 ± 0.4 for cadexomer iodine and -2.47 ± 0.3 with saline with a $p = 0.96$.			(1 RCT)	⊕⊕○○ LOW [d,e]

GRADE Working Group grades of evidence
⊕⊕⊕⊕ **High certainty**: We are very confident that the true effect lies close to that of the estimate of the effect.
⊕⊕⊕ **Moderate certainty**: We are moderately confident in the effect estimate: The true effect is likely to be close to the estimate of the effect, but there is a possibility that it is substantially different.
⊕⊕ **Low certainty**: Our confidence in the effect estimate is limited: The true effect may be substantially different from the estimate of the effect.
⊕ **Very low certainty**: We have very little confidence in the effect estimate: The true effect is likely to be substantially different from the estimate of effect.

* The risk in the intervention group (and 95% confidence interval) is based on the assumed risk in the comparison group and the relative effect of the intervention (and 95% CI). CI: confidence interval; RR: risk ratio; RCT: randomised controlled trial. (a) Missing outcome data, around 30% of patient losses during follow-up. (b) One study with a high risk of bias due to methodology risk and both studies at risk to outcome reporting. (c) Different results in the two studies. (d) Selective outcome report. (e) Study with an overall high risk of bias.

Among the three studies, two reported the healing rate (Table 2) [31,32]. Raju et al. presented the healing rate as reduction percentage in ulcer size from baseline to 12 weeks. They observed a significant reduction on wounds treated with both formulations of iodine (94.3% and 90.4%) compared to saline (67.8%) ($p < 0.001$) [32]. Holloway et al. defined the healing rate as the ulcer size reduction per week per baseline size. They noted no significant difference between patients treated with cadexomer iodine versus saline (0.04 $cm^2/week/cm^2$ ± 0.01 vs. 0.03 ± 0.01, respectively; $p = 0.079$) [31]. Finally, Gwak et al. reported three indirect markers of healing rate: the percentage rate of change of wound length, width, and area [28]. They found no difference between the two treatments. The heterogeneity of the data prevented pooling. The quality of the evidence was very low.

Pain was only evaluated in one study [31]. No statistical difference was observed between the two treatments (cadexomer iodine versus saline) ($p = 0.96$) (Table 2). The quality of the evidence was low.

Finally, no studies evaluated the bacterial bioburden.

All three studies reported AEs [28,31,32]. The pooled data showed no significant difference between AE incidence in the iodine group compared to the saline group (RR 1.44 (95%CI (0.77 to 2.68), $n = 3$ studies, moderate quality evidence) (Table 2).

2.3. *Polyhexanide Compared to Saline (2 RCT, 334 Patients)*

Two studies, with a total of 334 patients, compared polyhexanide to saline [30,33], both with unclear risk of bias. In 2011, Sibbald et al. studied the efficacy of polyhexanide solution in patients with chronic wounds over 4 weeks [33]. In 2016, Bellingeri et al. studied the efficacy of polyhexanide solution in patients with PUs or mixed aetiologies of chronic ulcers, over 4 weeks [30]. Neither study reported wound healing, and thus the primary outcome could not be evaluated.

The healing rate is described in Table 3 [30,33]. Bellingeri et al. evaluated healing rate as wound improvement on the 13 item BWAT scale (Bates Jensen Wound Assessment Tool) [30]. They observed a significantly improved healing rate in the polyhexanide group ($p = 0.072$), and a significantly better score at 4 weeks compared to the first week in the experimental group ($p = 0.025$). Sibbald et al. calculated the percentage decrease of the wound surface area by planimetry measurement and comparison [33]. They noted no significant difference between the median reduction of the wound surface in the polyhexanide group versus the saline group (35% vs. 28%; $p = 0.85$). Due to the heterogeneity of the measurement tools, we could not pool the data. The quality of the evidence was low.

Table 3. Summary of findings table with comparison between polyhexanide and saline solution in chronic wound care.

Comparison Two: Polyhexanide Compared to Saline for Chronic Wound Care					
Patient or Population: Chronic Wound Care Intervention: Polyhexanide Comparison: Saline					
Outcomes	Anticipated Absolute Effects * (95% CI)		Relative Effect (95% CI)	№ of Participants (Studies)	Certainty of the Evidence (GRADE)
	Risk with Saline	Risk with Polyhexanide			
Wound healing follow-up: mean 4 weeks Not measured	Not reported			(0 studies)	-
Adverse events assessed with: report follow-up: mean 4 days	12 per 1000	2 per 1000 (0 to 50)	RR 0.2024 (0.0098 to 4.1813)	334 (2 RCTs)	⊕⊕○○ LOW [a,b]
Healing rate assessed with: planimetry follow-up: median 4 weeks	Bellingeri et al. [30] found a significantly better progression of wounds in the polyhexanide group ($p = 0.0025$) using the BWAT wound assessment scale. Sibbald et al. [33] found no significant difference ($p = 0.85$) between the two groups by comparing wound surface reduction (35% vs. 28%).			334 (2 RCTs)	⊕⊕○○ LOW [a,b]
Pain assessment assessed with: Pain scales follow-up: mean 4 weeks	Bellingeri et al. [30] found similar pain scores with no significant difference in the two groups (average score = 3 with minimal or no change during follow up). Sibbald et al. [33] reported significant pain reduction in the polyhexanide group compared to the saline control group (73.1% vs. 38.1%; $p = 0.02$).			(2 RCTs)	⊕⊕○○ LOW [a,b]

GRADE Working Group grades of evidence
⊕⊕⊕⊕ **High certainty:** We are very confident that the true effect lies close to that of the estimate of the effect.
⊕⊕⊕ **Moderate certainty:** We are moderately confident in the effect estimate: The true effect is likely to be close to the estimate of the effect, but there is a possibility that it is substantially different.
⊕⊕ **Low certainty:** Our confidence in the effect estimate is limited: The true effect may be substantially different from the estimate of the effect.
⊕ **Very low certainty:** We have very little confidence in the effect estimate: The true effect is likely to be substantially different from the estimate of effect.

* The risk in the intervention group (and 95% confidence interval) is based on the assumed risk in the comparison group and the relative effect of the intervention (and 95% CI). CI: confidence interval; RR: risk ratio; RCT: randomised controlled trial. (a) Two studies with different results. (b) Very few participants in one study.

Concerning pain, Bellingeri et al. observed no significant difference between groups [30], whereas Sibbald et al. reported a significant pain reduction in the polyhexanide group compared to the saline group (73.1% vs. 38.1%, $p = 0.02$) [33]. We graded the quality of the evidence low.

Finally, Sibbald et al. assessed the reduction of the bacterial bioburden [33]. They noted polymicrobial microorganisms in 5.3% of the polyhexanide group wounds versus 33% in the control group ($p = 0.016$).

Bellingeri et al. reported no AEs in either group [30], but Sibbald et al. recorded two AEs as infection in the saline control group [33]. The pooled data showed a significant difference between the incidence of AEs in the polyhexanide group compared to the saline group (RR 0.2 (95%CI (0.01 to 4.18), $n = 2$ studies, low-quality evidence) (Table 3).

2.4. Octenidine Compared to Saline (1 RCT, 126 Patients)

One study, with 126 patients, compared octenidine to saline [29]. Vanscheidt et al. assessed the efficacy of octenidine in patients with locally infected chronic VLU that was at least 1 month old and who had had no previous or concomitant drug therapy for 12 weeks. Their study was categorised at unclear risk of bias.

Wound healing was assessed as time to complete wound closure and the proportion of patients with complete wound closure (Table S1). The time to complete wound healing was not significantly different between the groups (92 days for octenidine vs. 87 days for saline; $p = 0.952$) (Table 4). Accordingly, the proportion of patients with complete wound healing was similar (30.6% vs. 32%; $p = 0.882$). Interestingly, Vanscheidt et al. reported a significant proportion of healing for patients with ulcers larger than 6 cm^2 and older than 6 months in the octenidine group versus the saline group (33.3% vs. 0%, respectively p=0.022) [29]. We graded the quality of the evidence as high.

Table 4. Summary of findings table with comparison between octenidine and saline solution in chronic wound care.

Comparison Three: Octenidine Compared to Saline for Chronic Wound Care					
Patient or Population: Chronic Wound Care **Intervention: Octenidine** **Comparison: Saline**					
Outcomes	**Anticipated Absolute Effects * (95% CI)**		**Relative Effect (95% CI)**	**№ of Participants (Studies)**	**Certainty of the Evidence (GRADE)**
	Risk with Saline	**Risk with Octenidine**			
Wound healing assessed with: Proportion of patients with complete wound healing follow-up: mean 12 weeks	242 per 1000	250 per 1000 (136 to 461)	RR 1.0313 (0.5595 to 1.9009)	126 (1 RCT)	⊕⊕⊕⊕ HIGH
Adverse events assessed with: AE report follow-up: mean 12 weeks	317 per 1000	178 per 1000 (90 to 351)	RR 0.5614 (0.2844 to 1.1081)	120 (1 RCT)	⊕⊕⊕⊕ HIGH
Healing rate assessed with: planimetry follow-up: mean 12 weeks	No difference in the healing rate of the patients in the octenidine group, compared to patients in the saline group (37.9% vs. 40.3%; $p = 0.769$) [29].			(1 RCT)	⊕⊕⊕⊕ HIGH
Pain assessment—not measured	Not reported			-	-

GRADE Working Group grades of evidence
⊕⊕⊕⊕ **High certainty**: We are very confident that the true effect lies close to that of the estimate of the effect.
⊕⊕⊕ **Moderate certainty**: We are moderately confident in the effect estimate: The true effect is likely to be close to the estimate of the effect, but there is a possibility that it is substantially different.
⊕⊕ **Low certainty**: Our confidence in the effect estimate is limited: The true effect may be substantially different from the estimate of the effect.
⊕ **Very low certainty**: We have very little confidence in the effect estimate: The true effect is likely to be substantially different from the estimate of effect.

* The risk in the intervention group (and 95% confidence interval) is based on the assumed risk in the comparison group and the relative effect of the intervention (and 95% CI). CI: confidence interval; RR: risk ratio; RCT: randomised controlled trial.

No difference in the healing rate was noted in the octenidine group compared to the saline group (37.9% vs. 40.3%, $p = 0.769$). The octenidine group achieved better results from 5 weeks. We graded the quality of the evidence high.

The study did not assess pain or bacterial bioburden.

AEs were reported in 10 patients in the octenidine group and 19 patients in the saline group, without significant difference between the groups [29]. We graded the quality of the evidence high.

3. Discussion

We reviewed the RCT evidence for the use of antiseptic agents in chronic wound care in adult patients. Although saline is the main recommended product used in chronic wound cleansing, numerous clinical studies described the benefit of antiseptic agents in this situation [7,17,21,35–37]. A limited number of studies are available on the efficacy of antiseptic agents on chronic wound healing. More limited studies are available on the efficacy of antiseptics on pain. The trials are small, clinically heterogeneous, without clearly defined outcomes, and at high or unclear risk of bias. Of the 838 RCT identified, only 6 studies were included, representing a total of 725 patients.

Our review established a better wound healing with iodine compared to saline (2 RCT, 195 patients, RR 1.85 (1.27 to 2.69)), although the quality of the evidence was moderate. In contrast, no statistical efficacy of octenidine on healing rate (compared to saline) was seen with a high-quality evidence grade (1 RCT, 126 patients RR 1.03 (0.56 to 1.90)). Interestingly, none of the antiseptic agents influenced AE occurrence compared to saline. Notably, over half of the clinical trials have never been published. Most studies had unclear risk of bias, as previously described [38,39].

Of the 838 studies, most of them did not evaluate clinical signs of infection, and mainly focused on bacterial load reduction, a measure long deemed unsuitable [2–4]. Furthermore, the six included studies ignored the effect of biofilm in delaying the healing process. Of the two studies assessing microbiological impact on infection, none assessed biofilm reduction [30,31,33]. For future research, we suggest the use of dynamic models mimicking the wound environment instead of the traditional quantitative microbial load in in vitro studies [40]. This includes non-static models and the consideration of multispecies biofilm reduction over a clinically relevant time (>1 month).

Although most guidelines recommend against the use of antiseptic agents [2,4,25,27], a recent consensus suggested using hypochlorite and polyhexanide in chronic wound care [21]. We found no study demonstrating a significant effect of hypochlorite on the healing of chronic wounds. We could not assess the efficacy of polyhexanide due to the heterogeneity of outcomes between studies. However, this consensus included other types of non-healing wounds such as post-surgical or burn wounds and made no distinction between WAR score categories (colonised and infected wounds). Finally, it also included non-randomised trials, which provide lower evidence than RCTs and different systematic bias are encountered [41]. The main limitation of this guideline is the extrapolation of recommendations from various studied wounds to specific chronic wound care. Another key problem is the attribution of effect to antiseptic agents when antibiotics were systematically used in case of infected wounds.

Following the diverse interpretations of study results in recommendations, future investigations in primary research must focus on value to patients and healthcare professionals, particularly treatment choice. The design of future trials should be driven by high-priority questions. Moreover, good practice guidelines must be followed at each step (e.g., design, implementation, reporting). Assessment of complete wound healing instead of wound healing rate would be more relevant, and time to complete wound healing should be reported as the main endpoint. Future research should be controlled at least against saline, and preferably with another or multiple other antiseptic agents. Another fruitful area of research would be the impact of antiseptic agents according to wound size. Two of our included studies on two different antiseptic agents reported increased healing rate for wounds larger than 6 cm^2 versus smaller wounds [29,31]. Further good quality evidence studies may aid decision making about the use of topical antiseptics in the management of chronic wounds.

The main limitation of this review was the great heterogeneity in study designs, methodologies, and outcomes. Overall, the six studies included in this review were too heterogeneous for pooling. We could not compare the four antiseptic agents with one another. Sensitivity analysis could not be performed due to the low number of results nor could subgroup analysis by age or type of chronic wound.

Despite its limitations, this review assesses the quality of the available data on the four selected antiseptic agents on chronic wound healing, using a well-known and robust methodology. It focuses the topic on chronic wound, reducing the bias of specific cares required for the other types of wounds.

The relative effects of topical antiseptic treatment on chronic wounds are unclear. There is insufficient evidence to determine the superiority of one antiseptic agent over the others. We could not assess the effect of hypochlorous agents on chronic wound healing. Moderate evidence suggests an improvement of wound healing with iodine compared to saline. Currently, there is not enough evidence to recommend one antiseptic agent over another in this clinical situation.

4. Materials and Methods

4.1. Selection Criteria and Search Strategy

The study is registered at PROSPERO (CRD42020213494).

The review protocol was based on the Cochrane Handbook for Systematic Reviews of Intervention (version 6.2) [42] and devised in accordance with the Preferred Reporting Items for Systematic Review and Meta-Analyses (PRISMA) statement guidelines [43]. A meta-analysis was initially planned. Due to heterogeneity in included studies, a systematic review with a summary of effect estimates was performed.

Published and unpublished RCT were eligible. Two reviewers (K.B.B., J.O.) independently screened titles and abstracts to determine eligibility and assessed the full text of retained articles. Disagreements were solved by discussion or arbitration by three independent reviewers (A.S., P.L., J.P.L.). Exclusion reasons were recorded. This study followed this PICO strategy:

1. **Population**: Our population was adult patients ($\geq$18 years) with chronic wounds as previously defined [2]. We included studies from primary, secondary, and tertiary clinical settings. We included different types of chronic wounds (leg ulcer, DFU, PU, eschar). We excluded studies containing patients with wounds requiring specific care (acute wounds, burn wounds, systemic infected wound, postsurgical wounds).
2. **Interventions** of interest were those including antiseptics as cleansing method or within a dressing with at least weekly application.
3. **Comparative** regimens included saline solution or another antiseptic. We anticipated that interventions would consist of povidone-iodine, hypochlorite or hypochlorous acid, iodine, polyhexanide, and octenidine in the form of creams, ointments, powders, sprays, or impregnated into dressings. We included intervention schedules applying concurrent therapies (e.g., negative pressure wound therapy) if the therapy was common across study arms. We excluded (i) interventions where the antiseptic agent was not the only systematic difference between treatment groups; (ii) physical and biological therapies with antimicrobial properties, such as heat or larval therapy; (iii) studies evaluating topical antiseptics in prevention of chronical wounds or those using antiseptics as preparation for surgical treatment of ulcers; (iv) studies evaluating non-recommended antiseptics in chronic wound care and those evaluating antiseptic agents alongside antibiotic agents.
4. **Outcome**: The primary outcome was wound healing, evaluated as the proportion of patients with complete healing during follow-up and/or the time to complete wound healing (analysed using survival, time-to-event approaches). An adjustment

for relevant covariates such as baseline wound area or duration were ideally used to evaluate the outcome. The secondary outcomes were healing rate (described as changes or rate of change in wound size, with adjustment for baseline size); mean pain scores; bacterial bioburden reduction; and AEs, including infection.

We searched PubMed (NLM database), MEDLINE (OvidSP), Web of Science (Thomson Reuters), Google Scholar, and Cochrane library databases, as well as 3 clinical trial registries (ClinicalTrials.gov (www.clinicaltrials.gov, accessed on 17 January 2022)), EU Clinical Trials Register (https://www.clinicaltrialsregister.eu, accessed on 17 January 2022), and World Health Organisation (WHO) International Clinical Trial Registry Platform (https://apps.who.int/trialsearch/, accessed on 17 January 2022) up to 30 June 2021 without restrictions for language, study status, date of publication, or country, using a MeSH terms string chain (Figure 1). Furthermore, we searched the reference lists of reviewed studies for relevant studies. The search strategy for all databases is presented in Appendix A.1.

4.2. Data Collection and Analysis

One reviewer (K.B.B.) performed data extraction and quality assessment for the included studies, validated by a second author (J.O.). Disagreements were resolved by arbitration by three independent review authors (A.S., P.L., J.P.L.). We contacted study authors for additional data if necessary. We performed data extraction using a standardised sheet, as recommended by the Cochrane Collaboration's handbook for systematic review (trial authors, year of publication, patient population characteristics, duration of follow-up, trial design, measured outcomes, including assessment methods, objectives, results, country where trials were performed, number of participants randomly assigned to each treatment group, clinical setting, detail of interventions in each group, details of comparators in each group, source of funding, number of withdrawals, outcomes). Data are presented in Table S2.

4.3. Risk of Bias and Certainty of the Evidence

Two reviewers (K.B.B., J.O.) independently assessed the risk of bias of eligible studies; any disagreements were resolved by arbitration by three independent review authors (A.S., P.L., J.P.L.). Risk of bias was assessed using the bias excel tool (RoB 2 checklist, 2019) [44] as recommended by the Cochrane Handbook for Systematic Reviews of Intervention (version 6.2) [42]. The overall bias risk was rated as low, moderate, high, or unclear (also some concerns). The 7 domains of bias were also assessed for each trial, following the recommendation of the Cochrane collaboration, 2011 [34].

The overall quality of evidence of included studies was assessed using Grading of Recommendations Assessment, Development and Evaluation (GRADE) as recommended by the Cochrane Handbook for Systematic Reviews of Intervention (version 6.2) [44]. Summary tables for each antiseptic were produced with GRADEPro software considering five outcomes: complete wound healing (healing rate or proportion of patients with complete wound healing), rate of change in wound size, pain assessment, bacterial bioburden reduction, and AEs. We calculated the risk ratio (RR) for dichotomous outcomes (wound healing, AEs, infection) with 95% confidence interval (CI).

4.4. Role of the Funding Source

There was no funding source for this study.

5. Conclusions

Iodine compounds showed a better effect on chronic wound healing compared to saline. Octenidine and polyhexanide did not show any difference in this healing compared to saline. Currently, there is not enough evidence to recommend one antiseptic over another in this clinical situation. Future clinical trials assessing antiseptic agents in chronic wounds management should pay attention to include several antiseptic agents in their trial for comparison, respecting the double-blind trial and with a well-defined study population.

They should establish the main efficiency criteria as complete wound healing and not only wound size reduction. A sub-group analysis based on the size of the wound would be relevant to the matter. Finally, following good practice guidelines is mandatory in every step of trials in order to avoid the numerous biases found in the assessed studies.

Supplementary Materials: The following supporting information can be downloaded at https://www.mdpi.com/article/10.3390/antibiotics11030350/s1, Figure S1: Plot of the percentage of risk of bias assessments at each level of risk of bias per domain. Table S1: Summary of the included studies. Table S2: Risk of bias assessment among included studies. Table S3: Review authors' judgements about each risk of bias item for each included study. Table S4: Different outcomes reported among the studies.

Author Contributions: Conceptualisation, A.S., J.-P.L. and P.L.; methodology, K.B.-B., F.S. and P.L.; software, F.S.; validation, J.O., A.S., J.-P.L. and P.L.; formal analysis, K.B.-B. and J.O.; investigation, K.B.-B. and J.O.; writing—original draft preparation, K.B.-B., J.O. and P.L.; writing—review and editing, A.S., F.S. and J.-P.L.; supervision, A.S. and J.-P.L.; project administration, P.L.; funding acquisition, A.S. and J.-P.L. All authors have read and agreed to the published version of the manuscript.

Funding: This research received no external funding.

Acknowledgments: We thank Sarah Kabani for editing the article. We thank the Nîmes University hospital for its structural, human, and financial support through the award obtained by our team during the internal call for tenders "Thématiques phares". The authors belong to the FHU InCh (Federation Hospitalo Universitaire Infections Chroniques, Aviesan.

Conflicts of Interest: The authors declare no conflict of interest.

Appendix A Appendix

Appendix A.1 Medline and Cochrane Library Search

Our search strategy basis was to focus on the four main antiseptic agents recommended or used in practice: chlorous compounds, octenidine, iodophor compounds, and polyhexanide. Four different searches were conducted then combined:

- (((((chronic wound) OR diabetic foot) OR leg ulcer) OR pressure ulcer) OR eschar) AND (((iodine) OR povidone iodine OR betadine) OR cadexomer iodine).
- (((((chronic wound) OR diabetic foot) OR leg ulcer) OR pressure ulcer) OR eschar) AND ((((hypochlorite) OR hypochlorous) OR Dakin) OR Javel).
- (((((chronic wound) OR diabetic foot) OR leg ulcer) OR pressure ulcer) OR eschar) AND ((((PHMB) OR polyhexanide) OR Polyhexamethylene biguanide) OR Polyhexamethylene).
- (((((chronic wound) OR diabetic foot) OR leg ulcer) OR pressure ulcer) OR eschar) AND (Octenidine).

References

1. Järbrink, K.; Ni, G.; Sönnergren, H.; Schmidtchen, A.; Pang, C.; Bajpai, R.; Car, J. Prevalence and incidence of chronic wounds and related complications: A protocol for a systematic review. *Syst. Rev.* **2016**, *5*, 152. [CrossRef] [PubMed]
2. Kirsner, R. The Wound Healing Society chronic wound ulcer healing guidelines update of the 2006 guidelines—Blending old with new. *Wound Repair Regen.* **2016**, *24*, 110–111. [CrossRef] [PubMed]
3. Lipsky, B.A.; Berendt, A.R.; Deery, H.G.; Embil, J.M.; Joseph, W.S.; Karchmer, A.W.; LeFrock, J.L.; Lew, D.P.; Mader, J.T.; Norden, C.; et al. Infectious Diseases Society of America. Diagnosis and treatment of diabetic foot infections. *Clin. Infect. Dis.* **2004**, *39*, 885–910. [CrossRef] [PubMed]
4. Société de Pathologie Infectieuse de Langue Française. Management of diabetic foot infections. Long text. Société de Pathologie Infectieuse de Langue Française. *Med. Mal. Infect.* **2007**, *37*, 26–50.
5. Bjarnsholt, T. The role of bacterial biofilms in chronic infections. *APMIS* **2013**, *121*(s136), 1–58. [CrossRef]
6. Pouget, C.; Dunyach-Remy, C.; Pantel, A.; Schuldiner, S.; Sotto, A.; Lavigne, J.P. Biofilms in diabetic foot ulcers: Significance & clinical relevance. *Microorganisms* **2020**, *8*, 1580.
7. Schultz, G.; Bjarnsholt, T.; James, G.A.; Leaper, D.J.; McBain, A.J.; Malone, M.; Stoodley, P.; Swanson, T.; Tachi, M.; Wolcott, R.D. Global Wound Biofilm Expert Panel. Consensus guidelines for the identification and treatment of biofilms in chronic nonhealing wounds. *Wound Repair Regen.* **2017**, *25*, 744–757. [CrossRef]

8. Dissemond, J.; Assadian, O.; Gerber, V.; Kingsley, A.; Kramer, A.; Leaper, D.J.; Mosti, G.; Piatkowski de Grzymala, A.; Riepe, G.; Risse, A.; et al. Classification of wounds at risk and their antimicrobial treatment with polyhexanide: A practice-oriented expert recommendation. *Skin Pharmacol. Physiol.* **2011**, *24*, 245–255. [CrossRef]
9. Bakker, K.; Apelqvist, J.; Lipsky, B.A.; Van Netten, J.J. International Working Group on the Diabetic Foot. The 2015 IWGDF guidance documents on prevention and management of foot problems in diabetes: Development of an evidence-based global consensus. *Diabetes Metab. Res. Rev.* **2016**, *32*, 2–6. [CrossRef]
10. Wolcott, R.D.; Kennedy, J.P.; Dowd, S.E. Regular debridement is the main tool for maintaining a healthy wound bed in most chronic wounds. *J. Wound Care.* **2009**, *18*, 54–56. [CrossRef]
11. Sibbald, R.G.; Elliott, J.A.; Persaud-Jaimangal, R.; Goodman, L.; Armstrong, D.G.; Harley, C.; Coelho, S.; Xi, N.; Evans, R.; Mayer, D.O.; et al. Wound Bed Preparation 2021. *Adv. Skin Wound Care.* **2021**, *34*, 183–195. [CrossRef]
12. Sun, X.; Chen, J.; Zhang, J.; Wang, W.; Sun, J.; Wang, A. Maggot debridement therapy promotes diabetic foot wound healing by up-regulating endothelial cell activity. *J. Diabetes Complicat.* **2016**, *30*, 318–322. [CrossRef] [PubMed]
13. Ayello, E.A.; Cuddigan, J.E. Debridement: Controlling the necrotic/cellular burden. *Adv. Skin Wound Care* **2004**, *17*, 66–75. [CrossRef] [PubMed]
14. Kravitz, S.R.; McGuire, J.; Zinszer, K. Management of skin ulcers: Understanding the mechanism and selection of enzymatic debriding agents. *Adv. Skin Wound Care* **2008**, *21*, 72–74. [CrossRef] [PubMed]
15. De Francesco, F.; De Francesco, M.; Riccio, M. Hyaluronic Acid/Collagenase Ointment in the Treatment of Chronic Hard-to-Heal Wounds: An Observational and Retrospective Study. *J. Clin. Med.* **2022**, *11*, 537. [CrossRef] [PubMed]
16. Agarwal, P.; Kukrele, R.; Sharma, D. Vacuum assisted closure (VAC)/negative pressure wound therapy (NPWT) for difficult wounds: A review. *J. Clin. Orthop. Trauma.* **2019**, *10*, 845–848. [CrossRef]
17. O'Meara, S.; Cullum, N.; Nelson, E.A.; Dumville, J.C. Compression for venous leg ulcers. *Cochrane Database Syst. Rev.* **2012**, *11*, CD000265. [CrossRef] [PubMed]
18. Vouillarmet, J.; Bourron, O.; Gaudric, J.; Lermusiaux, P.; Millon, A.; Hartemann, A. Lower-extremity arterial revascularization: Is there any evidence for diabetic foot ulcer-healing? *Diabetes Metab.* **2016**, *42*, 4–15. [CrossRef]
19. Schaper, N.C.; van Netten, J.J.; Apelqvist, J.; Bus, S.A.; Hinchliffe, R.J.; Lipsky, B.A. IWGDF Editorial Board. Practical guidelines on the prevention and management of diabetic foot diseases (IWGDF 2019 update). *Diabetes Metab. Res. Rev.* **2020**, *36*, e3266. [CrossRef]
20. Bowers, S.; Franco, E. Chronic wounds: Evaluation and management. *Am. Fam. Physician.* **2020**, *101*, 159–166.
21. Kramer, A.; Dissemond, J.; Kim, S.; Willy, C.; Mayer, D.; Papke, R.; Tuchmann, F.; Assadian, O. Consensus on wound antisepsis: Update 2018. *Skin Pharmacol. Physiol.* **2018**, *31*, 28–58. [CrossRef] [PubMed]
22. Babalska, Z.Ł.; Korbecka-Paczkowska, M.; Karpiński, T.M. Wound Antiseptics and European Guidelines for Antiseptic Application in Wound Treatment. *Pharmaceuticals* **2021**, *14*, 1253. [CrossRef] [PubMed]
23. Pouget, C.; Dunyach-Remy, C.; Pantel, A.; Boutet-Dubois, A.; Schuldiner, S.; Sotto, A.; Lavigne, J.P.; Loubet, P. Alternative approaches for the management of diabetic foot ulcers. *Front. Microbiol.* **2021**, *12*, 747618. [CrossRef] [PubMed]
24. Rayman, G.; Vas, P.; Dhatariya, K.; Driver, V.; Hartemann, A.; Londahl, M.; Piagessi, A.; Apelqvist, J.; Attinger, C.; Game, F. International Working Group on the Diabetic Foot (IWGDF). Guidelines on use of interventions to enhance healing of chronic foot ulcers in diabetes (IWGDF 2019 update). *Diabetes Metab. Res. Rev.* **2020**, *36*, e3283. [CrossRef]
25. Norman, G.; Dumville, J.C.; Moore, Z.E.H.; Tanner, J.; Christie, J.; Goto, S. Antibiotics and antiseptics for pressure ulcers. *Cochrane Database Syst. Rev.* **2016**, *4*, CD011586. [CrossRef]
26. Dogra, S.; Sarangal, R. Summary of recommendations for leg ulcers. *Indian Dermatol. Online J.* **2014**, *5*, 400–407. [CrossRef]
27. Lipsky, B.A.; Senneville, E.; Abbas, Z.G.; Aragón-Sánchez, J.; Diggle, M.; Embil, J.M.; Kono, S.; Lavery, L.A.; Malone, M.; van Asten, S.A.; et al. International Working Group on the Diabetic Foot (IWGDF). Guidelines on the diagnostic and treatment of foot infection in persons with diabetes (IWGDF 2019 update). *Diabetes Metab. Res. Rev.* **2020**, *36*, e3280. [CrossRef]
28. Gwak, H.C.; Han, S.H.; Lee, J.; Park, S.; Sung, K.S.; Kim, H.J.; Chun, D.; Lee, K.; Ahn, J.H.; Kwak, K.; et al. Efficacy of a povidone-iodine foam dressing (Betafoam) on diabetic foot ulcer. *Int. Wound J.* **2020**, *17*, 91–99. [CrossRef]
29. Vanscheidt, W.; Harding, K.; Téot, L.; Siebert, J. Effectiveness and tissue compatibility of a 12-week treatment of chronic venous leg ulcers with an octenidine based antiseptic—A randomized, double-blind controlled study. *Int. Wound J.* **2012**, *9*, 316–323. [CrossRef]
30. Bellingeri, A.; Falciani, F.; Traspedini, P.; Moscatelli, A.; Russo, A.; Tino, G.; Chiari, P.; Peghetti, A. Effect of a wound cleansing solution on wound bed preparation and inflammation in chronic wounds: A single-blind RCT. *J. Wound Care* **2016**, *25*, 160–168. [CrossRef]
31. Holloway, G.A.; Johansen, K.H.; Barnes, R.W.; Pierce, G.E. Multicenter trial of cadexomer iodine to treat venous stasis ulcer. *West J. Med.* **1989**, *151*, 35–38. [PubMed]
32. Raju, R.; Kethavath, S.N.; Sangavarapu, S.M.; Kanjarla, P. Efficacy of cadexomer iodine in the treatment of chronic ulcers: A randomized, multicenter, controlled trial. *Wounds* **2019**, *31*, 85–90. [PubMed]
33. Sibbald, R.G.; Coutts, P.; Woo, K.Y. Reduction of bacterial burden and pain in chronic wounds using a new polyhexamethylene biguanide antimicrobial foam dressing-clinical trial results. *Adv. Skin Wound Care* **2011**, *24*, 78–84. [CrossRef] [PubMed]

34. Higgins, J.P.T.; Altman, D.G.; Gøtzsche, P.C.; Jüni, P.; Moher, D.; Oxman, A.D.; Savonic, J.; Schulz, K.F.; Weeks, L.; Sterne, J.A.C. Cochrane Bias Methods Group; Cochrane Statistical Methods Group. The Cochrane Collaboration's tool for assessing risk of bias in randomised trials. *BMJ* **2011**, *343*, d5928. [CrossRef]
35. O'Meara, S.M.; Cullum, N.A.; Majid, M.; Sheldon, T.A. Systematic review of antimicrobial agents used for chronic wounds. *Br. J. Surg.* **2001**, *88*, 4–21. [CrossRef]
36. Hämmerle, G.; Strohal, R. Efficacy and cost-effectiveness of octenidine wound gel in the treatment of chronic venous leg ulcers in comparison to modern wound dressings. *Int. Wound J.* **2016**, *13*, 182–188. [CrossRef]
37. To, E.; Dyck, R.; Gerber, S.; Kadavil, S.; Woo, K.Y. The effectiveness of topical polyhexamethylene biguanide (PHMB) agents for the treatment of chronic wounds: A systematic review. *Surg. Technol. Int.* **2016**, *29*, 45–51.
38. Dwan, K.; Altman, D.G.; Arnaiz, J.A.; Bloom, J.; Chan, A.W.; Cronin, E.; Decullier, E.; Easterbrook, P.J.; Von Elm, E.; Gamble, C.; et al. Systematic review of the empirical evidence of study publication bias and outcome reporting bias. *PLoS One* **2008**, *3*, e3081. [CrossRef]
39. Chan, A.W.; Hróbjartsson, A.; Haahr, M.T.; Gøtzsche, P.C.; Altman, D.G. Empirical evidence for selective reporting of outcomes in randomized trials: Comparison of protocols to published articles. *JAMA* **2004**, *291*, 2457–2465. [CrossRef]
40. Bahamondez-Canas, T.F.; Heersema, L.A.; Smyth, H.D.C. Current status of in vitro models and assays for susceptibility testing for wound biofilm infections. *Biomedicines* **2019**, *7*, 34. [CrossRef]
41. Schmidt, W.P. Randomised and non-randomised studies to estimate the effect of community-level public health interventions: Definitions and methodological considerations. *Emerg. Themes Epidemiol.* **2017**, *14*, 9. [CrossRef] [PubMed]
42. Higgings, J.P.T.; Thomas, J.; Chandler, J.; Cumpston, M.; Li, T.; Page, W.J.; Welch, V.A. Cochrane Handbook for Systematic Reviews of Interventions Version 6.2 (Updated February 2021). Cochrane. 2021. Available online: www.training.cochrane.org/handbook (accessed on 17 January 2022).
43. Moher, D.; Liberati, A.; Tetzlaff, J.; Altman, D.G. Preferred reporting items for systematic reviews and meta-analyses: The PRISMA statement. *Int. J. Surg.* **2010**, *8*, 336–341. [CrossRef] [PubMed]
44. Higgins, J.P.T.; Savović, J.; Page, M.J.; Elbers, R.G.; Sterne, J.A.C. Chapter 8: Assessing risk of bias in a randomized trial. In *Cochrane Handbook for Systematic Reviews of interventions v6.0*; Higgins, J.P.T., Higgins, J.P.T., Thomas, J., Chandler, J., Cumpston, M., Li, T., Page, M.J., Welch, V.A., Eds.; Cochrane: London, UK, 2019; Available online: www.training.cochrane.org/handbook/current/chapter-08 (accessed on 17 January 2022).

MDPI
St. Alban-Anlage 66
4052 Basel
Switzerland
Tel. +41 61 683 77 34
Fax +41 61 302 89 18
www.mdpi.com

Antibiotics Editorial Office
E-mail: antibiotics@mdpi.com
www.mdpi.com/journal/antibiotics

www.ingramcontent.com/pod-product-compliance
Lightning Source LLC
Chambersburg PA
CBHW061441210926
51850CB00008B/1570
9783036582382